BIONANOTECHNOLOGY

GLOBAL PROSPECTS

BIONANOTECHNOLOGY

GLOBAL PROSPECTS

Edited by

David E. Reisner

CRC Press
Taylor & Francis Group
Boca Raton London New York

CRC Press is an imprint of the
Taylor & Francis Group, an **informa** business

CRC Press
Taylor & Francis Group
6000 Broken Sound Parkway NW, Suite 300
Boca Raton, FL 33487-2742

First issued in paperback 2019

ISBN-13: 978-0-8493-7528-6 (hbk)
ISBN-13: 978-0-367-38705-1 (pbk)

Library of Congress Cataloging-in-Publication Data

Bionanotechnology: Global prospects / editor, David E. Reisner.
    p. ; cm.
    "A CRC title."
    Includes bibliographical references and index.
    ISBN 978-0-8493-7528-6 (hardback : alk. paper)
    1. Nanotechnology. 2. Biotechnology. I. Reisner, David Evans
    [DNLM: 1. Nanotechnology. 2. Biotechnology. QT 36.5 B6156 2008]

    TP248.25.N35B565 2008
    660.6--dc22   2008012855

Visit the Taylor & Francis Web site at
http://www.taylorandfrancis.com

and the CRC Press Web site at
http://www.crcpress.com

Contents

Preface

Nature has been engaged in its own unfathomable and uncanny nanotechnology project since the dawn of life, billions of years ago. It is only recently that humans have developed their own tools to observe Nature as she assembles and manipulates structures so complex and purposeful so as to defy the imagination. No one would argue that all molecular biology is based on nanotechnology. After all, these structural building blocks composed of ordered elements are well within the 100-nanometer scale that is generally agreed upon as the physical dimensional ceiling below which nanotechnology processes occur. It is no wonder that man is now attempting to mimic Nature by building analogous structures from the bottom up.

A few words about the book title: The temptation to consider "nanobiotechnology" as a subset of biotechnology fails to pay homage to the gargantuan impact of the burgeoning nanotechnology field—a field in the throes of revolutionary growth. The word *nanobiotechnology* feels redundant, a bromide. In distinction, the term *bionanotechnology* connotes a rapidly evolving sector of the nanotechnology field that deals strictly with biological processes and structures. Many refer to this synthesis as "convergence." As will be demonstrated in this monograph, the seeds of bionanotechnology development have been planted. Commercial products will likely be on the marketplace well before the next edition appears. Many nanotech soothsayers predict that as time goes on, this convergence of biotechnology and nanotechnology will become a dominant focus area for technological innovation worldwide and will impact all of our lives on a daily basis.

Naturally, this is also an engineering book. One need not stretch the imagination very far to appreciate that Nature has fundamentally engineered life as we know it, culminating in our own species. This fact has not gone unnoticed on the part of nanotechnologists, who have begun in earnest to mimic Nature's fundamental engineering processes through invoking precise controls over her building blocks. Self-assembly, a key construct of nanotechnology, forms the backbone of biological processes. For example, exploiting DNA as scaffolding for the engineering of DNA-templated molecular electronic devices is an inspiring example of our newfound ability to insinuate our own design skills at the nanoscale level in living systems. Using this approach, it is possible to create self-assembling electronic circuits or devices in solution. Directed evolution based on repeated mutagenesis experiments can be conducted at the nanoscale level. Along these lines, the use of the solar energy conversion properties of bacteriorhodopsin for making thin-film memories, photovoltaic convertors, holographic processors, artificial memories, logic gates, and protein-semiconductor hybrid devices is astounding.

Quantum dots are tiny light-emitting particles on the nanoscale. They have been developed as a new class of biological fluorophore. Once rendered hydrophilic with appropriate functional groups, quantum dots can act as biosensors that can detect biomolecular targets on a real-time or continuous basis. Different colors of quantum dots could be combined into a larger structure to yield an optical bar code. Gold nanoparticles can be functionalized to serve as biological tags.

Nanomedicine is a burgeoning area of development, encompassing drug delivery by nanoparticulates, including fullerenes, as well as new enabling opportunities in medical diagnostics, labeling, and imaging. Quantum dots will certainly play a large role in nanomedicine. Years from now, we will laugh at the archaic approach to treating disease we presently take for granted, carried over from the twentieth century, relying on a single drug formulation to treat a specific disease in all people without acknowledging each individual's unique genetic makeup. Nanocoatings also play an important near-term role in the lifetime of medical devices, especially orthopedic prosthetics. Nanocrystalline hydroxyapatite is far less soluble in human body fluid than conventional amorphous material, thereby anticipating great increases in its service life.

It is not the intention to provide a comprehensive treatise on bionanotechnology, rather I hope to provide representative reporting on a wide variety of activity in the field from all corners of the planet (now that the "world is flat" it has corners). I have attempted to assemble chapters that are relevant to looming product opportunities and instructional for those readers interested in developing the bionanotech products of the future. To that end, I felt it appropriate to conclude the discussion with a chapter that reviews the patent landscape for bionanotechnology, which is presently in a state of great flux. Now more than ever, intellectual property is relevant to both the academic and corporate sectors, and as such, patents are being ascribed greater value and importance. Bionanotechnology commercialization will be driven by the increases in government funding as well as the expiration of more traditional drug patents.

Accumulating author contributions from experts scattered across the globe acquired a life of its own in the evolution of this book. As a Technology Pioneer at the Annual Meeting of the World Economic Forum (WEF) in Davos, Switzerland, I was privy to a worldview that few technologists are able to enjoy. Klaus Schwab, WEF's driving force, has observed that everywhere in society and business, the power is moving from the center to the periphery. This monograph is a testimonial to that paradigm shift. Authors have contributed from 15 different countries in cities from as far as Florianópolis, Mumbai, Ramat-Gan, Pretoria, Havana, Tehran, Glasgow, Shenyang, and Kiev, just to name a few. Of course, this diaspora of academic excellence is largely enabled by the most pervasive technological innovation of our time, the Web.

Chris Anderson has postulated a compelling new economics of culture and commerce, dubbed the "Long Tail," so named because in statistics, the tail of a $1/x$ power law curve is very long relative to the head. Long Tail economics is about the economics of abundance (not scarcity), and we now see quantum shifts in customer buying habits at Amazon, Netflix, and eBay, as well as shifts in content distribution at Wikipedia, Google, and the emerging "Blogosphere." This phenomenon is also playing out in scientific research across the globe, where the Long Tail has now made possible world-class creative technology advances that not long ago were unimaginable. This monograph is proof in spades of this paradigm shift. I dedicate this book to all the authors who gave their valuable time to create the contributions that fill this volume. Many of those authors delivered expert chapters in the face of severe obstacles, some even endured personal hardship and loss over the course of their writing. They know who they are, and I thank them. I dedicate this book to the singer, not the song.

David E. Reisner
The Nano Group, Inc.
Farmington, Connecticut, USA

The Editor

David E. Reisner, Ph.D., is a well known entrepreneur in the burgeoning field of nanotechnology, having cofounded in 1996 two nanotech companies in Connecticut, Inframat® and US Nanocorp®. He has been the Chief Executive Officer of both companies since founding, which were recognized in Y2002–Y2005 for their fast revenue growth as Deloitte & Touche *Connecticut Technology Fast50 Award* recipients. In 2004, The Nano Group, Inc. was formed as a parent holding company for investment. Dr. Reisner and the cofounders were featured in *Forbes* magazine in 2004. He is also a Managing Director in Delta Capital Group.

Dr. Reisner has more than 175 publications and is an inventor on 10 issued patents. He is the editor for the "Bionanotechnology" section of the 3rd Edition of *The BioMedical Engineering Handbook*. He has written articles on the business of nanotechnology in *Nanotechnology Law & Business* as well as the Chinese publication *Science & Culture Review*.

Dr. Reisner served a 3-year term as a Technology Pioneer for the World Economic Forum and was a panelist at the 2004 Annual Meeting in Davos. He is on the Board of the Connecticut Venture Group and is Chairman of the Board of the Connecticut Technology Council. He was a National Aeronautics and Space Administration (NASA) *NanoTech Briefs* Nano50 awardee in 2006. For his efforts in the field of medical implantable devices, Reisner won the 1st Annual BEACON Award for Medical Technology in 2004. He is a member of the Connecticut Academy of Science and Engineering.

Reisner is a 1978 University Honors graduate from Wesleyan University and received his Ph.D. at MIT in 1983 in the field of chemical physics. He was recognized for his historic preservation efforts in 1994 when he received the Volunteer Recognition Award from the Connecticut Historical Commission and the Connecticut Trust for Historic Preservation. Dr. Reisner is known nationally for his expertise in vintage Corvette restoration and documentation.

Contributors

Reinaldo Acevedo
Finlay Institute
Havana, Cuba

Fred Allen
Always Ready, Inc.
Little Falls, New Jersey, USA

André Avelino Pasa
Departamento de Física
Universidade Federal de Santa Catarina
Florianópolis, Brazil

Hossein Baharvand
Department of Stem Cells
Royan Institute
Tehran, Iran

Ramón Barberá
Finlay Institute
Havana, Cuba

Raj Bawa
Biology Department and Office of Tech
  Commercialization
Rensselaer Polytechnic Institute
Troy, New York, USA

Tânia Beatriz Creczynski-Pasa
Departamento de Ciências Farmacêuticas
Universidade Federal de Santa Catarina
Florianópolis, Brazil

Robert R. Birge
University of Connecticut
Storrs, Connecticut, USA

Gustavo Bracho
Finlay Institute
Havana, Cuba

Concepción Campa
Finlay Institute
Havana, Cuba

Shawn D. Carrigan
McGill University
Montreal, Quebec, Canada

Paul K. Chelule
Polymers and Bioceramics
Council for Scientific and Industrial Research
  (CSIR)
Pretoria, South Africa

Michael Connolly
Integrated Nano-Technologies
Henrietta, New York, USA

Michelle Critchley
Nanotechnology Victoria Ltd.
Victoria, Australia

Remy Cromer
Oxonica, Inc.
Mountain View, California, USA

Matthew J. Dalby
University of Glasgow
Glasgow, Scotland, United Kingdom

Glenn Davis
Oxonica, Inc.
Mountain View, California

Judith del Campo
Finlay Institute
Havana, Cuba

William E. Doering
Oxonica, Inc.
Mountain View, California, USA

Kimberly L. Douglas
McGill University
Montreal, Quebec, Canada

Rebekah A. Drezek
Rice University
Houston, Texas, USA

Dror Fixler
School of Engineering
Bar-Ilan University
Ramat-Gan, Israel

R. Griffith Freeman
Oxonica, Inc.
Mountain View, California, USA

Peter Gammel
SiGe Semiconductor
Andover, Massachusetts, USA

Javier García
Departamento de Optica
Universitat de Valencia
Burjassot, Spain

Magnus Gittins
Advance Nanotech, Inc.
New York, New York, USA

Domingo González
Finlay Institute
Havana, Cuba

David Guagliardo
Oxonica, Inc.
Mountain View, California, USA

Naomi J. Halas
Rice University
Houston, Texas, USA

Jie Han
NASA Ames Research Center
Moffett Federal Airfield, California, USA

Yulin Hao
Institute of Metal Research
Chinese Academy of Sciences
Shenyang, Liaoning, China

Michael N. Helmus
Medical Devices, Biomaterials, and
  Nanotechnology
Worcester, Massachusetts, USA

Jason R. Hillebrecht
University of Connecticut
Storrs, Connecticut, USA

Theodore R. Holford
Yale University School of Medicine
New Haven, Connecticut, USA

Bob Irving
Nanotechnology Victoria Ltd.
Victoria, Australia

Manoj Joshi
Nanoelectronics Centre, Department of
  Electrical Engineering
Indian Institute of Technology—Bombay
Mumbai, India

Nitin Kale
Nanoelectronics Centre, Department of
  Electrical Engineering
Indian Institute of Technology—Bombay
Mumbai, India

Lonji Kalombo
Polymers and Bioceramics
Council for Scientific and Industrial Research
  (CSIR)
Pretoria, South Africa

Frans W.H. Kampers
BioNT, Wageningen Bionanotechnology Centre
Wageningen, The Netherlands

Jeremy F. Koscielecki
University of Connecticut
Storrs, Connecticut, USA

Alexis Labrada
Finlay Institute
Havana, Cuba

R. Lal
Nanoelectronics Centre, Department of
  Electrical Engineering
Indian Institute of Technology—Bombay
Mumbai, India

Miriam Lastre
Finlay Institute
Havana, Cuba

Shujun Li
Institute of Metal Research
Chinese Academy of Sciences
Shenyang, Liaoning, China

V.A. Lifton
mPhase Technologies, Inc.
Little Falls, New Jersey, USA

Rogerio S. Lima
National Research Council of Canada
Industrial Materials Institute
Boucherville, Quebec, Canada

Basil R. Marple
National Research Council of Canada
Industrial Materials Institute
Boucherville, Quebec, Canada

Narges Zare Mehrjardi
Department of Stem Cells
Royan Institute
Tehran, Iran

Vicente Mico
AIDO, Technological Institute of Optics,
   Colour and Imaging
Parc Tecnològic
Valencia, Spain

Sarah Morgan
Nanotechnology Victoria Ltd.
Victoria, Australia

S. Mukherji
Nanoelectronics Centre and School of
   Biosciences and Bioengineering
Indian Institute of Technology—Bombay
Mumbai, India

Vijay M. Naik
Hindustan Unilever Research Centre
Bangalore, India

Michael J. Natan
Oxonica, Inc.
Mountain View, California, USA

Shuming Nie
Emory University
Georgia Institute of Technology
Atlanta, Georgia, USA

Joo L. Ong
University of Tennessee
Knoxville, Tennessee, USA

Richard O.C. Oreffo
University of Glasgow
Glasgow, Scotland, United Kingdom

Oliver Pérez
Finlay Institute
Havana, Cuba

Jeanette Pritchard
Nanotechnology Victoria Ltd.
Victoria, Australia

V. Ramgopal Rao
Nanoelectronics Centre and Department of
   Electrical Engineering
Indian Institute of Technology—Bombay
Mumbai, India

Alexandra A. Revina
A.N. Frumkin Institute of Physical Chemistry
   and Electrochemistry
Russian Academy of Sciences
Moscow, Russia

Gualberto Ruaño
Genomas, Inc.
Hartford, Connecticut, USA

Boitumelo Semete
Polymers and Bioceramics
Council for Scientific and Industrial Research
   (CSIR)
Pretoria, South Africa

Michael Y. Sha
Oxonica, Inc.
Mountain View, California, USA

Elena Shembel
Inter-Intel, Inc.
Coral Springs, Florida, USA

Alexandra Shmyryeva
National Technical University of Ukraine
Kiev Polytechnic Institute
Kiev, Ukraine

Gustavo Sierra
Finlay Institute
Havana, Cuba

S. Simon
mPhase Technologies, Inc.
Little Falls, New Jersey, USA

Andrew Michael Smith
Emory University
Georgia Institute of Technology
Atlanta, Georgia, USA

Hulda S. Swai
Polymers and Bioceramics
Council for Scientific and Industrial Research
  (CSIR)
Pretoria, South Africa

Maryam Tabrizian
McGill University
Montreal, Quebec, Canada

Weihong Tan
University of Florida
Gainesville, Florida, USA

Lin Wang
University of Florida
Gainesville, Florida, USA

Jennifer L. West
Rice University
Houston, Texas, USA

Andreas Windemuth
Yale University School of Medicine
New Haven, Connecticut, USA

Rui Yang
Institute of Metal Research
Chinese Academy of Sciences
Shenyang, Liaoning, China

Yunzhi Yang
University of Tennessee
Knoxville, Tennessee, USA

Zeev Zalevsky
School of Engineering
Bar-Ilan University
Ramat-Gan, Israel

Caridad Zayas
Finlay Institute
Havana, Cuba

Zongtao Zhang
Inframat Corporation
Farmington, Connecticut, USA

1 Nanotechnology in Stem Cell Biology and Technology[*]

Hossein Baharvand and Narges Zare Mehrjardi
Department of Stem Cells, Royan Institute, Tehran, Iran

CONTENTS

1.1 INTRODUCTION

Nanotechnology is the science and engineering concerned with the design, synthesis, characterization, and application of materials and devices that have a functional organization in at least one dimension on the nanometer (nm) scale, ranging from a few to about 100 nm. Nanotechnology is beginning to help advance the equally pioneering field of stem-cell research, with devices that can precisely control stem cells (SCs) and provide nanoscaled-biodegradable scaffolds and magnetic tracking systems. SCs are undifferentiated cells generally characterized by their functional capacity to both self-renew and to generate a large number of differentiated progeny cells. The characteristics of SCs indicate that these cells, in addition to use in developmental biology studies, have the potential to provide an unlimited supply of different cell types for tissue replacement, drug screening, and functional genomics applications. Tissue engineering at the nanoscale level is a potentially useful approach to develop viable substitutes, which can restore, maintain, or improve the function of human tissue. Regenerating tissue can be achieved by using nanobiomaterials to convey signals to surrounding tissues to recruit cells that promote inherent regeneration or by using cells and a nanobiomaterial scaffold to act as a framework for developing tissue. In this regard, nanomaterials

[*] The authors would like to dedicate this chapter to the memory of Dr. Saeid Kazemi Ashtiani. He was a wonderful colleague, a great stem cell biologist, and an inspirational advocate of human stem cell research in Iran.

such as nanofibers are of particular interest. Three different approaches toward the formation of nanofibrous materials have emerged: self-assembly, electrospinning, and phase separation [1]. Each of these approaches is unique with respect to its characteristics, and each could lead to the development of a scaffolding system. For example, self-assembly can generate small-diameter nanofibers in the lowest end of the range of natural extracellular matrix (ECM) collagen, while electrospinning is more useful for generating large-diameter nanofibers on the upper end of the range of natural ECM collagen. Phase separation, on the other hand, has generated nanofibers in the same range as natural ECM collagen and allows for the design of macropore structures. Specifically designed amphiphilic peptides that contain a carbon alkyl tail and several other functional peptide regions have been synthesized and shown to form nanofibers through a self-assembly process by mixing cell suspensions in media with dilute aqueous solutions of the peptide amphiphil (PA) [2,3]. The challenges with the techniques mentioned above are that electrospinning is typically limited to forming sheets of fibers and thus limiting the ability to create a designed three-dimensional (3D) pore scaffold, and self-assembling materials usually form hydrogels, limiting the geometric complexity and mechanical properties of the 3D structure. Another class of nanomaterials includes carbon nanotubes (CNTs), which are a macromolecular form of carbon with high potential for biological applications due in part to their unique mechanical, physical, and chemical properties [4,5]. CNTs are strong, flexible, conduct electrical current [6], and can be functionalized with different molecules [7], properties that may be useful in basic and applied biological research (for review see [8]). Single-walled carbon nanotubes (SWNTs) have an average diameter of 1.5 nm, and their length varies from several hundred nanometers to several micrometers. Multiwalled carbon nanotube (MWNT) diameters typically range between 10 and 30 nm. The diameters of SWNTs are close to the size of the triple helix collagen fibers, which makes them ideal candidates for substrates for bone growth. As prepared CNTs are insoluble in most solvents, chemical modifications are aimed at increasing their solubility in water and organic solvents.

In this chapter, we aim to offer a basic understanding of this emerging field of SC nanoengineering based on the fusion of SCs, tissue engineering, and nanotechnology.

1.2 STEM CELLS AND TYPES

Although most cells of the body, such as heart cells or skin cells, are committed to conduct a specific function, a SC is an uncommitted cell that has the ability to self-renew and differentiate into a functional cell type [9–11]. Conventionally, SCs are classified as those derived either from embryo or adult tissues (Figure 1.1). Embryonic SCs, embryonic carcinoma cells, and embryonic germ cells are derived from the inner cell mass of blastocysts, teratocarcinoms, and primordial germ cells, respectively. These cells are pluripotent, because they have the ability to entirely colonize an organism and give rise to almost all cell types, except extracellular tissues (e.g., placenta). SCs found in adult organisms are referred to as adult SCs, and are present in most, if not all, adult organs [12]. They are considered multipotent, because they can originate mature cell types of one or more lineages but cannot reconstitute the organism as a whole. What determines SC potency is dependent to a large extent on the genetic makeup of the cell and whether it contains the appropriate genetic circuitry to differentiate to a specific cell type. However, the decision to differentiate or self-renew is often regulated by the SC microenvironment, also known as the SC niche. For example, changes in cytokine gradients, cell–cell, and cell–matrix contacts are important in switching "on" and "off" genes and gene pathways, thereby controlling the type of cell generated.

1.2.1 EMBRYONIC STEM CELLS

Embryonic stem cells (ESCs) from mice were first derived in 1981 from the inner cell mass (ICM) of developing mouse blastocysts [13,14]. Human ESCs were established by Thomson and

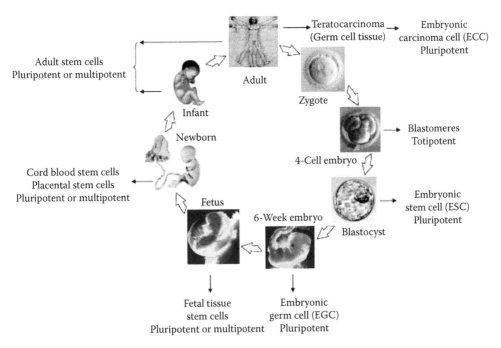

FIGURE 1.1 Origin of different stem cells. Stem cells at different developmental stages appear to have different capacities for self-renewal and differentiation.

colleagues in 1998 [15]. ESCs can be stably propagated indefinitely and maintain a normal karyotype without undergoing cell senescence *in vitro* when cultured in the presence of leukemia inhibitory factor (LIF) (in the case of the mouse) or over a layer of mitotically inactivated mouse embryonic fibroblasts (MEFs), in the monkey and human systems (Figure 1.2). Upon injection of mouse ESCs into blastocysts [16], their progeny is present in all tissues and organs, including the germ line of a chimeric individual (not shown in human ESC due to ethics) and can contribute in the formation of functional gametes [17]. The transmission of genetically manipulated ESCs *in vitro* can thus be passed into chimeric murine offspring and provide a useful approach for studying varying genetic aspects related to ESCs. Homologous recombination has become a useful transgenic approach for introducing selected mutations into the mouse germ line [16,18]. These mutant mice are useful animal models for studying gene function *in vivo* and for clarifying the roles of specific genes in all aspects of mammalian development, metabolic pathways, and immunologic functions.

Upon removal of ESCs from feeder layers and subsequent transfer to suspension cultures, ESCs begin to differentiate into 3D, multicellular aggregates, forming both differentiated and undifferentiated cells, termed embryoid bodies (EBs). Initiation of differentiation may also be induced following the addition of cells into two-dimensional (2D) cultures (i.e., on a differentiation inducing layer such as a matrix or feeder cells). EBs can spontaneously differentiate into different cells and the type of voluntary cells increased by addition of inducing substances or growth factors in their medium, including rhythmically contracting cardiomyocytes, pigmented and nonpigmented epithelial cells, neural cells with outgrowths of axons and dendrites, and mesenchymal cells (Figure 1.2) [19]. Recent studies have also demonstrated ESC differentiation into germ cells and more mature gametes, although significant unanswered questions remain about the functionality of these cells [20]. The derivation of germ cells from ESCs *in vitro* provides an invaluable assay both for the genetic dissection of germ cell development and for epigenetic reprogramming, and may one day facilitate nuclear transfer technology and infertility treatments.

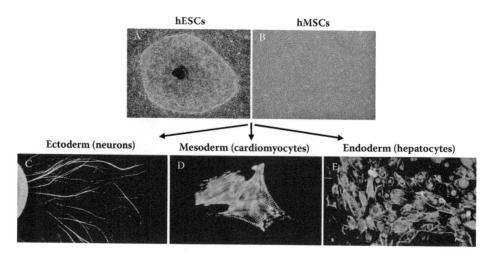

FIGURE 1.2 Morphology and derivatives of embryonic and adult stem cells. Phase-contrast microscopy of (A) a human embryonic stem cell (hESC) (Royan H5) colony cultured on mouse embryonic fibroblast feeder cells (see Baharvand, H., et al., *Dev. Growth. Differ.* 48, 117–128, 2006). (B) Human bone marrow mesenchymal stem cells (hMSCs). Immunocytochemistry of differentiated ESCs with (C) antineuron-specific tubulin III, (D) antialpha actinin, and (E) anticytokeratin 18. (**See color insert following page 112.**)

1.2.2 ADULT STEM CELLS

The ability of adult tissue such as skin, hemopoietic system, bone, and liver to repair or renew indicates the presence of stem or progenitor cells. The use of autologous or allogeneic cells taken from adult patients might provide a less difficult route to regenerative-cell therapies. In adult soma, SCs generally have been thought of as tissue specific and able to be lineage restricted and therefore only able to differentiate into cell types of the tissue of origin. However, several recent studies suggest that these cells might be able to break the barriers of germ layer commitment and differentiate *in vitro* and *in vivo* into cells of different tissues. For example, when bone marrow is extracted and the cells are placed in a plastic dish, the populations of cells that float are blood-forming SCs (hemopoietic SCs [HSCs]), and those that adhere are referred to as stromal cells [21], including mesenchymal stem or progenitor cells (MSCs, Figure 1.2) [22]. These cells can replicate as undifferentiated forms and have the potential to differentiate to lineages of mesodermal tissues, including bone, cartilage, fat, and muscle [23,24]. Moreover, transplanted bone marrow cells contribute to endothelium [25] and skeletal muscle myoblasts [26] and acquire properties of hepatic and biliary duct cells [27], lung, gut, and skin epithelia [28] as well as neuroectodermal cells [29]. Recently, bone marrow was shown as a potential source of germ cells that could sustain oocyte production in adulthood [30]. Furthermore, neural SCs (NSCs) may repopulate the hematopoietic system [31], and muscle cells may differentiate into hematopoietic cells [32].

Jiang and coworkers recently demonstrated a rare multipotent adult progenitor cell (MAPC) within MSC cultures from rodent bone marrow [33,34]. This cell type differentiates not only into mesenchymal lineage cells but also into endothelium and endoderm. Mouse MAPCs injected in the blastocyst contribute to most, if not all, somatic cell lineages including brain [33]. Furthermore, mouse MAPCs can also be induced to differentiate *in vitro* using a coculture system with astrocytes into cells with biochemical, anatomical, and electrophysiological characteristics of neuronal cells [35].

Umbilical cord blood (UCB) is a source of a population of pluripotent, mesenchymal-like SCs [36] and HSCs for transplantation. There are several reports of MSCs or somatic SCs with pluripotent differentiation potential from various sites in the umbilical cord [36–38]. For example, Buzanska and colleagues [39] reported recently that human UCB-derived cells attain neuronal and glial features *in vitro*. Thus, this tissue is a rich source of SCs that may be useful for a variety of therapeutic purposes. This has led to the establishment of cord blood banks and the increased use of UCB for transplantation [40,41].

1.2.3 Differentiation of Stem Cells In Vitro

SCs are a useful tool for investigating methods relating to the extraction of specific cell types from mixed cell populations or heterogeneous teratomas and to perhaps study the differentiation events of precursor cells toward a particular cell lineage. Feasible methods that may help to achieve these include the addition of specific combinations of growth factors or chemical morphogens; changes in physical and geometrical properties of the microenvironment; coculture or transplantation of SCs with inducer tissues or cells; implantation of SCs into specific organs or regions; and overexpression of transcription factors associated with the development of specific tissues. However, to date, these strategies have not yielded a 100% pure population of mature progeny. Therefore, efficient protocols for purifying cell populations are required. Methods such as fluorescence-activated cell sorting (FACS) or magnetic-activated cell sorting (MACS) allow such purification but are dependent on the cell type of interest expressing a surface marker that can be recognized by a fluorescent or magnetic microbead-tagged antibody, and to be fully effective, the marker needs to be cell-type specific. In most cases, these cell markers are not commercially available. Thus, sorting methods are reliant on genetic modification of the SCs, especially the ESCs, by tagging a lineage-specific promoter to a fluorescent marker. Alternatively, cells could be transduced with a drug-resistance gene instead of a marker to allow for preferential selection of subpopulations [19].

1.3 BEHAVIOR OF CELLS ON NANOBIOMATERIALS

Studies of the interactions between substrate topography and cells have encompassed a wide variety of cell types and substratum features, including grooves, ridges, steps, pores, wells, nodes, and adsorbed protein fibers. Grooves are the most common feature type employed in the study of the effects of surface structure on cells [42–45]. Typically, the grooves are arrayed in regular, repeating patterns, often with equal groove and ridge width. The cross sections of the groove are often of the square wave, V-shape, or truncated V-shape [46]. In general, investigations of grooved surfaces have revealed that the cells aligned to the long axis of the grooves [44,47] often with organization of actin and other cytoskeletal elements in an orientation parallel to the grooves [48,49]. The organization of cytoskeletal elements was observed to occur in some cases with actin and microtubules aligned along walls and edges [48,50]. Many studies have found that the depth of the grooves was more important than their width in determining cell orientation [51], because the orientation often increased with increasing depth but decreased with increasing groove width. Repeat spacing also played a role, with orientation decreasing at higher repeat spacing [52]. There are some studies investigating the behavior of cells on other synthetic features. Green and coworkers found that nodes of 2 μm and 5 μm resulted in increased cell proliferation compared to 10 μm nodes and smooth surfaces [53]. Campbell and von Recum found that pore size played a larger role than material hydrophobicity in determining tissue response [54]. The behavior of cells on sandblasted surfaces has been studied, although the observed trends seem less clear than those on controlled morphologies, such as grooves. In general, adhesion, migration, and ECM production were greater on rougher surfaces than on surfaces sandblasted with larger grain sizes [55,56]. Studies have also been performed in which protein patterns were used to guide cues for several cell types, including neural cells [57,58]. Isolated tracks were found to provide stronger guidance than repeated tracks [57]. Goodman et al. used polymer casting to replicate the topographical features of the ECM [59] and observed endothelial cells cultured on the ECM textured replicas spread faster and had appearance more like cells in their native arteries than did cells grown on untextured surfaces [59,60]. Advancements in electron beam lithography technology have allowed engineers to fabricate well-defined nanostructures down to a possible lateral feature size of 12 nm [61]. The ability to fabricate these nanofeatures has enabled biologists to look at the effects of such features (which are of a similar size to those that surround a cell, for example, the 66 nm repeat of collagen) on SCs.

When a cell interacts with a biomaterial, it senses the surface topography and will respond accordingly. If a suitable site for adhesion is detected, focal adhesions and actin stress fibers are formed; later, microtubules are recruited, which stabilize the contact [61]. Recently, it was reported that regular nanotopography significantly reduces cell adhesion [62]. Gallagher et al. cultured fibroblasts onto nanopatterned ε-PCL (polycaprolactone) surfaces and showed that cell spreading is reduced compared with that on a planar substrate. Furthermore, cytoskeletal organization is disrupted as indicated by a marked decrease in the number and size of focal contacts [63]. Focal adhesion contacts are of great importance in signal transductive pathways [64]. The signal transductive events originating from focal contacts can affect the long-term cell differentiation in response to materials [61,65].

For scaffolds, it is generally agreed that a highly porous microstructure with interconnected pores and a large surface area is conducive to tissue ingrowth. For bone regeneration, pore sizes between 100 μm and 350 μm and porosities of more than 90% are preferred. For example, rat MSCs that were cultured on highly porous electrospun (PCL) nanofiber scaffolds migrated more rapidly and differentiated into osteoblasts in rotating bioreactors [66]. It is also believed that small fiber diameter and the overall porous structure aid in the adhesion and migration of cells into the scaffold [66].

Nanofibrous scaffolds formed by electrospinning are highly porous, have a high surface-area-to-volume ratio, and have morphological properties that are similar to collagen fibrils [67]. These physical characteristics promote favorable biological responses of seeded cells within these scaffolds, including enhanced cell attachment, proliferation, and maintenance of the chondrocytic phenotype [68,69].

Nanoparticles can also affect ECM properties and cell behavior. For example, carboxylated SWNT can be incorporated into type I collagen scaffolds. Living smooth muscle cells were also incorporated at the time of collagen gelation to produce cell-seeded collagen–CNT composite matrices. These cell-seeded collagen matrices can be further aligned through constrained gelation and compaction, as well as through the application of external mechanical strain [70].

1.4   EXTRACELLULAR MATRIX ENHANCEMENT

Tissues are complex and are typically organized into a well-defined, 3D structure in our bodies. This architecture contributes significantly to the biological functions in the tissues. Furthermore, it provides oxygen and nutrient support and spatial environment for the cells to grow [71]. In this respect, there are three key factors to be considered for the success of tissue regeneration: cells, scaffold, and cell–matrix (scaffold) interaction. The scaffold plays a pivotal role in accommodating the cells. An ideal scaffold for tissue engineering application should mimic the natural microenvironment of the natural tissue and present the appropriate biochemical and topographical cues in a spatially controlled manner for cell proliferation and differentiation. When a cell comes into contact with biomaterial, it will perceive the chemistry of a surface using integrin transmembrane proteins to find suitable sites for adhesion, growth, and maturation. *In vitro*, cells will readily translocate on the material surface to the sites of preferential attachment, and cells will produce distinct morphologies when motile and when adhered and entering the S-phase [72].

Tissue structure and function depend greatly on the arrangement of cellular and noncellular components at the micro- and nanoscale levels—featuring a higher specific surface and thus a higher interface area—in ECM [73]. In addition to providing a physical support for cells, the native ECM also provides a substrate with specific ligands for cell adhesion and migration, and regulates cellular proliferation and function by providing various growth factors. A well-known feature of native ECM is the nanoscaled dimensions of its physical structure. In a typical connective tissue, structural protein fibers such as collagen and elastin fibers, have diameters that range from several tens to several hundred nanometers. The nanoscaled protein fibers entangle with each other to form a nonwoven mesh that provides tensile strength and elasticity, and laminin, which provides a specific binding site for cell adhesion, also exists as nanoscaled fibers in the ECM. The scaffold should therefore mimic the structure and biological function of native ECM as much as possible, both in terms of chemical composition and physical structure. It is reasonable to expect that an ECM-mimicking tissue-engineered scaffold will play a similar role to promote tissue regeneration *in vitro* as native ECM does *in vivo*. Accordingly, the design of nanofeatured tissue scaffolds is novel and exciting, opening a new area in tissue engineering. Work with ECM components has demonstrated that the physical presentation of these molecules affects morphology, proliferation, and morphogenesis of differentiated cells [74–76]. Culture on or within 3D as opposed to 2D arrays of matrix molecules promotes cellular phenotypes that display more *in vivo*-like structure and function [74,77,78].

These observations were made in the absence of added ECM, suggesting that the geometry of the ECM can influence cellular phenotype and function even in the absence of chemistry. The understanding of how the microenvironment can influence the cell behavior will aid the development of the next generation of scaffolds for tissue engineering and SC applications.

1.4.1 Proliferation of Stem Cells

To provide a more topologically accurate and reproducible representation of the geometry and porosity of the ECM/basement membrane for SC culture [79], Ultra-Web™ (Corning, New York), a commercially available 3D nanofibrillar and nanoporous matrix produced by depositing electrospun nanofibers composed of polyamide onto the surface of glass or plastic coverslips, was used. Within these scaffolds, mouse ESCs had enhanced proliferation with self-renewal in comparison with tissue culture surfaces independent of soluble factors such as LIF. Significantly, cells did not adhere to 2D films composed of polyamide, demonstrating the importance of the nanofibrillar geometry for SC proliferation. It is important to note that these proliferation measurements were performed in the presence of less than 5% of the original feeder MEFs, which remained during passage. Because MEFs normally provide cues that promote SC proliferation, these results suggest that the 3D nanofibrillar surfaces can compensate, at least in part, for the absence of MEFs, but standard tissue culture surfaces cannot perform the same synergistic or replacement function.

This was the first report in which ESCs were cultured on a defined synthetic 3D nanofibrillar surface that resembles the geometry of the basement membrane and in which a relationship is demonstrated between the 3D nanotopology, proliferation with self-renewal, upregulation of Nanog, a homeoprotein shown to be required for maintenance of pluripotency [80], the activation of the small GTPase Rac, and the activation of the phosphoinositide 3-kinase (PI3K)/AKT, components of the PI3K signaling pathway. SCs cultured on the 3D nanofibrillar surface maintained their ability to differentiate in the presence of differentiating factors such as retinoic acid. Because nanofibers influence cellular parameters such as cell shape, actin cytoskeleton, and fibronectin deposition [77,78], it is possible that they influence SCs both directly and indirectly by altering the phenotype of the feeder cells. Moreover, Ultra-Web was used to culture NIH-3T3 fibroblasts and normal rat kidney cells and observed dramatic changes in cellular morphology [77,78]. These observations more closely resembled their *in vivo* counterparts [77,78]. SCs cultured on the 3D nanofibrillar surface maintained their ability to differentiate in the presence of differentiating factors such as retinoic acid. Because nanofibers influence cellular parameters such as cell shape, actin cytoskeleton, and fibronectin deposition [77,78], it is possible that they influence SCs both directly and indirectly by altering the phenotype of the feeder cells. Moreover, Ultra-Web was used to culture NIH-3T3 fibroblasts and normal rat kidney cells, and dramatic changes were observed in cellular morphology that more closely resembled their *in vivo* counterparts [77,78].

Recent experiments using MSCs demonstrated an important role for mechanical cues in regulating SC fate [81]. Li et al. [68] were the first to examine the ability of an electrospun nanofibrous scaffold to support MSC proliferation. Jin et al. also reported that the nanofibrous scaffold supported MSC adhesion and proliferation [82]. Recently, Kommireddy et al. [83] reported that the proliferation, spreading, and attachment of mouse MSCs increased after deposition of layer-by-layer (LbL) assembled titanium dioxide (TiO_2) nanoparticle thin films and a higher number of cells attached on increasing numbers of layers of TiO_2 nanoparticle thin films. The spreading of cells was found to be faster on surfaces with an increasing number of layers of TiO_2 nanoparticles. Therefore, one topic for future work is directed toward promoting osteogenesis and chondrogenesis of MSC on TiO_2 nanoparticle thin-film surfaces for possible applications for soft and hard tissue repair and reconstruction using the LbL nanoassembly technique. In a related study, TiO_2 thin films were shown to be the optimal surface for the faster attachment and spreading of cells compared with other kinds of nanoparticle thin films [84]. Investigations demonstrating that mechanical signals are transduced to the cell cytoskeleton through the activation of Rho, a small GTPase, and Rho kinase [81] has provided important evidence that SC fate may not only depend on soluble factors but may require attachment to 3D ECM/bone marrow surfaces whose physical, mechanical, and chemical properties can directly or indirectly regulate the pathways controlling SC fate [85–88]. Recently, Engler et al. reported that microenvironments appear important in SC lineage specification but can be difficult to adequately characterize or control in soft tissues [89]. Naive MSCs are shown here to specify lineage and commit to phenotypes with extreme sensitivity to tissue-level elasticity. Soft matrices that mimic brain are neurogenic, stiffer matrices that mimic muscle are myogenic, and comparatively rigid matrices that mimic collagenous bone prove osteogenic. During the initial week in culture, reprogramming of these lineages is possible with the addition of soluble induction factors, but after several weeks in culture, the cells commit to the lineage specified by matrix elasticity, consistent with the elasticity-insensitive commitment of differentiated cell types. Inhibition of nonmuscle myosin II blocks all elasticity-directed lineage specification without strongly perturbing many other aspects of cell function and shape. The results have significant implications for understanding physical effects of the *in vivo* microenvironment and also for the therapeutic uses of SCs.

A growing body of evidence also suggests the importance of surface chemistry as well as topographical features on the rate of HSC proliferation and $CD34^+$ cell expansion [90–94]. For example, Laluppa et al. [90] have shown that the type of substrate used in culture, ranging from polymers (polystyrene, polysulfone, polytetrafluoroethylene, cellulose acetate) to metals (titanium, stainless-steel) and glasses, can significantly affect the outcome of *ex vivo* expansion of HSCs [90]. Li et al. have shown that culture in 3D nonwoven polyester matrices enhanced cell–cell and cell–matrix interactions and expansion of stromal and hematopoietic cells [91]. Recently, it was reported that covalent surface immobilization of ECM proteins such as fibronectin [92] and adhesion peptides (CS-1 and arginine-glycine-aspartic acid [RGD]) [94] mediate HSC adhesion to the substrate and increase HSC expansion. RGD is the best-known peptide sequence for prompting cell adhesion on synthetic material surfaces. The RGD sequence is a cell recognition motif found in many ECM proteins, such as collagen, laminin, fibrinogen, and vitronectin, that bind to integrin receptors [67]. These results suggest that biochemical as well as topographical cues could be actively involved in dictating the proliferation and differentiation of cultured HSCs. Recently, Chua et al. examined human umbilical cord HSC expansion on surface-functionalized polyethersulfone (PES) nanofiber meshes and PES films [95]. Among the carboxylated, hydroxylated, and aminated PES substrates and tissue culture polystyrene surface (TCPS), aminated PES substrates mediated the highest expansion efficiency of $CD34^+ CD45^+$ cells (3.5 times) and colony-forming unit (CFU) potential. Aminated nanofiber mesh could further enhance the HSC-substrate adhesion and expansion of multilineage colonies (CFU–granulocyte erythroid mieloid [GEMM]) forming progenitor cells. This study demonstrates the importance of culture substrate in influencing the proliferation and differentiation of HSCs. One possible mechanism for the observed effects is that the aminated substrate, being positively charged, could selectively enrich certain protein components from the medium, which then contribute to the expansion outcome [96,97]. These studies have shown that the functional presentation of adsorbed fibronectin was different on hydroxylated, methylated, aminated, and carboxylated surfaces, which consequently led to variations in cell adhesion and differentiation [96]. It is possible that aminated

PES surface mediated HSC proliferation by a similar mechanism by binding critical cytokines and growth factors from the medium, and presenting them in a more effective immobilized form, thereby mimicking a salient feature of the bone marrow HSC niche [98,99]. Another possible mechanism by which aminated surface enhanced HSC expansion and CD34+ phenotype maintenance is by direct interaction with the HSCs through CD34 receptors. CD34 is a highly sialylated and negatively charged glycophosphoprotein, and its expression decreases as HSCs become differentiated [100,101]. Chua et al. therefore postulate that a positively charged "ligand"—in this case, the surface-bound amine groups—could bind and engage the negatively charged CD34 antigen, and the engagement of CD34 antigens on HSCs might activate downstream signaling pathways that subsequently influence fate choices upon proliferation [95,102]. It was also shown that stimulation of undifferentiated hematopoietic (myeloblastic leukemia cell line) KG1a cells with anti-CD34 antibody induces homotypic cytoadhesion [102]. Binding of aggregating antibody to CD34 antigens induced tyrosine phosphorylation, cell polarization and adhesion, and perhaps cell motility. Interestingly, being colocalized with F-actin, the cross-linked CD34 "cap" is quite stable and persists on the cell surface for at least 2 days after stimulation, whereas many other cell-surface molecules are rapidly internalized for degradation or recycling, upon stimulation. It is possible that the aminated PES surface serves the same role by engaging cell surface CD34 antigen. Moreover, it is generally accepted that the native bone marrow microenvironment provides a complex 3D meshwork of ECM that serves as a SC niche to regulate hematopoietic stem/progenitor cell functions such as self-renewal, proliferation, fate choice, and homing [98,99]. It has also been demonstrated that integrin–ECM interactions within a specific tissue niche play a critical role in the self-renewal of NSCs [103].

Nanotechnology can be used to design intelligent bioreactors to direct the SC behavior with high efficiency for regenerative medicine. The ability of bioreactors to support the development of multicellular, multilayer, 3D tissue-like constructs has been attributed to the low-shear environment, randomization of the gravity vector, and colocalization of cells, cell aggregate, and microcarrier beads within the bioreactor [104]. Future bioreactors therefore will incorporate appropriate fluid physics, nanobiomaterials, nanogeometry of environment, and automated control systems into its tissue engineering program with the goal of generating novel bioreactor designs with enhanced transport and appropriate fluid shear properties to support the growth of complex tissue constructs.

1.4.2 DIFFERENTIATION INTO OSTEOBLASTS

Bioengineered scaffold materials for the growth of bone have been the focus of intense research. Ideally, a suitable scaffold material for bone formation should provide the adequate microenvironment for proliferation and differentiation of bone cells as well as formation of a mineralized matrix, therefore acting as a template. In the bioengineered bone, bone formation and remodeling is a dynamic process that involves production and resorption within the scaffold material. It has been demonstrated that osteoblast adhesion, proliferation, alkaline phosphatase (ALP) activity, and ECM secretion on carbon nanofibers increased with decreasing fiber diameter in the range of 60 to 200 nm, whereas the adhesion of other kinds of cells such as chondrocytes, fibroblasts, and smooth muscle cells was not influenced [105,106]. It has been suggested that the nanostructured surfaces affect the conformation of adsorbed adhesion proteins such as vitronectin and thus affect the subsequent cell behavior [107]. In addition, the nanoscaled dimensions of cell membrane receptors such as integrins should be considered. Yoshimoto et al. [66] reported osteogenic differentiation of MSCs in the nanofibrous scaffold, and Li et al. [67] described the support of chondrogenic differentiation of MSCs in the nanofibrous scaffold. Similarly, it was shown that the level of MSC chondrogenesis in the nanofibrous scaffolds was equivalent to, or higher than, that seen in high-density pellet cultures of MSCs [87]. Nanofibrous scaffolds characteristically have a high surface-to-volume ratio, thus providing a favorable 3D, porous space to accommodate MSCs at a high density, similar to that in high-density cell pellet cultures, and promote cellular condensation required for chondrogenesis. Microscopic images of the nanofibrous scaffolds cultures revealed the formation of a thick, cartilage-like tissue upon transforming growth factor-β1 (TGF-β1) treatment, also seen in similarly treated cell pellet cultures. Interestingly, this engineered cartilage produced within the nanofibrous scaffolds displays zonal morphologies (that is, a superficial zone at the surface of the culture consisting of flat cells, a middle layer of round cells, and the bottom zone containing small flat cells surrounded by abundant ECM). This zonal architecture is thus reminiscent of the superficial, middle, and deep zones of native articular cartilage [67].

Recently, Hosseinkhani et al. investigated the proliferation and differentiation of MSC in a 3D network of nanofibers formed by self-assembly of peptide-amphiphile (PA) molecules [108]. PA molecules—that is, peptide-based molecules with a hydrophobic tail and hydrophilic head group—were designed to self-assemble into a network of nanofiber scaffolds only when present in physiological ionic conditions [2,109,110]. The surface of the nanofibers, made up of the hydrophilic head groups of aligned PA molecules, consisted of physiologically active peptide sequences designed to engage in cell signaling by acting as ligands for cell surface receptors. The PA molecules existed in a solution of water until they encountered physiological concentrations of cations such as calcium, which triggered their self-assembly into nanofibers that "held" the water molecules in place, macroscopically forming a gel-like substrate. PA was synthesized by standard solid-phase chemistry that ends with the alkylation of the NH2 terminus of the peptide. The sequence of RGD was included in peptide design as well. A 3D network of nanofibers was formed by mixing cell suspensions in media with dilute aqueous solution of PA. When rat MSCs were seeded into the PA nanofibers with or without RGD, a larger number of cells attached to the PA nanofibers that contained RGD. In addition, the osteogenic differentiation of MSC as measured by ALP activity and osteocalcin content was higher for the PA nanofibers that contained RGD. In another study, a combination of MSC-seeded hybrid scaffold and the perfusion method was used to enhance *in vitro* osteogenic differentiation of MSCs and *in vivo* ectopic bone formation. The hybrid scaffold consists of two biomaterials: a hydrogel formed through self-assembly of PA with cell suspensions in media, and a collagen sponge reinforced with poly-(glycolic acid) (PGA) fiber incorporation.

In an interesting experiment, Woo et al. compared osteoblastic properties for osteoblasts seeded on nanofibrous 3D scaffolds with those on solid-walled 3D scaffolds, both of which were made from the same material (poly-L-lactic acid [PLLA]) and had the same macropore structures (interconnected spherical pores) [111]. The most notable differences between the nanofibrous and the solid-walled scaffolds were in the extent of mineralization and in changes of BSP gene expression, a marker for the osteoblast phenotype. Collagen fibers are known to mediate mineralization [112], and bone sialoprotein (BSP) is considered to promote mineralization *in vitro* and *in vivo* [113,114]. Similar to collagen fibers, nanofibers of the PLLA scaffolds were associated with mineralization. Consistent with this finding was a much higher level of BSP expression in cells cultured on nanofibrous scaffolds versus solid-walled scaffolds. These findings suggest that synthetic nanofibers act, at least to certain degree, in a manner similar to natural collagen fibers, and that nanofibrous scaffolds hold promise for bone tissue engineering. Moreover, CNTs may be used not only to stimulate bone regeneration but also to serve a permanent mechanical role. Zhao et al. reported the mineralization of chemically functionalized SWNTs, a class of nanomaterials with hydroxyapatite (HA) as a scaffold for bone growth [115]. With the purpose of finding the type of CNT that best supports bone formation, Zanello et al. [116] cultured rat osteosarcoma ROS 17/2.8 cells on SWNTs and MWNTs, with and without the introduction of chemical modifications. They found CNTs carrying neutral electric charge sustained the highest cell growth and production of plate-shaped crystals. There was a dramatic change in cell morphology in osteoblasts cultured on MWNTs, which correlated with changes in plasma membrane functions. Application of CNTs to bone therapy may lead to the development of new bone graft materials and techniques.

Supronowicz et al. demonstrated that novel nanocomposites consisting of blends of polylactic acid and carbon nanotubes can effectively be used to expose cells to electrical stimulation [117]. When osteoblasts cultured on the surfaces of these nanocomposites were exposed to electric stimulation, there was a 46% increase in cell proliferation after 2 days, a 307% increase in the concentration of extracellular calcium after 21 days, and an upregulation of mRNA expression for collagen type 1. These results provide evidence that electrical stimulation delivered through novel, current-conducting polymer/nanophase composites promotes osteoblast functions that are responsible for the chemical composition of the organic and inorganic phases of bone. Moreover, collagen–CNT composite materials sustained high smooth muscle cell viability [70].

1.4.3 DIFFERENTIATION INTO NEURONS

When human MSCs were cultured on nanopatterns, the MSCs, as well as their nuclei, aligned along the direction of the patterns. Gene profiling with reverse transcriptase-polymerase chain reaction (RT-PCR), microarray analysis, and real-time polymerase chain reaction (PCR) showed significant upregulation of neuronal markers compared to nonpatterned controls [118]. The combination of nanotopography and biochemical cues such as retinoic acid created a synergistic condition for the neuronal differentiation, but nanotopography showed a stronger effect compared to retinoic acid alone on nonpatterned surfaces when human ESCs were cultured on the nanogratings [118]. With the aid of nanotechnology, Yang et al. explored the possibilities of the fabrication of a 3D PLLA nanostructured porous scaffold by a liquid–liquid phase separation method [119]. The *in vitro* cell culture studies showed that the 3D porous scaffold seeded with NSCs plays a significant role in nerve tissue engineering as evidenced by the support of NSC differentiation and neurite outgrowth. These findings provide the feasibility of using nanostructured scaffold in nerve tissue engineering for better cell adhesion and differentiation *in vitro*. The suitability of fabricating aligned PLLA nano-/microfibrous scaffolds for the NSC culture was studied in terms of their fiber alignment and dimension [120]. It was found that the NSCs elongated, and their neurites grew along the fiber direction for the aligned scaffolds, whereas the fiber diameter did not show any significant effect on the cell orientation. On the contrary, the NSC differentiation rate was higher for nanofibers than for microfibers but was independent with respect to the fiber alignment. As expected, the aligned nanofibers highly supported the NSC culture and improved the neurite outgrowth. Based on these experimental results, this study suggested that the aligned PLLA nanofibrous scaffold may be of great value as a cell carrier in neural tissue engineering.

On the other hand, neural progenitor cells or neural retinal cells are able to be encapsulated in PA gels by mixing cell culture suspensions with PA solutions, trapping the cells in the interior of the gels [121]. Because the functional peptide sequence formed the outer surface of the nanofibers, it allowed the functional signaling of encapsulated cells in 3D. Encapsulated neural progenitor cells differentiated faster and more robustly into mature neuronal phenotypes compared with controls. By days 1 and 7 *in vitro*, 30% and 50%, respectively, of the neural progenitor cells expressed the mature neuron marker β-tubulin III [122]. Moreover, it was also reported that few astrocytes were found in these cultures (less than 1% and 5% at 1 and 7 days *in vitro*, respectively) [122] despite the multipotent nature of the progenitor cells. This suggests novel approaches for limiting the effects of reactive gliosis and glial scarring after traumatic or degenerative events by transplanting donor cells *in vivo* with the peptide amphiphile substrates. Similar results were observed with encapsulated retinal cells. The ultimate vision is for the peptide amphiphile solutions to be (minimally invasively) injected stereotaxically in combination with donor cells, have the nanofiber network form in situ and *in vivo*, and then provide functional cellular signaling to both donor and host cells while at the same time limiting the effects of reactive gliosis. This nanofiber system is currently being explored for spinal cord injury, stroke, and degenerative retinal disorders, including age-related macular degeneration [121]. Moreover, chemically functionalized CNTs have been used successfully as potential devices improve neural signal transfer while supporting dendrite elongation and cell adhesion [123–126]. The results strongly suggest that the growth of neuronal circuits on a CNT grid is accompanied by a significant increase in network activity. The increase in the efficacy of neural signal transmission may be related to the specific properties of CNT materials, such as the high electrical conductivity.

One of the limitations of cell culture including SC culture is the presence of chemical species such as superoxide (O2-), hydroxyl (·OH), peroxynitrite (ONOO-), and peroxide (H_2O_2) that produce a host of oxidative-mediated deleterious changes in cells, including DNA fragmentation, peroxidation of cell membrane lipids, decreased mitochondrial energy production, and transporter protein inactivation [127,128]. A potential approach to remedy this problem is the use of carbon-60 fullerene-based neuroprotective compounds [127,129–131]. Fullerenes are molecules composed of large 3D arrays of evenly spaced carbon atoms, similar to the pattern produced by the rhombuses on a soccer ball. Fullerenols are hydroxyl (i.e., OH) functionalized fullerene derivatives and have been shown to possess antioxidant and free-radical scavenger properties that are able to reduce induced excitotoxic and apoptotic cell death [127,129–131]. The mechanism of fullerenol-mediated neuroprotection is due at least in part to inhibition of glutamate channels, because neither GABA (A) nor taurine receptors were affected. In addition, fullerenols appear to lower glutamate-induced increases of intracellular calcium concentrations, a critical mechanism of excitotoxicity in neurons [127,129–131].

1.5 STEM CELL LABELING AND TRACKING

During the last few years, SC biology has experienced a rapid advancement, leading to rising hopes that SCs, migrating toward the lesion target area, will contribute to functional improvement. There is therefore a need to develop noninvasive methods for visualizing transplanted SCs [132,133] for a better understanding of their migrational dynamics and differentiation processes and of their regeneration potential [134]. Single-cell sensitivity is especially important in a new field such as that of SCs, because the pattern of migration of SCs, even after local injection, is unknown, and there is a distinct possibility that single SCs scattered diffusely throughout the body might be effective therapeutics for certain disease states. Because no single contrast agent/detector pair will satisfy all needs of SC clinical trials, dual- and multimodality contrast agents, which combine the best features of each technology, have been developed.

An ideal imaging technology for SC tracking during clinical trials has the following characteristics: biocompatible, safe, and nontoxic; no genetic modification or perturbation to the SC; single-cell detection at any anatomic location; quantification of cell number; minimal or no dilution with cell division; minimal or no transfer of contrast agent to nonstem cells; noninvasive imaging in the living subject over months to years; and no requirement for injectable contrast agent. At present, no imaging technology fulfills the criteria presented above [135]. Examination of the literature thus far shows nanoparticles are feasible as imaging contrast agents for optical and magnetic resonance.

1.5.1 MAGNETIC RESONANCE IMAGING (MRI) CONTRAST AGENTS

Given its extraordinary 3D capabilities and high safety profile, MRI is the imaging modality used by most research studies to track SCs *in vivo* [133,135]. Magnetic resonance (MR) molecular imaging contrast agents take many forms, but nanoparticulate systems have emerged as the most successful genre to date. In general, nanoparticulates provide enormous surface area, which can be functionalized to serve as a scaffold for targeting ligands and magnetic labels. In regard to the biophysical and metabolic properties of iron oxide, there is growing interest in using MRI to track cells after they are labeled with iron oxide contrast agents because these loaded cells have long-term viability and growth rate, and apoptotic indexes are unaffected, with the iron payload disappearing after five to eight cell divisions [136]. A wide variety of iron oxide-based nanoparticles have been developed that differ in hydrodynamic particle size and surface coating material (dextran, starch, albumin, silicones) [137]. In general terms, these particles are categorized based on nominal diameter into superparamagnetic iron oxides (SPIOs) (50 to 500 nm) and ultrasmall, superparamagnetic iron oxides (USPIOs) (<50 nm), which dictates their physicochemical and pharmacokinetic properties. Iron oxide particles have been used for imaging the gastrointestinal tract, liver, spleen, and lymph nodes; however, USPIOs are also useful as blood-pool agents for angiography and perfusion imaging in myocardial and neurological disease. Ultrasmall dextran-coated iron oxide particles were developed for uptake into cells to allow MRI tracking. However, dextran-free anionic magnetic nanoparticles are stable in colloidal suspension and are adsorbed through nonspecific electrostatic interactions with the membranes of most cell types, followed by spontaneous cell internalization [138]. This label gives magnetic properties to the entire cell body, properties that potentially allow for the detection of cells. Fluorophore-labeled monocrystalline iron oxide nanocrystals (MIONs) can be used to detect SCs using optical imaging and image single cells in pathological specimens *ex vivo*. Similarly, because of the magnification effect, MIONs can be used to track small numbers of SCs, on the order of thousands, at high field strengths, for up to several weeks [132]. Frank and colleague labeled MSCs and other mammalian cells with SPIO (nanoparticles) MRI contrast agents. They showed magnetic labeling of mammalian cells with use of ferumoxides is possible and may enable cellular MR imaging and tracking in experimental and clinical settings [139,140]. Hoehn et al. implanted 30,000 ESCs labeled with green fluorescent protein (GFP) and with the dextran-coated iron oxide nanoparticles of Sinerem™ (Guerbet, France) stereotactically both into the cortex (next to the corpus callosum) and into the striatum of the normal contralateral hemisphere in the rat with brain stroke. After 3 weeks, they observed cell migration *in vivo* and monitored cell dynamics within individual animals, and for a prolonged time. Cells migrated along the corpus callosum to the ventricular walls, and massively populated the border zone of the damaged brain tissue on the hemisphere opposite to the implantation sites. Their results indicated that ESCs have higher migrational dynamics, targeted to the cerebral lesion area, and this imaging approach is ideally suited for the noninvasive observation of cell migration, engraftment, and morphological differentiation at high spatial and temporal resolution [134].

In another study, implanted rat bone marrow MSCs and mouse ESCs labeled with iron-oxide nanoparticles and human HSCs (CD34+) labeled with magnetic MicroBeads in rats with a cortical or spinal cord lesion were studied [141,142]. Cells were grafted intracerebrally, contralaterally to a cortical photochemical lesion, or injected intravenously. During the first week posttransplantation, transplanted cells migrated to the lesion. About 3% of MSCs and ESCs differentiated into neurons, while no MSCs, but 75% of ESCs differentiated into astrocytes. Labeled MSCs, ESCs, and CD34+ cells were visible in the lesion on MR images as a hypointensive signal, persisting for more than 50 days. In rats with a balloon-induced spinal cord compression lesion, intravenously injected MSCs migrated to the lesion, leading to a hypointensive MRI signal. Their studies demonstrate that grafted adult as well as ESCs labeled with iron-oxide nanoparticles migrate into a lesion site in brain as well as in spinal cord [141,142].

NSCs from an adult rat's spinal cord were loaded with superparamagnetic gold-coated monocrystalline iron oxide nanoparticles (Au-ION) as contrast agent for MRI. A dose-dependent attenuation of MRI signals was observed for Au-MION down to 0.001 µg Fe/µl and for nanoparticle-loaded clusters of only 20 cells. The labeled cells were infused into the spinal cord of anesthetized rats and tracked by MRI at 1 hour, 48 hours, and 1 month postinjection. Histological analysis revealed that MRI signals correlated well with gold-positive staining of transplanted cells. They showed that Au-MION exerts powerful contrast-enhancing properties and may represent novel MRI labels for labeling and tracking the transplanted cells *in vivo* [143,144].

Huang et al. reported a cellular labeling approach with a novel vector composed of mesoporous silica nanoparticles (MSNs) conjugated with fluorescein isothiocyanate (FITC) in human bone marrow MSCs and 3T3-L1 cells. FITC-MSNs were efficiently internalized into MSCs and 3T3-L1 cells even in short-term incubation. The internalization of FITC-MSNs did not affect the cell viability, proliferation, immunophenotype, and differentiation potential of the MSCs, or the 3T3-L1 cells. Finally, FITC-MSNs could escape from endolysosomal vesicles and retained the architectonic integrity after internalization. They concluded that the advantages of biocompatibility, durability, and higher efficiency in internalization make MSNs a better vector for SC tracking than others that are currently used [145].

SCs can also be labeled with antibody-bound magnetic nanoparticles and tracked [146]. To examine the potential of ESCs to differentiate into dopamine neurons and integrate within the host brain, Björklund et al. transplanted ESCs in a Parkinson rat model. For tracking they used the increased relaxivity of the iron-based nanoparticles [147] technique with monocrystalline iron oxide nanocolloid. Using a long-half-life iron oxide, relative cerebral blood volume (CBV) can be estimated easily, because at steady state, changes in local tissue relativity are directly proportional to changes in local tissue blood volume [148]. The use of such agents increases the signal-to-noise ratio such that at low and intermediate magnetic field strength, CBV provides temporal and spatial resolution sufficient to detect slowly evolving activation induced by pharmacological challenges [149].

Rather than nonspecific surface coating, Lewin et al. [150] labeled NSCs and HSCs (CD34$^+$) with short HIV-Tat peptides to derivatize superparamagnetic nanoparticles. The particles are efficiently internalized into hematopoietic and neural progenitor cells in quantities of up to 10 to 30 pg of superparamagnetic iron per cell. Iron incorporation did not affect cell viability, differentiation, or proliferation of CD34$^+$ cells. Following intravenous injection into immunodeficient mice, 4% of magnetically CD34$^+$ cells homed to the bone marrow per gram of tissue, and single cells could be detected by MR imaging in tissue samples. Localization and retrieval of cell populations *in vivo* enabled detailed analysis of specific SC and organ interactions critical for advancing the therapeutic use of SCs [150]. The attachment of TAT and related translocation sequences to dextran cross-linked iron oxide (CLIO) nanoparticles has been improved and shown to increase uptake by over 100-fold. In addition, magnetism-based interaction capture (MAGIC) identifies molecular targets on the basis of induced movement of superparamagnetic nanoparticles inside living cells. Efficient intracellular uptake of superparamagnetic nanoparticles (coated with a small molecule of interest) was mediated by a transducible fusogenic peptide. These nanoprobes captured the small molecule's labeled target protein and were translocated in a direction specified by the magnetic field. MAGIC is also used to monitor signal-dependent modification and multiple interactions of proteins as well as for imaging signaling pathways in SCs [151].

1.5.2 OPTICAL LABELING

Biological probes are indispensable tools for studying biological samples, cells in culture, and animal models. Exogenous probes are frequently multifunctional, having one component that can detect a biological molecule or event, and another component that reports the presence of the probe. Fluorescent dyes are used routinely for determining the presence and location of biological molecules in cultured cells and tissue sections. Concise evaluation, accurate quantification, and precise localization with ultimate sensitivity in the analysis of biomolecules can be achieved when single molecule microscopy is used [152]. Classically, studies of biomolecules at the single molecule level often involve large teams of specialized biophysicists, sophisticated and finely tuned equipment, and long periods of time to record and analyze relatively few events. Moreover, these studies also suffer from the limitations of certain organic dyes, such as photobleaching, instability, and low quantum efficiency [153,154].

Recently, semiconductor particles or quantum dots (QDs) have been developed as a new class of biological fluorophore with easily tunable properties and significant spectral advantages over conventional fluorophores. The QD imaging technique has a higher resolution compared to many other imaging techniques such as fluorescence. QDs for biological environments have opened the doors to an expanding variety of biological applications, such as serving as specific markers for cellular structures and molecules, tracing cell lineage, monitoring physiological events in live cells, measuring cell motility, and tracking cells *in vivo* [155].

QDs are nanocrystals of inorganic semiconductors, typically with a diameter of 2 nm to 8 nm (on the order of 200 to 10,000 atoms). Bulk-phase semiconductors are characterized by valence electrons that can be excited to a higher-energy conduction band. The energy difference between the valence band and the conduction band is the bandgap energy of the semiconductor. An excited electron may then relax to its ground state through the emission of a photon with energy equal to that of the bandgap. When a semiconductor is of nanoscale dimensions, the bandgap is dependent on the size of the nanocrystal. As the size of a semiconductor nanocrystal decreases, the bandgap increases, resulting in shorter wavelengths of light emission [152]. Dots of slightly different sizes fluoresce at different wavelengths. For example, larger dots shine red, and smaller dots shine blue, with a rainbow of colors in between. Researchers can create up to 40,000 different labels by mixing QDs of different colors and intensities as an artist would mix paint. In addition to coming in a vast array of colors, the dots also are brighter and more versatile than more traditional fluorescent dyes. For example, they can be used to visualize individual molecules. Indeed, the unique properties of QDs not only provide biologists the opportunity to explore advanced imaging techniques, such as single molecule or lifetime imaging, but also to revisit traditional fluorescence imaging methodologies and extract yet unobserved or inaccessible information *in vitro* or *in vivo*. Investigators have performed a variety of experiments in which QDs have been used to localize molecules in cells and tissues, both in live and fixed specimens [156].

In 1998, QDs were first used for cell labeling. Bruchez et al. demonstrated dual-color labeling of fixed mouse fibroblasts by staining the nucleus with green QDs and labeling the F-actin filaments in the cytoplasm with red QDs [157]. In 2003, QDs were used for the first time to visualize cellular structures at high resolution [152]. Kaul and colleagues [158] reported immunofluorescence labeling of the heat shock 70 protein, mortalin, using QDs to show different staining patterns in normal and transformed cells. Yeh et al. [159] also described the application of QD probes for the fluorescent detection of DNA sequences. The process can further our knowledge concerning the impact of DNA mutation, damage and repair information, and mechanisms in the development of cancer or other genetic diseases. It might be possible to use this method as a standard technique for protein, RNA, and DNA quantification, because new QD developments will draw on such unique biophysical properties to strengthen the quantitative data obtained and reduce the amount of processing analysis required to compensate for photobleaching [160]. This procedure has been used in fluorescence *in situ* hybridization (FISH) and can help follow transplanted cells. QDs are ideal probes for probing structure and locating signal transduction-related molecules. Rosenthal et al. used serotonin-linked QDs to target the neurotransmitter receptor on the cell surface. The QD probes not only recognized and labeled serotonin-specific neurotransmitters on cell membranes, but also inhibited the serotonin transportation in a dose-dependent manner [161]. Although one to two orders of magnitude less potent at inhibiting the receptor than free serotonin, the behavior of the QD conjugates was similar to that of free serotonin, making QDs a valuable probe for exploring the serotonin transportation mechanism [155].

Antibody-conjugated QDs have been used for labeling and tracking single receptors. In addition, ligand-conjugated QDs retain their activity and bind to receptors. nerve growth factor (NGF)-QD ligand conjugates as well as antibody against tyrosin kinase A (anti-TrkA)-QD antibody conjugates can be used to visualize NGF-TrkA spatiotemporal dynamics in the neuronal PC12 cell line [162]. Other signaling pathways, such as epidermal growth factor receptor B/homologous receptor (erbB/HER) receptor-mediated signal transduction, have also been examined using QDs [163]. This procedure can similarly be used for studying the signaling pathways in SCs.

Although the imaging of fixed cells is useful and sufficient for many applications, live cell microscopy is ideal for visualizing cellular processes but is considerably more difficult [152]. Two-photon imaging of QDs can be used to more efficiently image thick specimens. In 2003, Larson et al. injected QDs intravenously into mice and demonstrated that these QDs were detectable through intact skin at the base of the dermis (~100 micron) using an excitation wavelength of 900 nm [164]. To optimize the conditions of *in vivo* experiments [165], QDs with different polymer coatings were tested *in vivo* using various imaging techniques, including light and electron microscopy, on tissue sections, and in noninvasive whole-body fluorescence imaging [166]. Amazingly, these QDs maintained their fluorescence even after 4 months *in vivo*. Although the QDs showed no deleterious effects upon the animals in these studies, a more detailed evaluation of potential QD toxicity in the body is warranted prior to their long-term usage in humans.

As mentioned before, QDs are extremely bright and photostable, and they can be used as cell markers for long-term studies such as cell–cell interactions, cell differentiation, and cell lineage tracking. Most experiments conducted to date have shown that QDs do not interfere with normal cell physiology and cell differentiation [167]. In regard to properties, QDs seem to be suitable for studying SCs. For example, NSCs divide to generate all of the required cell types of the mature brain, but the mechanisms controlling this process are only partially understood. One reason for this lack of knowledge is the inability to label and track NSCs in their intact environment for long periods of time. Haydar and colleagues used highly fluorescent QD nanocrystals as novel labeling tools to better understand NSC behavior *in vivo*. They demonstrated that NSCs loaded with QDs *in vivo* retained the capacity to differentiate into neurons [168].

1.6   EVALUATION OF TOXICITY OF NANOMATERIALS WITH STEM CELLS

Despite the wide application of nanomaterials, there is a serious lack of information concerning the impact of manufactured nanomaterials on human health and the environment. Typically, after systemic administration, the nanoparticles are small enough to penetrate even very small capillaries throughout the body; therefore, they offer the most effective approach to distribution in certain tissues. Because nanoparticles can pass through biological membranes, they can affect the physiology of any cell in an animal body [169]. SWNTs in suspension in the culture medium were incorporated into the cell cytoplasm by macrophages and leukemia cells without affecting the cell population growth [170,171]. It has been shown that CNT substrates decreased keratinocyte [172], glial [173], and HEK293 cell survival significantly [174], raising important concerns about the biocompatibility of the nanomaterial. In addition, SWNTs have been shown to block potassium channel activities in heterologous mammalian cell systems when applied externally to the cell surface [175], which suggests a degree of cytotoxicity. This consideration is of importance for SCs, where the effects of nanoparticles on their potential for self-renewal and differentiation are unknown. Data available from toxicity studies of nanoparticles, in particular in ASCs, are limited. Recently, Braydich-Stolle et al., by using mouse spermatogonial SC line as a model to assess nanotoxicity in the male germline *in vitro*, demonstrated a concentration-dependent toxicity for all types of particles tested, whereas the corresponding soluble salts had no significant effect [176]. Silver nanoparticles were the most toxic, and molybdenum trioxide (MoO_3) nanoparticles were less toxic. However, the increase of interfacial bonding and the introduction of surfactants are likely to have an impact on degradation kinetics and cytotoxicity of the composite. These effects are largely unknown and remain to be investigated.

1.7   CONCLUSIONS AND FUTURE OUTLOOK

Tissue engineering at the nanoscale level is leading to the development of viable substitutes that can restore, maintain, or improve the function of human tissue. Regenerating tissue can be achieved by using biomaterials to convey signals to surrounding tissues to recruit cells that promote inherent regeneration or by using SCs and a biomaterial scaffold to act as a framework for developing tissue. Nanobiomaterials may serve as an important component to imparting novel properties to the biomatrix for directing SC proliferation, differentiation, functionalization, and transplantation. SC nano-engineering is beginning to help advance the equally pioneering field of stem-cell research, with devices that can precisely control SCs and provide bioscaffolds and labeling and tracking systems to create significant advances in *in vivo* monitoring of engineered tissues. While new uses of nano-materials for biomedical applications are being developed, concerns about cytotoxicity may be mitigated by chemical functionalization. However, there will be some limitations to this nanomaterial, because it is not biodegradable. As such, in order for nanotechnology applications to develop to their fullest potential, it will be important for SC scientists and physicians to participate and contribute to the scientific process alongside physical and chemical science and engineering colleagues.

ACKNOWLEDGMENTS

We gratefully thank Dr. Ali Khademhosseini, Massachusetts Institute of Technology, and Professor Klaus Ingo Matthaei, Australian National University, for his critical reading and helpful comments on the manuscript. This work was supported by grants from the Royan Institute, Iran. We also acknowledge the assistance of Adeleh Taei, Mohammad Pakzad, Saeid Mahmood Hashemi, and Zahra Maghari.

REFERENCES

1. Smith, L.A., and Ma, P.X. (2004) Nano-fibrous scaffolds for tissue engineering. *Colloids and Surfaces* 39, 125–131.
2. Hartgerink, J.D., et al. (2001) Self-assembly and mineralization of peptide-amphiphile nanofibers. *Science* 294, 1684–1688.
3. Silva, G.A., et al. (2001) Selective differentiation of neural progenitor cells by high epitope density nanofibers. *Science* 303, 1352–1355.
4. Ajayan, P.M. (1999) Nanotubes from carbon. *Chem Rev* 99, 1787–1800.
5. Yu, M.F., et al. (2000) Tensile loading of ropes of single wall carbon nanotubes and their mechanical properties. *Phys Rev Lett* 84, 5552–5555.
6. Wong, E.W., et al. (1997) Nanobeam mechanics: Elasticity, strength, and toughness of nanorods and nanotubes. *Science* 277, 1971–1975.
7. Wong, S.S., et al. (1998) Covalently functionalized nanotubes as nanometre-sized probes in chemistry and biology. *Nature* 394, 52–55.
8. Harrison, B.S., and Atala, A. (2007) Carbon nanotube applications for tissue engineering. *Biomaterials* 28, 344–353.
9. McKay, R. (1997) Stem cells in the central nervous system. *Science* 276, 66–71.
10. Gordon, M.Y., and Blackett, N.M. (1998) Reconstruction of the hematopoietic system after stem cell transplantation. *Cell Transplantation* 7, 339–344.
11. Scheffler, B., et al. (1999) Marrow-mindedness: A perspective on neuropoiesis. *Trends in Neurosciences* 22, 348–357.
12. Preston, S.L., et al. (2003) The new stem cell biology: Something for everyone. *Mol Pathol* 56, 86–96.
13. Evans, M.J., and Kaufman, M.H. (1981) Establishment in culture of pluripotential cells from mouse embryos. *Nature* 292, 154–156.
14. Martin, G.R. (1981) Isolation of a pluripotent cell line from early mouse embryos cultured in medium conditioned by teratocarcinoma stem cells. *Proc Natl Acad Sci USA* 78, 7634–7638.
15. Thomson, J.A., et al. (1998) Embryonic stem cell lines derived from human blastocysts. *Science* 282, 1145–1147.
16. Bradley, A., et al. (1984) Formation of germ-line chimaeras from embryo-derived teratocarcinoma cell lines. *Nature* 309, 255–256.
17. Labosky, P.A., et al. (1994) Mouse embryonic germ (EG) cell lines: Transmission through the germline and differences in the methylation imprint of insulin-like growth factor 2 receptor (Igf2r) gene compared with embryonic stem (ES) cell lines. *Development* 120, 3197–3204.
18. Koller, B.H., and Smithies, O. (1992) Altering genes in animals by gene targeting. *Annu Rev Immunol* 10, 705–730.
19. Odorico, J.S., et al. (2001) Multilineage differentiation from human embryonic stem cell lines. *Stem Cells* 19, 193–204.
20. West, J.A., and Daley, G.Q. (2004) *In vitro* gametogenesis from embryonic stem cells. *Curr Opinion in Cell Biol* 16, 688–692.
21. Javazon, E.H., et al. (2001) Rat marrow stromal cells are more sensitive to plating density and expand more rapidly from single-cell-derived colonies than human marrow stromal cells. *Stem Cells* 19, 219–225.
22. Dennis, J.E., et al. (1999) A quadripotential mesenchymal progenitor cell isolated from the marrow of an adult mouse. *J Bone Miner Res* 14, 700–709.
23. Orlic, D., et al. (2001) Bone marrow cells regenerate infarcted myocardium. *Nature* 410, 701–705.
24. Sekiya, I., et al. (2002) Expansion of human adult stem cells from bone marrow stroma: Conditions that maximize the yields of early progenitors and evaluate their quality. *Stem Cells* 20, 530–541.

25. Orlic, D., et al. (2003) Bone marrow stem cells regenerate infarcted myocardium. *Pediatric Transplantation* 7 Suppl 3, 86–88.
26. Gussoni, E., et al. (1999) Dystrophin expression in the mdx mouse restored by stem cell transplantation. *Nature* 401, 390–394.
27. Lagasse, E., et al. (2000) Purified hematopoietic stem cells can differentiate into hepatocytes *in vivo*. *Nature Med* 6, 1229–1234.
28. Krause, D.S., et al. (2001) Multi-organ, multi-lineage engraftment by a single bone marrow-derived stem cell. *Cell* 105, 369–377.
29. Mezey, E., et al. (2003) Transplanted bone marrow generates new neurons in human brains. *Proc Natl Acad Sci USA* 100, 1364–1369.
30. Johnson, J., et al. (2005) Oocyte generation in adult mammalian ovaries by putative germ cells in bone marrow and peripheral blood. *Cell* 122, 303–315.
31. Shih, C.C., et al. (2002) Hematopoietic potential of neural stem cells. *Nature Med* 8, 535–536; author reply 536–537.
32. Jackson, K.A., et al. (1999) Hematopoietic potential of stem cells isolated from murine skeletal muscle. *Proc Natl Acad Sci USA* 96, 14482–14486.
33. Jiang, Y., et al. (2002) Pluripotency of mesenchymal stem cells derived from adult marrow. *Nature* 418, 41–49.
34. Jiang, Y., et al. (2002) Multipotent progenitor cells can be isolated from postnatal murine bone marrow, muscle, and brain. *Exp Hematol* 30, 896–904.
35. Jiang, Y., et al. (2003) Neuroectodermal differentiation from mouse multipotent adult progenitor cells. *Proc Natl Acad Sci USA* 100 Suppl 1, 11854–11860.
36. Kogler, G., et al. (2004) A new human somatic stem cell from placental cord blood with intrinsic pluripotent differentiation potential. *J Exp Med* 200, 123–135.
37. Wang, H.S., et al. (2004) Mesenchymal stem cells in the Wharton's jelly of the human umbilical cord. *Stem Cells* 22, 1330–1337.
38. Sarugaser, R., et al. (2005) Human umbilical cord perivascular (HUCPV) cells: A source of mesenchymal progenitors. *Stem Cells* 23, 220–229.
39. Buzanska, L., et al. (2002) Human cord blood-derived cells attain neuronal and glial features *in vitro*. *J Cell Sci* 115, 2131–2138.
40. Cohen, Y., and Nagler, A. (2004) Cord blood biology and transplantation. *Isr Med Assoc J* 6, 39–46.
41. Warwick, R., and Armitage, S. (2004) Cord blood banking. *Best Pract and Res* 18, 995–1011.
42. Meyle, J., et al. (1993) Surface micromorphology and cellular interactions. *J Biomater Appl* 7, 362–374.
43. Rajnicek, A., and McCaig, C. (1997) Guidance of CNS growth cones by substratum grooves and ridges: effects of inhibitors of the cytoskeleton, calcium channels and signal transduction pathways. *J Cell Sci* 110 (Pt 23), 2915–2924.
44. Chou, L., et al. (1998) Effects of titanium substratum and grooved surface topography on metalloproteinase-2 expression in human fibroblasts. *J Biomed Mater Res* 39, 437–445.
45. van Kooten, T.G., et al. (1998) Influence of silicone (PDMS) surface texture on human skin fibroblast proliferation as determined by cell cycle analysis. *J Biomed Mater Res* 43, 1–14.
46. Wojciak-Stothard, B., et al. (1996) Guidance and activation of murine macrophages by nanometric scale topography. *Exp Cell Res* 223, 426–435.
47. Meyle, J., et al. (1995) Variation in contact guidance by human cells on a microstructured surface. *J Biomed Mater Res* 29, 81–88.
48. Wojciak-Stothard, B., et al. (1995) Role of the cytoskeleton in the reaction of fibroblasts to multiple grooved substrata. *Cell Motility and the Cytoskeleton* 31, 147–158.
49. den Braber, E.T., et al. (1998) Orientation of ECM protein deposition, fibroblast cytoskeleton, and attachment complex components on silicone microgrooved surfaces. *J Biomed Mater Res* 40, 291–300.
50. Oakley, C., and Brunette, D.M. (1995) Response of single, pairs, and clusters of epithelial cells to substratum topography. *Biochemistry and Cell Biology. Biochimie et Biologie Cellulaire* 73, 473–489.
51. Webb, A., et al. (1995) Guidance of oligodendrocytes and their progenitors by substratum topography. *J Cell Sci* 108 (Pt 8), 2747–2760.
52. Brunette, D.M. (1986) Spreading and orientation of epithelial cells on grooved substrata. *Exp Cell Res* 167, 203–217.
53. Green, A.M., et al. (1994) Fibroblast response to microtextured silicone surfaces: Texture orientation into or out of the surface. *J Biomed Mater Res* 28, 647–653.

54. Campbell, C.E., and von Recum, A.F. (1989) Microtopography and soft tissue response. *J Invest Surg* 2, 51–74.

55. Martin, J.Y., et al. (1995) Effect of titanium surface roughness on proliferation, differentiation, and protein synthesis of human osteoblast-like cells (MG63). *J Biomed Mater Res* 29, 389–401.

56. Lampin, M., et al. (1997) Correlation between substratum roughness and wettability, cell adhesion, and cell migration. *J Biomed Mater Res* 36, 99–108.

57. Clark, P., et al. (1993) Growth cone guidance and neuron morphology on micropatterned laminin surfaces. *J Cell Sci* 105 (Pt 1), 203–212.

58. Britland, S., and McCaig, C. (1996) Embryonic Xenopus neurites integrate and respond to simultaneous electrical and adhesive guidance cues. *Exp Cell Res* 226, 31–38.

59. Goodman, S.L., et al. (1996) Three-dimensional extracellular matrix textured biomaterials. *Biomaterials* 17, 2087–2095.

60. Flemming, R.G., et al. (1999) Effects of synthetic micro- and nano-structured surfaces on cell behavior. *Biomaterials* 20, 573–588.

61. Dalby, M.J., et al. (2002) Polymer-demixed nanotopography: Control of fibroblast spreading and proliferation. *Tissue Eng* 8, 1099–1108.

62. Curtis, A.S. (2001) Cell reactions with biomaterials: The microscopies. *Eur Cells and Mater [electronic resource]* 1, 59–65.

63. Gallagher, J.O., et al. (2002) Interaction of animal cells with ordered nanotopography. *IEEE Trans on Nanobiosci* 1, 24–28.

64. Burridge, K., and Chrzanowska-Wodnicka, M. (1996) Focal adhesions, contractility, and signaling. *Annu Rev Cell Dev Biol* 12, 463–518.

65. Cary, L.A., et al. (1999) Integrin-mediated signal transduction pathways. *Histology and Histopathology* 14, 1001–1009.

66. Yoshimoto, H., et al. (2003) A biodegradable nanofiber scaffold by electrospinning and its potential for bone tissue engineering. *Biomaterials* 24, 2077–2082.

67. Li, W.J., et al. (2005) A three-dimensional nanofibrous scaffold for cartilage tissue engineering using human mesenchymal stem cells. *Biomaterials* 26, 599–609.

68. Li, W.J., et al. (2002) Electrospun nanofibrous structure: A novel scaffold for tissue engineering. *J Biomed Mater Res* 60, 613–621.

69. Li, W.J., et al. (2003) Biological response of chondrocytes cultured in three-dimensional nanofibrous poly(epsilon-caprolactone) scaffolds. *J Biomed Mater Res* 67A, 1105–1114.

70. MacDonald, R.A., et al. (2005) Collagen-carbon nanotube composite materials as scaffolds in tissue engineering. *J Biomed Mater Res A* 74, 489–496.

71. Bissell, M.J., and Barcellos-Hoff, M.H. (1987) The influence of extracellular matrix on gene expression: Is structure the message? *J Cell Sci Suppl* 8, 327–343.

72. Dalby, M.J., et al. (2004) Changes in fibroblast morphology in response to nano-columns produced by colloidal lithography. *Biomaterials* 25, 5415–5422.

73. Langer, R., and Vacanti, J.P. (1993) Tissue engineering. *Science* 260, 920–926.

74. Cukierman, E., et al. (2001) Taking cell-matrix adhesions to the third dimension. *Science* 294, 1708–1712.

75. Schmeichel, K.L., and Bissell, M.J. (2003) Modeling tissue-specific signaling and organ function in three dimensions. *J Cell Sci* 116, 2377–2388.

76. Wang, Y.K., et al. (2003) Rigidity of collagen fibrils controls collagen gel-induced down-regulation of focal adhesion complex proteins mediated by $\alpha 2\beta 1$ integrin. *J Biol Chem* 278, 21886–21892.

77. Nur, E.K.A., et al. (2005) Three dimensional nanofibrillar surfaces induce activation of Rac. *Biochem Biophys Res Commun* 331, 428–434.

78. Schindler, M., et al. (2005) A synthetic nanofibrillar matrix promotes *in vivo*-like organization and morphogenesis for cells in culture. *Biomaterials* 26, 5624–5631.

79. Nur, E.K.A., et al. (2006) Three-dimensional nanofibrillar surfaces promote self-renewal in mouse embryonic stem cells. *Stem Cells* 24, 426–433.

80. Chambers, I., et al. (2003) Functional expression cloning of Nanog, a pluripotency sustaining factor in embryonic stem cells. *Cell* 113, 643–655.

81. McBeath, R., et al. (2004) Cell shape, cytoskeletal tension, and RhoA regulate stem cell lineage commitment. *Dev Cell* 6, 483–495.

82. Jin, H.J., et al. (2004) Human bone marrow stromal cell responses on electrospun silk fibroin mats. *Biomaterials* 25, 1039–1047.

83. Kommireddy, D.S., et al. (2006) Stem cell attachment to layer-by-layer assembled TiO2 nanoparticle thin films. *Biomaterials* 27, 4296–4303.

84. Kommireddy, D.S., et al. (2005) Nanoparticle thin films: Surface modification for cell attachment and growth. *J Biomed Nanotechnol* 3, 286–290.

85. Levenberg, S., et al. (2003) Differentiation of human embryonic stem cells on three-dimensional polymer scaffolds. *Proc Natl Acad Sci USA* 100, 12741–12746.

86. Chaudhry, G.R., et al. (2004) Osteogenic cells derived from embryonic stem cells produced bone nodules in three-dimensional scaffolds. *J Biomed Biotechnol* 2004, 203–210.

87. Li, W.J., et al. (2005) Multilineage differentiation of human mesenchymal stem cells in a three-dimensional nanofibrous scaffold. *Biomaterials* 26, 5158–5166.

88. Liu, H., and Roy, K. (2005) Biomimetic three-dimensional cultures significantly increase hematopoietic differentiation efficacy of embryonic stem cells. *Tissue Eng* 11, 319–330.

89. Engler, A.J., et al. (2006) Matrix elasticity directs stem cell lineage specification. *Cell* 126, 677–689.

90. LaIuppa, J.A., et al. (1997) Culture materials affect *ex vivo* expansion of hematopoietic progenitor cells. *J Biomed Mater Res* 36, 347–359.

91. Li, Y., et al. (2001) Human cord cell hematopoiesis in three-dimensional nonwoven fibrous matrices: *In vitro* simulation of the marrow microenvironment. *J Hematother Stem Cell Res* 10, 355–368.

92. Feng, Q., et al. (2006) Expansion of engrafting human hematopoietic stem/progenitor cells in three-dimensional scaffolds with surface-immobilized fibronectin. *J Biomed Mater Res A* 78, 781–791.

93. Ehring, B., et al. (2003) Expansion of HPCs from cord blood in a novel 3D matrix. *Cytotherapy* 5, 490–499.

94. Jiang, X.S., et al. (2006) Surface-immobilization of adhesion peptides on substrate for *ex vivo* expansion of cryopreserved umbilical cord blood CD34+ cells. *Biomaterials* 27, 2723–2732.

95. Chua, K.N., et al. (2006) Surface-aminated electrospun nanofibers enhance adhesion and expansion of human umbilical cord blood hematopoietic stem/progenitor cells. *Biomaterials* 27, 6043–6051.

96. Keselowsky, B.G., et al. (2004) Surface chemistry modulates focal adhesion composition and signaling through changes in integrin binding. *Biomaterials* 25, 5947–5954.

97. Wilson, C.J., et al. (2005) Mediation of biomaterial-cell interactions by adsorbed proteins: A review. *Tissue Eng* 11, 1–18.

98. Calvi, L.M., et al. (2003) Osteoblastic cells regulate the haematopoietic stem cell niche. *Nature* 425, 841–846.

99. Zhang, J., et al. (2003) Identification of the haematopoietic stem cell niche and control of the niche size. *Nature* 425, 836–841.

100. Krause, D.S., et al. (1996) CD34: structure, biology, and clinical utility. *Blood* 87, 1–13.

101. Lanza, F., et al. (2001) Structural and functional features of the CD34 antigen: An update. *J Biol Regul Homeost Agents* 15, 1–13.

102. Tada, J., et al. (1999) A common signaling pathway via Syk and Lyn tyrosine kinases generated from capping of the sialomucins CD34 and CD43 in immature hematopoietic cells. *Blood* 93, 3723–3735.

103. Yamagishi, H., et al. (1995) Development photochemical technique for modifying poly(arylsulfone) ultrafiltration membranes. *J Membr Sci* 105, 237–247.

104. Behravesh, E., et al. (2005) Comparison of genotoxic damage in monolayer cell cultures and three-dimensional tissue-like cell assemblies. *Adv Space Res* 35, 260–267.

105. Elias, K.L., et al. (2002) Enhanced functions of osteoblasts on nanometer diameter carbon fibers. *Biomaterials* 23, 3279–3287.

106. Pricea, R.L., et al. (2003) Selective bone cell adhesion on formulations containing carbon nanofibers. *Biomaterials* 24, 1877–1887.

107. Webster, T.J., et al. (2001) Mechanisms of enhanced osteoblast adhesion on nanophase alumina involve vitronectin. *Tissue Eng* 7, 291–301.

108. Hosseinkhani, H., et al. (2006) Osteogenic differentiation of mesenchymal stem cells in self-assembled peptide-amphiphile nanofibers. *Biomaterials* 27, 4079–4086.

109. Hartgerink, J.D., et al. (2002) Peptide-amphiphile nanofibers: A versatile scaffold for the preparation of self-assembling materials. *Proc Natl Acad Sci USA* 99, 5133–5138.

110. Niece, K.L., et al. (2003) Self-assembly combining two bioactive peptide-amphiphile molecules into nanofibers by electrostatic attraction. *J Am Chem Soc* 125, 7146–7147.

111. Woo, K.M., et al. (2007) Nano-fibrous scaffolding promotes osteoblast differentiation and biomineralization. *Biomaterials* 28, 335–343.

112. Bonucci, E. (1992) Role of collagen fibrils in calcification. In *Calcification in Biologic Systems* (Bonucci, E., Ed.), 19–39, Boca Raton: CRC Press.
113. Byers, B.A., et al. (2002) Cell-type-dependent up-regulation of *in vitro* mineralization after overexpression of the osteoblast-specific transcription factor Runx2/Cbfal. *J Bone Miner Res* 17, 1931–1944.
114. Midura, R.J., et al. (2004) Bone acidic glycoprotein-75 delineates the extracellular sites of future bone sialoprotein accumulation and apatite nucleation in osteoblastic cultures. *J Biol Chem* 279, 25464–25473.
115. Zhao, B., et al. (2005) Synthesis and characterization of water soluble single-walled carbon nanotube graft copolymers. *J Am Chem Soc* 127, 8197–8203.
116. Zanello, L.P., et al. (2006) Bone cell proliferation on carbon nanotubes. *Nano Lett* 6, 562–567.
117. Supronowicz, P.R., et al. (2002) Novel current-conducting composite substrates for exposing osteoblasts to alternating current stimulation. *J Biomed Mater Res* 59, 499–506.
118. Yim, E.K. (2006) Nanotopography-induced cell behavior and human stem cell differentiation. Bioengineering Seminar Series, Sept. 29.
119. Yang, F., et al. (2004) Fabrication of nano-structured porous PLLA scaffold intended for nerve tissue engineering. *Biomaterials* 25, 1891–1900.
120. Yang, S.H., et al. (2005) *N*-Myristoyltransferase 1 is essential in early mouse development. *J Biol Chem* 280, 18990–18995.
121. Silva, G.A. (2005) Nanotechnology approaches for the regeneration and neuroprotection of the central nervous system. *Surg Neurol* 63, 301–306.
122. Silva, G.A., et al. (2004) Selective differentiation of neural progenitor cells by high-epitope density nanofibers. *Science* 303, 1352–1355.
123. Mattson, M.P., et al. (2000) Molecular functionalization of carbon nanotubes and use as substrates for neuronal growth. *J Mol Neurosci* 14, 175–182.
124. Hu, H., et al. (2004) Chemically functionalized carbon nanotube as substrates for neuronal growth. *Nano Lett* 4, 507.
125. Hu, H., et al. (2005) Polyethyleneimine functionalized single-walled carbon nanotubes as a substrate for neuronal growth. *J Phys Chem* 109, 4285–4289.
126. Lovat, V., et al. (2005) Carbon nanotube substrates boost neuronal electrical signaling. *Nano Lett* 5, 1107–1110.
127. Dugan, L.L., et al. (2001) Fullerene-based antioxidants and neurodegenerative disorders. *Parkinsonism Relat Disord* 7, 243–246.
128. Wilson, J.X., and Gelb, A.W. (2002) Free radicals, antioxidants, and neurologic injury: Possible relationship to cerebral protection by anesthetics. *J Neurosurg Anesthesiol* 14, 66–79.
129. Dugan, L.L., et al. (1996) Buckminsterfullerenol free radical scavengers reduce excitotoxic and apoptotic death of cultured cortical neurons. *Neurobiol Dis* 3, 129–135.
130. Dugan, L.L., et al. (1997) Carboxyfullerenes as neuroprotective agents. *Proc Natl Acad Sci USA* 94, 9434–9439.
131. Jin, H., et al. (2000) Polyhydroxylated C(60), fullerenols, as glutamate receptor antagonists and neuroprotective agents. *J Neurosci Res* 62, 600–607.
132. Riviere, C., et al. (2005) Iron oxide nanoparticle-labeled rat smooth muscle cells: cardiac MR imaging for cell graft monitoring and quantitation. *Radiology* 235, 959–967.
133. Shapiro, E.M., et al. (2006) *In vivo* detection of single cells by MRI. *Magn Reson Med* 55, 242–249.
134. Hoehn, M., et al. (2002) Monitoring of implanted stem cell migration *in vivo*: A highly resolved *in vivo* magnetic resonance imaging investigation of experimental stroke in rat. *Proc Natl Acad Sci USA* 99, 16267–16272.
135. Frangioni, J.V., and Hajjar, R.J. (2004) *In vivo* tracking of stem cells for clinical trials in cardiovascular disease. *Circulation* 110, 3378–3383.
136. Arbab, A.S., et al. (2006) Cellular magnetic resonance imaging: Current status and future prospects. *Expert Rev Med Devices* 3, 427–439.
137. Bulte, J., and Kraitchman, D. (2004) Iron oxide MR contrast agents for molecular and cellular imaging. *NMR Biomed* 17, 484–499.
138. Wilhelm, C., et al. (2003) Rotational magnetic endosome microrheology: Viscoelastic architecture inside living cells. *Phys Rev E Stat Nonlin Soft Matter Phys* 67, 061908 (12 pages).
139. Frank, J.A., et al. (2003) Clinically applicable labeling of mammalian and stem cells by combining superparamagnetic iron oxides and transfection agents. *Radiology* 228, 480–487.
140. Frank, J.A., et al. (2004) Methods for magnetically labeling stem and other cells for detection by *in vivo* magnetic resonance imaging. *Cytotherapy* 6, 621–625.

141. Syková, E., et al. (2006) Bone marrow stem cells and polymer hydrogels—Two strategies for spinal cord injury repair. *Cell Mol Neurobiol* 26, 1111–1127.

142. Stroh, A., et al. (2005) In vivo detection limits of magnetically labeled embryonic stem cells in the rat brain using high-field (17.6 T) magnetic resonance imaging. *NeuroImage* 24, 635–645.

143. Zhu, J., et al. (2005) [Application of FTIR spectra fitting method in retrieving gas concentrations]. *Guang pu xue yu guang pu fen xi, Guang pu* 25, 1573–1576.

144. Wang, F.H., et al. (2006) Magnetic resonance tracking of nanoparticle labelled neural stem cells in a rat's spinal cord. *Nanotechnology* 17, 1911–1915.

145. Huang, D.M., et al. (2005) Highly efficient cellular labeling of mesoporous nanoparticles in human mesenchymal stem cells: Implication for stem cell tracking. *Faseb J* 19, 2014–2016.

146. Payne, A.G. (2004) Using immunomagnetic technology and other means to facilitate stem cell homing. *Medical Hypotheses* 62, 718–720.

147. Bjorklund, L.M., et al. (2002) Embryonic stem cells develop into functional dopaminergic neurons after transplantation in a Parkinson rat model. *Proc Natl Acad Sci USA* 99, 2344–2349.

148. Hamberg, L.M., et al. (1996) Continuous assessment of relative cerebral blood volume in transient ischemia using steady state susceptibility-contrast MRI. *Magn Reson Med* 35, 168–173.

149. Mandeville, J.B., et al. (2001) Regional sensitivity and coupling of BOLD and CBV changes during stimulation of rat brain. *Magn Reson Med* 45, 443–447.

150. Lewin, M., et al. (2000) Tat peptide-derivatized magnetic nanoparticles allow *in vivo* tracking and recovery of progenitor cells. *Nat Biotechnol* 18, 410–414.

151. Won, J., et al. (2005) A magnetic nanoprobe technology for detecting molecular interactions in live cells. *Science* 309, 121–125.

152. Smith, A.M., and Nie, S. (2005) Semiconductor quantum dots for molecular and cellular imaging. In *Tissue Engineering and Artificial Organs*, 22-21-22-10.

153. Guo, P. (2005) RNA nanotechnology: Engineering, assembly and applications in detection, gene delivery and therapy. *J Nanosci and Nanotechnol* 5, 1964–1982.

154. Vo-Dinh, T., et al. (2006) Nanoprobes and nanobiosensors for monitoring and imaging individual living cells. *Nanomed: Nanotechnol, Biol, and Med* 2, 22–30.

155. Alivisatos, A.P., et al. (2005) Quantum dots as cellular probes. *Annu Rev Biomed Eng* 7, 55–76.

156. Reisner, Y., et al. (2005) Hematopoietic stem cell transplantation across major genetic barriers: Tolerance induction by megadose CD34 cells and other veto cells. *Ann NY Acad Sci* 1044, 70–83.

157. Bruchez, M., Jr., et al. (1998) Semiconductor nanocrystals as fluorescent biological labels. *Science* 281, 2013–2016.

158. Kaul, Z., et al. (2003) Mortalin imaging in normal and cancer cells with quantum dot immuno-conjugates. *Cell Res* 13, 503–507.

159. Yeh, H.C., et al. (2005) Quantum dot-mediated biosensing assays for specific nucleic acid detection. *Nanomedicine* 1, 115–121.

160. Chan, W.C., and Nie, S. (1998) Quantum dot bioconjugates for ultrasensitive nonisotopic detection. *Science* 281, 2016–2018.

161. Rosenthal, S.J., et al. (2002) Targeting cell surface receptors with ligand-conjugated nanocrystals. *J Am Chem Soc* 124, 4586–4594.

162. Sundara Rajan, S., and Vu, T.Q. (2006) Quantum dots monitor TrkA receptor dynamics in the interior of neural PC12 cells. *Nano Lett* 6, 2049–2059.

163. Lidke, D.S., et al. (2004) Quantum dot ligands provide new insights into erbB/HER receptor-mediated signal transduction. *Nat Biotechnol* 22, 198–203.

164. Larson, D.R., et al. (2003) Water-soluble quantum dots for multiphoton fluorescence imaging *in vivo*. *Science* 300, 1434–1436.

165. Ballou, B. (2005) Quantum dot surfaces for use *in vivo* and *in vitro*. *Curr Top Dev Biol* 70, 103–120.

166. Hoshino, A., et al. (2004) Applications of T-lymphoma labeled with fluorescent quantum dots to cell tracing markers in mouse body. *Biochem Biophys Res Commun* 314, 46–53.

167. Mattheakis, L.C., et al. (2004) Optical coding of mammalian cells using semiconductor quantum dots. *Anal. Biochem.* 7, 200–208.

168. Haydar, T.F. (2005) Advanced microscopic imaging methods to investigate cortical development and the etiology of mental retardation. *Mental Retard and Dev Disabil Res Rev* 11, 303–316.

169. Brooking, J., et al. (2001) Transport of nanoparticles across the rat nasal mucosa. *J Drug Target* 9, 267–279.

170. Cherukuri, P., et al. (2004) Near-infrared fluorescence microscopy of single-walled carbon nanotubes in phagocytic cells. *J Am Chem Soc* 126, 15638–15639.

171. Shi Kam, N.W., et al. (2004) Nanotube molecular transporters: Internalization of carbon nanotube-protein conjugates into mammalian cells. *J Am Chem Soc* 126, 6850–6851.

172. Shvedova, A.A., et al. (2003) Exposure to carbon nanotube material: Assessment of nanotube cytotoxicity using human keratinocyte cells. *J Toxicol and Environ Health* 66, 1909–1926.

173. McKenzie, J.L., et al. (2004) Decreased functions of astrocytes on carbon nanofiber materials. *Biomaterials* 25, 1309–1317.

174. Cui, D., et al. (2005) Effect of single wall carbon nanotubes on human HEK293 cells. *Toxicol Lett* 155, 73–85.

175. Park, K.H., et al. (2003) Single-walled carbon nanotubes are a new class of ion channel blockers. *J Biol Chem* 278, 50212–50216.

176. Braydich-Stolle, L., et al. (2005) *In vitro* cytotoxicity of nanoparticles in mammalian germline stem cells. *Toxicol Sci* 88, 412–419.

177. Baharvand, H., et al. (2006) Generation of new human embryonic stem cell lines with diploid and triploid karyotypes. *Dev. Growth. Differ.* 48, 117–128.

2 Lipid Membranes in Biomimetic Systems

Tânia Beatriz Creczynski-Pasa and André Avelino Pasa
Departamento de Ciências Farmacêuticas, Universidade
Federal de Santa Catarina, Florianópolis, Brazil

CONTENTS

2.1 INTRODUCTION

Lipid membranes of living cells are the most important barriers to control the majority of cellular processes. They play a fundamental role in cell to cell communication involving the exchange of ions and biomolecules, including calcium [1], neurotransmitters [2], proteins [3], reactive species [4], and drugs [5], among others.

The biological membranes are very complex in composition, implying very complex investigations to determine their functions and properties. The biochemists choose to create models for cellular membranes, allowing for the reduction of the number of variables and better control of the experimental parameters.

Biomimetic systems are artificial structures conceptually designed to perform actions typical of natural biological systems, or in other words, to mimetize functions and structures of biological entities of living beings. Biomimetic systems are an achievement of the application of methods and systems found in nature to the study of fundamental phenomena and design of engineered structures for modern technology. In fundamental investigations, researchers need to decrease the complexity of the natural systems, as the mimetizing of cell membranes by liposomes. From the applied point of view, the development of devices named "biosensors" is based on the combination of inorganic materials with biological structures.

In this chapter, we will present an overview of lipid membranes, with special interest paid to specific preparation and characterization methods of immobilized phospholipid membranes. The preparation by vesicle fusion and the characterization by atomic force microscopy of the supported membranes will be described.

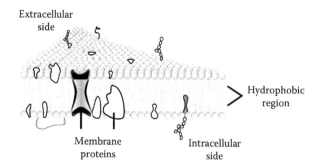

FIGURE 2.1   General scheme of a cell membrane structure.

2.2   LIPID MEMBRANES

The biological membranes are responsible for the structural and functional characteristics of eukaryotic cells. They are considered as the regulatory center of cellular activity and are fundamental for biological energy conservation. The principal membrane functions are to create chemical and electrical gradients to control the activity of membrane enzymes; to organize the proteins in a convenient manner so as to optimize their actions in a coordinated form; to provide substrates for metabolism; and to turn viable processes such as endocytose, exocytose, cell fusion, and division [6,7]. Being the anchor of these processes, the cellular membranes are key for cellular signal transduction [6–8].

Illustrated in Figure 2.1 is a scheme of a biological membrane known as the mosaic fluid model, which describes a three-dimensional structure consisting of a lipid bilayer associated with proteins [9]. However, it is unanimous in the literature that the cell membranes have a complex lateral structure. Lateral lipid domains with distinct physicochemical properties were already observed, corresponding to different kinds or phases of lipids [10]. Moreover, the presence and the biological role of the lipid domains or microdomains (called rafts) [11] (Figure 2.2) in the cell surface are still under investigation. It is challenging to describe how the lipids interact with each other and with the intercalated proteins in order to understand the dynamic of cellular membranes [12]. An interesting example is a domain called caveolae that is fundamental in cells by mediating the transport through the membrane of low-molecular solutes and macromolecules and infectious agents, such as viruses. The domain presents a particular lipid and protein composition that forms polyps through a membrane invagination. The caveolae has been seen as an efficient biological vector for drug transport and delivery [13].

The membranes are composed by a heterogeneous lipid matrix with the insertion of about 10^9 molecules, such as proteins and polypeptides [14–16]. Additionally, increasing the grade of complexity, there are more than 2000 kinds of lipids in mammal cell membranes, and the membrane lipid composition determines the physical properties, including free charge, dipole potential, elasticity, phase transition, and hydration [17].

The chemical composition and the physical state of the membranes allow the interaction between its constituents dictating their organization. The phospholipids, the major constituents of the cell membranes, are disposed structurally in symmetric bilayers in which the hydrophilic faces are formed by their polar side exposed to the aqueous environment of the cell (cytoplasm and extracellular side). Between the polar faces, the carbon chains of the fatty acids are self-organized, protecting them from the aqueous medium, as shown in Figure 2.1. The principal forces preserving

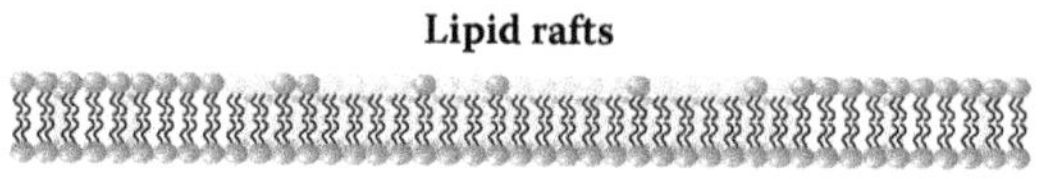

Lipid rafts

FIGURE 2.2   Scheme of lipid rafts.

the lipid bilayers are hydrogen bonds, others between the hydrophilic parts of the molecules, and van der Waals interactions (between the hydrophobic chains). There is also the additional interaction between the polar head and nonpolar tail with the aqueous medium [14]. More details regarding the phospholipids will be presented in the next section.

2.3 MEMBRANE LIPIDS

In general, the membrane lipids are amphipatic. The polar heads are characterized by the presence of phosphate, amine and alcohol groups, and the two nonpolar tails characterized by fatty acid chains. They are classified as phospholipids, sphingolipids, glycolipids, and sterols [6]. In this chapter, we will focus our attention on phospholipids, because the cell membranes and the membrane models are basically constituted by them.

The fatty acids of the phospholipids are bonded to the polar part through ester bindings either with the phosphate group or with the alcohol group [6,18]. Between the phospholipids, the phosphatidylcholine (PC) (structure illustrated in Figure 2.3) is the principal constituent of the cell membranes [6,18]. The natural PCs are con-

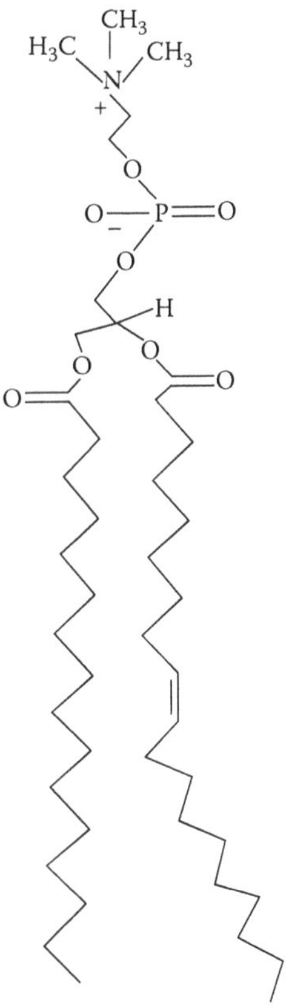

FIGURE 2.3   Phosphatidylcholine: Structural formula.

stituted by a zwitterion phosphocoline group, and the fatty acid chains present the same number of carbons. The phospholipids and glycolipids molecules of the cell membranes show fatty acid chains containing normally 14 to 24 carbon atoms. They are named according to the number of the carbon atoms and to the number of unsaturations; for example, the hexadecanoic acid presents C_{16} and two unsaturations—it is named 16:2 [6].

The nonpolar chain bonds to carbon 1 of the glycerol group (alcohol group), being normally saturated (presenting only simple bonds), and the chains bond to carbon 2, being normally unsaturated, presenting usually one to six double bonds in a *cis* conformation [19].

X-ray crystallographic measurements have shown that the presence of *cis* double bonds outcomes as a kink in the carbon chain, as illustrated in the Figure 2.3. The angle of a *cis* double bond is 123°, and the angle of a simple bond is 111° [20]. The packing, fluidity, lipid-phase transitions, as well as the thicknesses of the lipid membranes depend on the acyl chain saturation degree [21,22]. In general, membranes with high concentrations of unsaturated lipids present lower thicknesses. Similar to the size of the lipid bilayers, the unsaturations in the lipid acyl chains determine the phase transition temperatures of the membranes. The higher are the presence of unsaturations, the lower is the transition-phase temperature.

Additionally, the membranes in the gel phase present higher thicknesses than the membranes in the liquid crystalline phase. Figure 2.4 displays the gel to liquid phase transition of a lipid membrane mediated by temperature changes with the

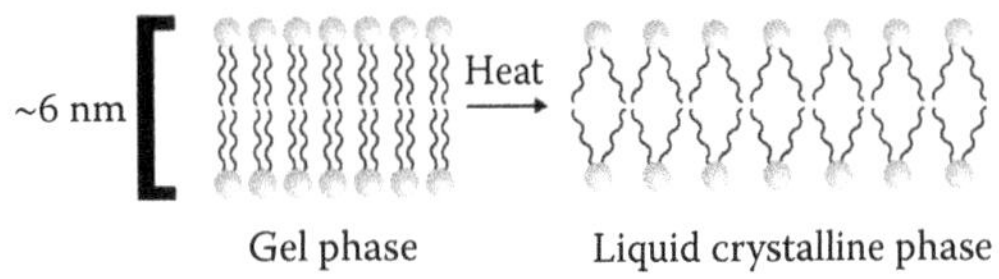

FIGURE 2.4   Scheme of the gel to liquid crystalline phase transition in lipid membranes, depicting the reduction of the thickness of the bilayers mediated by temperature changes.

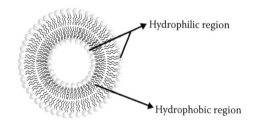
Hydrophilic region

Hydrophobic region

FIGURE 2.5 A unilamellar liposome.

consequent reduction of the height of the bilayer. The above description of the phospholipids is fundamental for the understanding of liposome formation and the characterization of supported bilayers through atomic force microscopy, as will be discussed later.

2.4 LIPOSOMES

Liposomes, in which phospholipid composition, structure, and dynamics can be fully controlled, are generally accepted as a suitable model for *in vitro* studies of cell membrane structures and properties [23,24]. It is a vesicle formed by a lipid bilayer which is structurally similar to the lipid matrix of cell membranes [23–25]. It is a vesicle formed by a lipid bilayer, as depicted in Figure 2.5, which is structurally similar to the lipid matrix of cell membranes [23–25].

Liposomes have been broadly studied because they are built with biocompatible material, have structural versatilities (i.e., size, composition, and bilayer fluidity), and present capabilities to incorporate a great variety of molecules despite their structure. Liposomes are particularly interesting for drug delivery and vaccine development, because they trespass natural biological drug barriers as the hematoencephalic one, and induce immunity, respectively. In this chapter, we are not going to discuss the liposomes in depth, although it is a biomimetic model, because a vast discussion would be warranted. The liposomes are presented briefly in this context, because one of the methods to obtain supported lipid membranes is based on the fusion of vesicles on solid surfaces.

There are several methods by which to prepare liposomes. According to the objective to be reached and the elected preparation method, it is possible to control the size and number of the vesicle layers and the manipulation of the lipid composition, which varies the fluidity and charge of the vesicles. The majority of preparation methods and the classification of the liposomes are resumed in Table 2.1.

2.5 SUPPORTED MEMBRANE LAYERS

Lipid monolayers and bilayers supported on solid substrates have attracted the interest of the researchers in the last two decades [33–35]. These models have been used in studies regarding the structure and properties of natural biological membranes and for the investigation of biological processes [36,37], including molecular reorganization and antigen–antibody interactions [38], enzymatic catalysis and membrane fusion [36–38], biosensors [39], inorganic solids biofunctionalization [38], and DNA immobilization [40]. Studies on morphology alterations in the cellular structure and protein behavior as well as controlled drug and genes delivery have also been done with the objective of developing biomaterials [40]. The clinical applications for such artificial biomaterials

TABLE 2.1
Classification of the Liposomes

Abbreviated Liposome Name	Name	Size	Obtaining Method
SUV	Small unilamellar vesicles	20–50 nm	Sonication of MLVs [26]
LUV	Large unilamellar vesicles	50–400 nm	Extrusion, reverse phase and dialysis [27–29]
MLV	Multilamellar vesicles	>400 nm	Dispersion in aqueous medium [30]
GUV	Giant unilamellar vesicles	(~ μm)	Rapid evaporation with CA [31] and with electrical field [32]

that mimetize biological structures are of great interest for modern medicine. An extra advantage of the supported lipid layers is that their morphology can be measured and manipulated in the range of nanometer scale with scanning probe techniques, such as atomic force microscopy (AFM) [41–45].

The preparation of supported lipid membranes on the surface of inorganic materials as gold and mica gives important results related to phase transition, stability, and morphology of the layers and are promising structures for the development of biosensors. However, it has to be considered that the mechanical properties of the lipid layers in these structures are significantly different from the membranes in fluid environments [41]. The choice of the model has to be done according to the objective of the study, and the model limitations have to be considered when the conclusions are elaborated.

Basically, there are two important methods to prepare membrane assemblies on surfaces that are Langmuir-Blodgett [46] and vesicle fusion [36]. These techniques allow for the preparation of high-quality lipid layers, with a controllable number of layers and large areas free of macroscopic defects. However, they are not easily available or require several experimental steps. An alternative procedure that also yields high-quality lipid layers, the solution spreading method, is simply based on the spreading of organic solutions on solid surfaces [47–49].

The results presented in session 7 were obtained with supported membranes prepared by the vesicle fusion method [36]. Briefly, it consists of dropping the vesicles onto a freshly cleaved mica surface and incubating for 90 minutes. During this time period, the vesicles simply adsorb and explode on the surface or interact with other vesicles by fusion and explode forming the bilayers. The vesicles were prepared by dissolving the lipids in chloroform and drying them under a stream of N_2, followed by a period under vacuum. After that, the dried lipid films were hydrated with HEPES buffer at pH 7.4 [30].

The choice of the solid surface is decided according to the objective to be reached. The majority of studies were done on mica substrates. However, it is possible to use glass for cell fixing and silicon or gold when conducting electrodes are required.

In the case of supported phospholipid layers on gold substrates, it is necessary to first bind thiols molecules on the surface, given that the polar heads of the phospholipids have no affinity for the mild hydrophobic gold surface. Self-assembled monolayers (SAMs) of thiols molecules on gold are very interesting in this case, because they can act as a bridge or a spacer between the inorganic surface and the assembled macromolecules. The reason to create a space between the solid surface and the macromolecules, such as the phospholipids immobilized on a solid surface, ordered in bilayers, is to provide a natural environment to accommodate hydrophilic domains of proteins and to make available the biomolecule transport from one side of the membrane to the other [49–51]. Moreover, the sulfur bridge formed by using substances such as 11-mercaptoundecanol [54] and thiopeptides [50,52,53], has been appropriate to self-organize lipid membranes on gold substrates. The interest on gold surfaces is to preserve this substrate as an electrode for detection of charge transfer processes for fundamental studies and biosensor applications.

2.6   ATOMIC FORCE MICROSCOPY: A BRIEF INTRODUCTION

AFM is a powerful tool for direct visualization of supported biological membranes [42,55]. The AFM uses a sharp probe (tip) on the free end of a flexible cantilever to measure the force between the apex of the tip and the surface of the sample. Depending on the local distance "d" between the tip and the surface, the atomic force will be attractive (van der Waals force, d < 100 nm) or repulsive (d < 1 nm). Different imaging modes were derived from the concept that the surface of the sample should be scanned by the tip in order to measure the surface topography. The usual operation modes are contact, noncontact, and intermittent. The contact mode is most used for hard samples, where the surface is not affected by the dragging forces of the tip. In this imaging mode, the tip contacts the surface at distances corresponding to short scale forces (1 nm or less), and the height (distance "d") is kept constant. The cantilever bending toward or away from the surface is measured by the deflection of a laser. The tip is always kept in contact by a feedback circuit that moves up and down the fixed end of the cantilever through a piezoelectric actuator (i.e., the cantilever bending is kept constant by the feedback system during the scanning of the tip). The piezoelectric material allows a vertical resolution well below 0.1 nm.

The noncontact mode is most used for soft samples. To minimize the tip–sample interaction, the tip does not touch the surface but hovers at long-range distances of about 10 nm. The attractive van der Waals forces acting between the tip and the sample are detected, and topographic images are constructed by scanning the tip above the surface. The sensitivity of this operation mode is increased by using modulation techniques (i.e., the tip oscillates at frequencies near the resonance frequency of the cantilever). During the scanning procedure to reveal the surface topography, the feedback loop maintains the tip to sample distance by keeping constant the amplitude of the oscillation or the resonant frequency.

The intermittent mode measures the topography by tapping the surface with the tip. The tip oscillates on top of the sample during the scanning procedure. This operation mode is applied for hard and soft samples, overcoming most of the problems faced by the other modes of operation, such as dragging forces and sticking of the tip to the surface. During the scanning procedure to determine the topography of the sample, the tip touches the surface at frequencies that are at or near the resonant frequency of the cantilever. The feedback loop maintains constant the amplitude of oscillation, at values of about 20 nm, allowing the measurement of the surface features.

2.7   *IN SITU* ATOMIC FORCE MICROSCOPY FOR LIPID MEMBRANES

For biological samples as supported lipid membranes, it is important to measure the surface topography inside liquid cells. The *in situ* measurements have the additional advantage in that they permit the study of morphological modifications due to biological processes in real time and under physiological conditions. In a previous work, we studied the morphology and stability of supported phospholipid layers prepared by solution spreading (casting) on mica [43]. The images were acquired in the contact or contact-intermittent modes, and the samples were analyzed *ex situ* just after solvent evaporation and after a hydration step and *in situ* with immersion in a buffer solution.

AFM contact mode is the straightforward way to measure samples immersed in a liquid. This operation mode avoids the necessity of having a cantilever oscillating in a resonant frequency damped by the fluid. However, as said above, contact-mode imaging is less suitable for soft or weakly attached materials, because the tip can often scrape or drag the membranes during scanning, a disadvantage that can be overcome by applying intermittent methods. On the other hand, studies have also demonstrated that by adjusting the operative force, it is possible to use contact mode to obtain AFM images of soft phospholipid layers [44].

Figure 2.6 illustrates the AFM setup for *in situ* measurements (i.e., with the tip immersed in the liquid contained in a fluid cell). The figure also shows the laser beam deflected on the free part of the cantilever and the light detector with two segments. This setup was applied to measure the topography of lipid layers grown by vesicle fusion on mica substrates, as will be shown in Figure 2.6 and Figure 2.7.

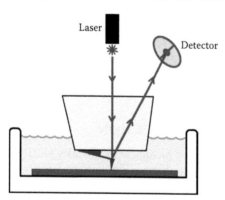

FIGURE 2.6 The atomic force microscopy (AFM) setup for *in situ* measurements. The tip is immersed in the liquid of the fluid cell, and the laser beam is deflected by the free part of the cantilever and detected by the segmented light detector.

FIGURE 2.7 Atomic force microscopy (AFM) topographic image (25 × 25 μm²) of a homogeneously covered mica surface with a 1,2-dimyristoyl-*sn*-glycero-3-phosphatitidylcholine (DMPC) layer prepared by vesicle fusion. The measurements were performed at a temperature of 24°C, showing the lipid phase transition from liquid-crystalline to gel. (**See color insert following page 112.**)

The phospholipid layers studied were 1,2-dimyristoyl-*sn*-glycero-3-phosphatitidylcholine (DMPC) and 1,2-dioleoyl-*sn*-glycero-3-phosphatidylcholine (DOPC), as well as a binary mixture of these phospholipids. The supported membranes were prepared on mica substrates by vesicle fusion method, and the lipid concentration was 0.3 mg/mL. AFM measurements were performed using a Molecular Imaging® PicoSPM atomic force microscope with oxide sharpened silicon nitride probes (nominal spring constant of 0.12 N/m). During the *in situ* measurements, the tip and cantilever were immersed completely in 0.9% of NaCl solution.

Figure 2.7 shows an AFM topographic image of a DMPC bilayer membrane on mica. The membrane covers the entire surface, 25 × 25 μm², homogeneously. The experimental conditions at a temperature of the system fluid cell at 24°C allowed observation of the presence of the liquid crystalline (dark regions) and gel (bright regions) phases of the DMPC, because the phase transition temperature of this lipid is 24°C. For these large lipid-covered surface areas, the height differences between the phases are ≈0.1 nm. The two phases coexist, and the domains were effectively measured *in situ* with the contact mode AFM.

Figure 2.8 shows the topographic image of a binary mixture of DMPC:DOPC in a (1:1) molar ratio. The DOPC has a transition temperature at –19°C and the experiments were performed at ≈19°C. The phase separation is easily observed, with gel DMPC coexisting with liquid-crystalline DOPC. Assuming that the higher domains (brighter regions) are formed by DMPC, step heights of about 1.0 nm to the surrounding DOPC membrane regions were measured, as showed in the profile of Figure 2.8. This approach provided a convenient way to examine the effect of lipid composition on the phase-separated morphology of the binary system, induced by the accentuated difference of transition temperature of the phospholipids.

2.8 CONCLUSION

In this chapter, we discussed lipid membranes as biomimetic systems. Emphasis was given to the preparation and characterization of immobilized lipid layers on solid surfaces and the characterization of these structures *in situ* with AFM. The examples given are DMPC and DMPC:DOPC supported on surfaces of mica substrates.

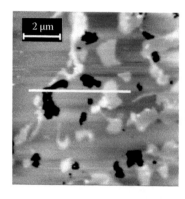
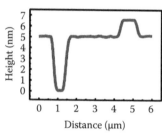

FIGURE 2.8 Contact mode atomic force microscopy (AFM) topographic image and line scan profile obtained *in situ*, at ≈19°C, of a (1:1) 1,2-dimyristoyl-*sn*-glycero-3-phosphatitidylcholine (DMPC):1,2-dioleoyl-*sn*-glycero-3-phosphatidylcholine (DOPC) mixture.

REFERENCES

1. Maldonado-Perez, D., Riccardi, D. *Recent Res. Develop. Physiol.*, v.2, p.403 (2004).
2. Millhorn, D.E., Bayliss, D.A., Erickson, J.T., Gallman, E.A., Szymeczek, C.L., Czyzyk-Krezeska, M., Dean, J.B. *Am. J. Physiol.*, v.257, 6, p.L289 (1989).
3. Bucking, W., Urban, G.A., Nann, T. *Sens. Actuators B*, v.104, p.111 (2005).
4. Halliwell, B., Gutteridge, M.C. *Free Rad. Biol. Med.* New York: Oxford University Press (2000).
5. Malheiros, S.V.P., Luciana Pinto, M.A., Gottardo, L., Yokaichiya, D.K., Fraceto, L.F., Meirelles, N.C., Paula, E. *Biophys. Chem.* v.110, p.213 (2004).
6. Yeagle, P.L. *The Membranes of Cells.* New York: Academic Press (1993).
7. Bloom, M., Evans, E., Mouritsen, O.G. *Q. Rev. Biophys.*, v.24, p.293 (1991).
8. Curtis, M.T., Gilfor, D., Farber, J.L. *Arch. Biochem. Biophys.*, v.235, p.644 (1984).
9. Singer, S.J., Nicholson, G.L. *Science*, v.175, p.720 (1972).
10. Kirchhausen, T. *Annu. Rev. Cell. Dev. Biol.*, v.15, p.705 (1999).
11. Hirabayashi, Y., Ichikawa, S. *Roles of Glycolipids and Sphingolipids in Biological Membrane: The Frontiers in Molecular Biology Series* (Eds., Fukuda, M., Hindsgaul, O.), IRL Press at Oxford Press (1999).
12. Anderson, R.G.W., Jacobson, K. *Science*, v.296, p.1821 (2002).
13. Galbiati, F., Razani, B., Lisanti, M.P. *Cell*, v.106, p.403, 2001.
14. Beney, L., Gervais, P. *Appl. Microbiol. Biotechnol.*, v. 57, p. 34 (2001).
15. Mouritsen, O.G. *Chem. Phys. Lipids*, v.57, p.179 (1991).
16. Raggers, R.J., Pomorski, T., Holthuis, J.C., Kalin, N., Van Meer, G. *Traffic*, v.1, p.226 (2000).
17. Kinnunen, P.K.J., Kôiv, A., Lehtonen, J.Y.A., Rytömaa, M., Mustonen, P. *Chem. Phys. Lipids*, v.73, p.181 (1994).
18. Ohvo-Rekilä, H., Ramstedt, B., Leppimäki, P., Slotte, J.P. *Prog. Lipid Res.*, v.41, p. 66 (2002).
19. Barenholz, Y., Thompson, T.E. *Chem. Phys. Lipids*, v.102, p. 29 (1999).
20. Hauser, H., Pascher, I., Pearson, R.H., Sundell, S. *Biochim. Biophys. Acta*, v.650, p.21 (1981).
21. Kates, M., Pugh, E.L., Ferrante, G. *Biomembranes*, v.12, p.379 (1984).
22. Cossins, A.R. *Temperature Adaptation of Biological Membranes.* London: Portland Press (1994).
23. Samuni, A.M., Lipman, A., Barenholz, Y. *Chem. Phys. Lipids*, v.105, p.121 (2000).
24. Casteli, F., Trombeta, D., Tomaiano, A., Bonina, F., Romeo, G., Uccella, N., Saija, A. *J. Pharmacol. Toxicol. Meth.*, v.37, p.135 (1997).
25. Lima, V.R., Morfim, M.P., Teixeira, A., Creczynski-Pasa, T.B. *Chem. Phys. Lipids*, v.132, p.197 (2004).
26. Rigaud, J.L., Bluzat, A., Büscheln, S. *Biochim. Biophys. Res. Commun.*, v.111, p.373 (1983).
27. Szoka, F., Papahadjopoulos, D. *Proc. Natl. Acad. Sci. USA*, v.75, p.4194 (1978).
28. Costa, E.J., Shida, C.S., Biaggi, M.H., Ito, A.S., Lamy-Freund, M.T. *FEBS Lett.*, v.416, p.103 (1997).
29. Sone, N., Yoshida, M., Hirata, H., Kagawa, Y. *J. Biochem.*, v.81, p.519 (1977).
30. Hope, M.J., Bally, M.B., Mayer, L.D., Janoff, A.S., Cullis, P.R. *Chem. Phys. Lipids*, v.40, p.89 (1986).

31. Moscho, A., Orwar, O., Chiu, D.T., Modi, B.P., Zare, R.N. *Proc. Natl. Acad. Sci. USA*, v.93, p.11443 (1996).
32. Estes, D.J., Mayer, M. *Biochim. Biophys. Acta*, v.1712, p.152 (2005).
33. Poglitsh, C.L., Thomson, N.L. *Biochemistry*, v.29, p.248 (1990).
34. Kumar, S., Hoh, J.H. *Langmuir*, v.16, p.9936 (2000).
35. Sackman, E. *Science,* v.271, p. 43 (1996).
36. Reviakine, I., Simon, A., Brisson, A. *Langmuir*, v.16, p.1473 (2000).
37. Mou, J., Czajkowsky, D.M., Zhang, Y., Shao, Z. *FEBS Lett.*, v.371, p.279 (1995).
38. Groves, J.T., Ulman, N., Boxer, S.G. *Science*, v.275, p.551 (1997).
39. Lawrie, G.A., Barnes, G.T., Gentle, I.R. *Colloids Surf. A*, v.155, p.69 (1999).
40. Muresan, A.S., Lee, K.Y.C. *J. Phys. Chem. B*, v.105, p.852 (2001).
41. Schneider, J., Dufrêne, Y.F., Barger, Jr., W.R., Lee, G.U. *Biophys. J.*, v.79, p.1107 (2000).
42. Tokumasu, F., Jin, A.J., Dvorak, J.A. *J. Electron. Micros.*, v.51, p.1 (2002).
43. Spangenberg, T., De Mello, N.F., Creczynski-Pasa, T.B., Pasa, A.A., Niehus, H. *Phys. Stat. Sol. A*, v.201, p.857 (2004).
44. Munford, L.M., Lima, V.R., Vieira, T.O., Heinzelmann, G., Creczynski-Pasa, T.B., Pasa, A.A. *Microsc. Microanal.*, v.11, Supp.3, p.154 (2005).
45. McConnell, H.M., Tamm, L.K., Weis, R.M. *Proc. Natl. Acad. Sci. USA*, v.81, p.3249 (1984).
46. Blodgett, K.A., Langmuir, I. *Phys. Rev.*, v.51, p.964 (1937).
47. McConnell, H.M., Tamm, L.K., Weis, R.M. *Proc. Natl. Acad. Sci. USA*, v.81, p.3249 (1984).
48. Seul, M., Sammon, M.J. *Thin Solid Films*, v.185, p.287 (1990).
49. Wang, L., Song, Y., Han, X., Zhang, B., Wang, E. *Chem. Phys. Lipids*, v.123, p.177 (2003).
50. Poirier, G.E. *Chem. Rev.*, v.97, p.1117 (1997).
51. Lang, H., Duchl, C., Vogel, H. *Langmuir*, v.10, p.197 (1994).
52. Naumann, R., Jonckzyc, A., Hampel, C., Ringdorf, H., Knoll, W., Bunjes, N., Graeber, P. *Bioelectrochem. Bioenerg.*, v.42, p.241 (1997).
53. Bunjes, N., Schmidt, E.K., Jonczyk, A., Rippmann, F., Beyer, D., Ringsdorf, H., Graeber, P., Knoll, W., Naumann, R. *Langmuir*, v.13, p.6188 (1997).
54. Bucking, W., Urban, G.A., Nann, T. *Sens. Actuators B*, v.104, p.111 (2005).
55. Alessandrini, A., Facci, P. *Meas. Sci. Technol.*, v.16, p.R65 (2005).

3 Mesenchymal Stem Cells and Controlled Nanotopography

Matthew J. Dalby and Richard O.C. Oreffo
University of Glasgow, Glasgow, Scotland, United Kingdom

CONTENTS

3.1 INTRODUCTION

The natural environment of the cell will contain topographical information. This may be on the microscale presented by large protein chains, other cells, and from other matrix constituents (for example, mineral in bone). The topography may also be on the nanoscale presented to the cells from protein folding and banding and nanocrystalline elements. It has been known for many years that cells will react to the shape of their environment [1], and this phenomenon was later termed *contact guidance* [2].

The borrowing of lithographical techniques from the microelectronics industry facilitated research into cell response to the topography of the cell environment. Initially, photolithography was used, which allowed fabrication of micron-scale (width, diameter) features such as grooves and pits with submicron depths [3]. Every cell type tested responded to the features by contact guidance [4–8]. Further examination revealed that contact guidance led to changes in cell adhesion, migration, cytoskeletal organization, and genomic regulation [9].

As fabrication techniques evolved, exploration into the cellular response to the nanoenvironment became possible. Microelectronics techniques such as electron beam lithography (EBL) have allowed fabrication of features with 10 nm X and Y dimensions, and theoretically, the Z dimension could be smaller if controlled. EBL is the most high-resolution top-down fabrication tool [10,11]. Other fabrication techniques include colloidal lithography, where monodispersed nanocolloids are used as an etch mask from which to create nanofeatures with controlled height and diameter but random placement [12], and polymer demixing, where spontaneous phase separation of polymers in a solvent creates nanotopography with reproducible height but random distribution and diameters [13].

We note at this point that it is envisaged that bottom-up fabrication, or the manipulation of individual molecules (or even atoms) to produce ultra-small structures, may play a critical role in the production of topographies for cells to react to, although this is presently largely in the realms of theory.

For top-down approaches, in the past few years, however, there has been considerable research effort into ascertaining the breadth of cellular response to nanoscale features. Again, contact guidance has been observed in many cell types. The physical contact guidance (that is, alignment), however, is on a different scale—that of the filopodia, which shall be discussed shortly.

3.2   MESENCHYMAL STEM CELLS

Before discussion of how mesenchymal stem cells may respond to nanotopography, a brief introduction to the cells is required. Bone is characterized by an extraordinary potential for growth, regeneration, and remodeling throughout life. This is largely due to the directed differentiation of mesenchymal cells into osteogenic cells, a process subject to exquisite regulation and complex interplay by a variety of hormones, differentiation factors, and environmental cues present within the bone matrix [14–16].

The osteoblast, the cell responsible for bone formation, is derived from pluripotent mesenchymal stem cells (MSCs). These MSCs can give rise to cells of the adipogenic, reticular, osteoblastic, myoblastic, and fibroblastic lineages and generate progenitors committed to one or more cell lines with an apparent degree of plasticity or interconversion [17–20]. Thus, the MSC gives rise to a hierarchy of bone cell populations artificially divided into a number of developmental stages, including MSC, determined osteoprogenitor cell (DOPC), preosteoblast, osteoblast, and ultimately, osteocyte [21–24].

3.3   CELL FILOPODIA

It seems likely that filopodia are one of the cells' main sensory tools. Gustafson and Wolpert first described filopodia in living cells in 1961 [25]. They observed mesenchymal cells migrating up the interior wall of the blastocoelic cavity in sea urchins and noted that the filopodia produced appeared to explore the substrate. This led them to speculate that they were being used to gather spatial information by the cells.

When considering filopodial sensing of topography, fibroblasts have been described as using filopodia to sense and align the cells to microgrooves [5]. Macrophages have been reported to sense grooves down to a depth of 71 nm by actively producing many filopodia and elongating in response to the shallow topography [7]. Although distinctly different from the filopodia of the aforementioned cell types, neuronal growth cone filopodia have been described as first sensing microgrooves and then aligning neurons to the grooves [26,27].

Cytoskeletal actin bundles drive the filopodia, and as the filopodia encounter a favorable guidance cue, they become stabilized following the recruitment of microtubules and accumulation of actin in a direction predictive of the future turn if a cell is to experience contact guidance.

Once cells locate a suitable feature using the filopodia presented on the cell's leading edge, lamellipodium are formed which move the cell to the desired site [28]. These actions require G-protein signaling and actin cytoskeleton. Specifically of interest are Rho, Rac, and Cdc42. Rho induces actin contractile stress fiber assembly to allow the cell to pull against the substrate, Rac induces lamellipodium formation, and Cdc42 activation is required for filopodial assembly [29]. Rho and Rac are both required for cell locomotion, but cells can translocate when Cdc42 is knocked out. Cells lacking Cdc42 cannot, however, sense chemotactic gradients and simply migrate in a random manner [30]. This, again, presents compelling evidence for filopodial involvement in cell sensing.

We recently produced evidence that MSCs also use filopodia to probe their nanoenvironment. In fact, it appears that the filopodia of these progenitor cells are more highly sensitive to nanotopography than with more mature cell types [31]. The smallest feature (thus far) that cells (fibroblasts) have been observed to respond to are 10 nm high-polymer demixed islands (Figure 3.1) [28]. MSCs

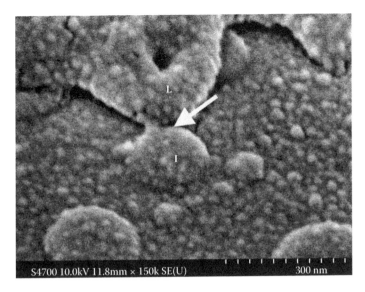

FIGURE 3.1 Fibroblast contact guiding to a 10-nm-high island (I). Filopodial interaction has caused lamellae to form in the direction of the island. The arrow shows the area of guidance.

have now been shown to produce filopodia in response to a variety of nanotopographies, the smallest being 40 nm high-polymer demixed islands [32]. It is clear that this is not the limit of their sensory capability, but rather the limit of the emerging cell testing so far.

On grooved substrata, the classical examples of contact guidance (i.e., alignment to the grooves) are quickly seen in MSCs after filopodial guidance (Figure 3.2) [33].

Filopodia seem to be likely candidates for nanotopographical sensing for another reason—that of size. MSCs are large cells, and when well spread can have lengths and widths of several hundreds of micrometers (μm). Some of the topographies tested, however, are several orders of magnitude smaller than the cell, and for many years, it seemed unlikely that the cells would notice them. However, the tips of filopodia are approximately 100 nm in diameter, thus putting them on a similar scale to the topographies now being fabricated.

Thus far, topographies that induce filopodial interactions and cell spreading have been discussed. A second mode of nanotopographical action on cells is reduction of adhesion and spreading. This is typically achieved using highly ordered pits fabricated by EBL. Such pits (e.g., 120

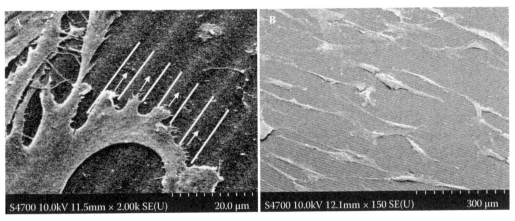

FIGURE 3.2 Contact guidance of mesenchymal stem cells to grooves. (A) Filopodial outgrowth along the ridges of 100-nm-high (5-μm-wide) grooves. (B) Full cell alignment along 300-nm-high (5-μm-wide) grooves.

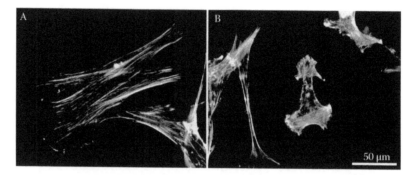

FIGURE 3.3 Mesenchymal stem cells stained for actin cytoskeleton after culture on (A) planar control and (B) electron beam lithography (EBL) nanopits. Note the well-spread cells on the control with well-defined actin stress fibers and compare to the small, stellate cells on the nanopits with puncatate actin.

nm diameter, 100 nm deep, and 300 nm center–center spacing in a square arrangement) have been shown to reduce adhesion of epitenon cells [34], fibroblasts [35] and now MSCs [36] (Figure 3.3).

3.4   CELL CYTOSKELETON AND CELLULAR ADHESIONS

The ability of a cell to form adhesions will affect the organization of the cytoskeletons and alter the cell's mechanotransductive pathways. It was recently published that nanotopography can change the mesenchymal stem cells' ability to form focal adhesions. (This has regularly been observed with differentiated cell types, and it has recently been shown with osteoblasts that nanotopography can change the type of adhesion formed, from very small and immature focal complexes to large fibrillar adhesions that align to endogenous matrix proteins [37].)

The cytoskeleton is anchored to the adhesions, with actin microfilaments linked directly. The cytoskeleton is a network of protein filaments extending through the cell cytoplasm within eukaryotic cells. The cytoskeleton is of fundamental importance in the control of many aspects of cell behavior, movement, and metabolism, including proliferation, intracellular signaling, movement, and cell attachment. The three main cytoskeletal components are microfilaments (actin), microtubules (tubulin), and intermediate filaments (for example, vimentin, cytokeratin, and desmin depending on cell type).

There are two types of microfilament bundles, involving different bundling proteins. One type of bundle, containing closely spaced actin filaments aligned in parallel, supports projections in the cell membrane (filopodia or mikrospikes involved in cell sensing, lamellipodia involved in cell crawling). The second type is composed of more loosely spaced filaments and is capable of contraction; these are the stress fibers. The ability of cells to crawl across substrate surfaces is a function of the actin cytoskeleton (by pulling [contracting] against focal adhesions).

The microtubules provide a system by which vesicles and other membrane-bound organelles may travel. They also help regulate cell shape, movement, and the plane of cell division. Microtubules, composed of tubulin, exist as single filaments that radiate outward through the cytoplasm from the centrosome, near the nucleus. Microtubules emanating from the centrosome (microtubule organizing center) act as a surveying device that is able to find the center of the cell.

Intermediate filaments are tough protein fibers in the cell cytoplasm. They extend from the nucleus in gently curving arrays to the cell periphery, and they are particularly dominant when the cells are subject to mechanical stress. The intermediate filaments are classified into four broad bands: type I are keratin-based proteins, type II includes vimentin and desmin, type III includes neurofilament proteins, and type IV are nuclear lamins. In cells of mesenchymal origin, vimentin is the main intermediate filament protein. The filaments provide mechanical support for the cell and nucleus [38–41].

As shown on the EBL pits, nanometer (nm) scale topographies will alter cytoskeletal organization [42–44]. Other topographies, such as grooves produced by photolithography (PL) and features produced by polymer demixing and colloidal lithography can increase cytoskeletal organization of MSCs [32,33].

3.5 MECHANOTRANSDUCTION

The ability to change cytoskeletal organization will lead to alterations in proliferation and differentiation, as many signaling cascades are regulated by the cytoskeleton. These mechanotransductive signaling events may be chemical (e.g., kinase based, linked to focal adhesions influenced by cytoskeletal contraction). An example would be integrin gathering as an adhesion is formed, which will activate myosin light-chain kinase (MCLK), which will generate actin–myosin sliding (the key event in stress fiber contraction), and in turn will change focal adhesion kinase (FAK) activity. Cytoskeletal involvement in contraction against adhesions will also alter calcium influx and G-protein events. These chemical signaling events are collectively known as indirect mechanotransduction [39].

Another form of mechanotransduction, direct, is considered to be transduced by the cytoskeletons as an integrated unit. The most accepted theory is that of cellular tensegrity, whereby the cell's mechanical structure is explained via tensional integrity [45–51]. Through this tensegrity structure, tensional forces from the extracellular environment (e.g., from tissue loading or changes in cell spreading) are possibly conferred to the nucleus and alter genome regulation [52–56].

We currently support the idea that the intermediate filaments of the cytoskeleton are linked to the lamin intermediate filaments of the nucleoskeleton. It is known that the telomeric ends of the interphase chromosomes are intimately linked to the peripheral lamins [57]. Thus, tension directed through the cytoskeleton may be passed directly to the chromosomes during gene transcription. Changes in chromosomal three-dimensional arrangement may affect transcriptional events such as access to the genes by transcription factors and polymerases. Also, changes in DNA tension can cause polymerase enzymes to slow down, speed up, or even stall [58]. These may be mechanisms by which changes in cell spreading, as seen by MSCs on nanotopographies, can change differentiation events.

At this point, it is important to mention another discrete theory of how nanotopography may alter cytoskeletal mechanotransductive events, which has recently been mooted—nanoimprinting into cells by nanofeatures [59,60]. This describes a phenomenon that has been clearly seen in platelets and for which some evidence has been provided in more complex cell types. For nanoimprinting to occur, the pattern of the topography must be transferred to the cytoskeletal filaments (i.e., the topographies produce a template that is favorable or unfavorable to condensation of cytoskeletal polymer chains through invagination of the basal membrane against the topography). As with filopodia, the dimensions of the cytoskeletal elements are small enough to warrant consideration of influence by nanoscale features. Figure 3.4 shows a fibroblast with colloidal nanocolumns impressing into the basal membrane. There is evidence that this leads to increased "attempted" endocytosis, that is, the cells recognize the features as being in the correct size range to try to endocytose and to form claterin-coated pits [61]. Such endocytotic vesicles are moved by actin cables,

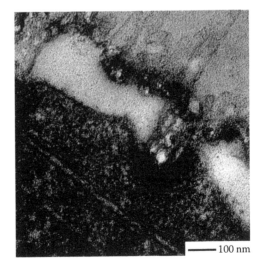

————— 100 nm

FIGURE 3.4 Nanoimprinting into the basal membrane of a fibroblast.

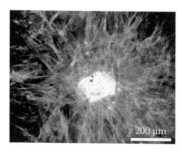

FIGURE 3.5 Osteocalcin stain showing nodule formation in a mesenchymal stem cell population cultured on 400-nm-deep pits with a 40 µm diameter.

and perhaps this mechanism causes the topography mimicking actin patterning described by Curtis and others [60].

3.6 DIFFERENTIATION

A key question arising from the above is can nanotopography alter cellular differentiation? The answer is yes. When MSCs are cultured *in vitro* on planar materials without chemical treatment (either in the material, such as hydroxyapatite, or in the media, such as dexamethasone), they are unlikely to become osteoblastic and produce osteocalcin-rich nodules as they start to produce bone mineral (apatite) and instead remain fibroblastic in appearance. This is a key problem with orthopedic biomaterials for load-bearing bone replacement. (Please see review by Balasundaram and Webster for in-depth orthopedic considerations of nanofeatures [62].) The materials come into contact with MSCs of the bone marrow and do not encourage the expression of an osteoblastic phenotype. This produces a weak repair with disorganized tissue formation and implant encapsulation in soft tissue (i.e., it appears that the cells need cues from the materials in order to commit to a specialized differentiation pathway).

It seems that topography can produce such cues. *In vitro* culture in basal media (without dexamethasone, etc.) on nanosurfaces can induce increased expression of osteocalcin and osteopontin (Figure 3.5) [32,33]. On planar control, however, no nodules tend to be observed. This demonstrates that the phenotype has been changed.

3.7 SUMMARY

Thus, it is seen that nanoscale topographies have strong effects on MSCs and can alter their commitment to different phenotypical lineages through mechanisms of spatial sensing and cytoskeletal organization. It is likely that stem cells, both adult and embryonic, will be very sensitive to their environment. All cells have the same genome, transcriptional machinery, and transcription factor pool, yet stem cells can take on different roles depending on their locations in the body [63].

Topography is likely to be one cue to which they will respond. Others will probably be chemistry (interleukins, hormones, surrounding matrix proteins, etc.) and possibly the modulus of their surroundings.

If you change the chemistry, Young's modulus, or topography of the cell environment, this changes the actual modulus of the cell (possibly from changes in cytoskeletal organization). This will lead to mechanotransductive changes. Certainly, changes in cytoskeletal tension cause changes in MSC commitment. Simply by changing the modulus of the environment from soft to hard, MSCs can form either adipocytes or osteoblasts, respectively [64].

Throughout, we have not considered nanoscale roughness. Due to its ease of production, nanoroughness will probably find crucial roles in biomaterials science. However, the results are unclear and can be conflicting (i.e., materials with similar measures of roughness, e.g., Ra, can produce very different results for, e.g., osseointegration). Thus, it is our opinion that controlled topographies are more valuable learning tools at this stage.

The potential for use of nanotopography and nanoroughness in medical science is huge. The creation of "next-generation" materials that elicit desired responses combined with multipotent stem cells is likely to underpin the regenerative field of tissue engineering. This will be aided by the stem cells' apparent desire to seek out a complex array of cues from their extracellular environment. Indeed, it is starting to look like stem cells are mechanically interconnected with their environment and are "hot-wired" to detect very subtle changes, almost as if part of the extracellular filamentous protein percolation network.

ACKNOWLEDGMENTS

Matthew Dalby is supported by a Biotechnology and Biological Sciences Research Council (BBSRC) David Phillips Fellowship. Richard Oreffo is supported by awards from the BBSRC and Engineering and Physical Science Research Council (EPSRC).

REFERENCES

1. Weiss P, Garber B. Shape and movement of mesenchyme cells as functions of the physical structure of the medium. *Proc. Natl. Acad. Sci. USA* 1952;38:264–280.
2. Curtis ASG, Varde M. Control of cell behaviour: Topological factors. *J. Nat. Cancer Res. Inst.* 1964;33:15–26.
3. Wilkinson CDW, Riehle M, Wood M, Gallagher J, Curtis ASG. The use of materials patterned on a nano- and micro-metric scale in cellular engineering. *Mater. Sci. and Eng.* 2002;19:263–269.
4. Clark P, Connolly P, Curtis AS, Dow JA, Wilkinson CD. Topographical control of cell behaviour. I. Simple step cues. *Development* 1987;99(3):439–448.
5. Clark P, Connolly P, Curtis AS, Dow JA, Wilkinson CD. Cell guidance by ultrafine topography *in vitro*. *J. Cell Sci.* 1991;99 (Pt 1):73–77.
6. Wojciak-Stothard B, Curtis ASG, Monaghan W, McGrath M, Sommer I, Wilkinson CDW. Role of the cytoskeleton in the reaction of fibroblasts to multiple grooved substrata. *Cell Motil. Cytoskeleton* 1995;31:147–158.
7. Wojciak-Stothard B, Madeja Z, Korohoda W, Curtis A, Wilkinson C. Activation of macrophage-like cells by multiple grooved substrata—Topographical control of cell behavior. *Cell Biol. Int.* 1995:485–490.
8. Britland S, Morgan H, Wojciak-Stothard B, Riehle M, Curtis A, Wilkinson C. Synergistic and hierarchical adhesive and topographic guidance of BHK cells. *Exp. Cell Res.* 1996;228:313–325.
9. Dalby MJ, Riehle MO, Yarwood SJ, Wilkinson CD, Curtis AS. Nucleus alignment and cell signaling in fibroblasts: Response to a micro-grooved topography. *Exp. Cell Res.* 2003;284(2):274–282.
10. Gadegaard N, Thoms S, MacIntyre DS, McGhee K, Gallagher J, Casey B, et al. Arrays of nano-dots for cellular engineering. *Microelectron. Eng.* 2003;67–68:162–168.
11. Vieu C, Carcenac F, Pepin A, Chen Y, Mejias M, Lebib A, et al. Electron beam lithography: Resolution limits and applications. *Appl. Surf. Sci.* 2000;164:111–117.
12. Denis FA, Hanarp P, Sutherland DS, Dufrene YF. Fabrication of nanostructured polymer surfaces using colloidal lithography and spin coating. *Nanoletters* 2002;2:1419–1425.
13. Affrossman S, Henn G, O'Neill SA, Pethrick RA, Stamm M. Surface topography and composition of deuterated polystyrene—Poly(bromostyrene) blends. *Macromolecules* 1996;29(14):5010–5016.
14. Bianco P, Riminucci M, Gronthos S, Robey PG. Bone marrow stromal stem cells: Nature, biology, and potential applications. *Stem Cells* 2001;19(3):180–192.
15. Bianco P, Robey PG. Stem cells in tissue engineering. *Nature* 2001;414:118–121.
16. Oreffo ROC. Growth factors for skeletal reconstruction and fracture repair. *Curr. Opin. Investig. Drugs* 2004;5(4):419–423.
17. Yang X, Tare RS, Partridge KA, Roach HI, Clarke NM, Howdle SM, et al. Induction of human osteoprogenitor chemotaxis, proliferation, differentiation, and bone formation by osteoblast stimulating factor-1/pleiotrophin: Osteoconductive biomimetic scaffolds for tissue engineering. *J. Bone Miner. Res.* 2003;18(1):47–57.
18. Rose FR, Oreffo RO. Bone tissue engineering: hope vs hype. *Biochem. Biophys. Res. Commun.* 2002;292(1):1–7.
19. Triffitt JT, Oreffo RO. Osteoblast Lineage. In: Zaidi M, editor. *Advances in Organ Biology: Molecular and Cellular Biology of Bone*. Greenwich, CT: JAI Press; 1998. pp. 475–498.
20. Friedenstein AJ. Precursor cells of mechanocytes. *Int. Rev. Cytol.* 1976;47:327–359.
21. Mirmalek-Sani SH, Roach HI, Wilson DI, Hanley NA, Oreffo ROC. Characterisation of human fetal populations: A comparative model for skeletal and stem cell differentiation. *J. Bone Miner. Res.* 2005;20:1292.

22. Locklin RM, Oreffo RO, Triffitt JT. Modulation of osteogenic differentiation in human skeletal cells in vitro by 5-azacytidine. *Cell Biol. Int.* 1998;22(3):207–215.

23. Oreffo RO, Bord S, Triffitt JT. Skeletal progenitor cells and ageing human populations. *Clin. Sci.* (London) 1998;94(5):549–555.

24. Oreffo RO, Triffitt JT. Future potentials for using osteogenic stem cells and biomaterials in orthopedics. *Bone* 1999;25(2 Suppl):5S–9S.

25. Gustafson T, Wolpert L. Studies on the cellular basis of morphogenesis in the sea urchin embryo. Directed movements of primary mesenchvme cells in normal and vegetalized larvae. *Exp. Cell Res.* 1961;24:64–79.

26. Rajnicek A, McCaig C. Guidance of CNS growth cones by substratum grooves and ridges: Effects of inhibitors of the cytoskeleton, calcium channels and signal transduction pathways. *J. Cell Sci.* 1997;110 (Pt 23):2915–2924.

27. Rajnicek A, Britland S, McCaig C. Contact guidance of CNS neurites on grooved quartz: Influence of groove dimensions, neuronal age and cell type. *J. Cell Sci.* 1997;110 (Pt 23):2905–2913.

28. Dalby MJ, Riehle MO, Johnstone H, Affrossman S, Curtis AS. Investigating the limits of filopodial sensing: A brief report using SEM to image the interaction between 10 nm high nano-topography and fibroblast filopodia. *Cell Biol. Int.* 2004;28(3):229–236.

29. Schmitz AA, Govek EE, Bottner B, Van Aelst L. Rho GTPases: Signaling, migration, and invasion. *Exp. Cell Res.* 2000;261(1):1–12.

30. Jones GE, Allen WE, Ridley AJ. The Rho GTPases in macrophage motility and chemotaxis. *Cell Adhes. Commun.* 1998;6(2–3):237–245.

31. Hart A, Gadegaard N, Wilkinson CDW, Oreffo ROC, Dalby MJ. Osteoprogenitor response to low-adhesion nanotopographies originally fabricated by electron beam lithography. *J. Mater. Sci. Mat. in Med.* 2007;18(6):1211–1218.

32. Dalby MJ, McCloy D, Robertson M, Agheli H, Sutherland D, Affrossman S, et al. Osteoprogenitor response to semi-ordered and random nanotopographies. *Biomaterials* 2006;27(15):2980–2987.

33. Dalby MJ, McCloy D, Robertson M, Wilkinson CDW, Oreffo ROC. Osteoprogenitor response to defined topographies with nanoscale depths. *Biomaterials* 2006;27:1306–1315.

34. Gallagher JO, McGhee KF, Wilkinson CDW, Riehle MO. Interaction of animal cells with ordered nanotopography. *IEEE Trans. on Nanobiosci.* 2002;1(1):24–28.

35. Dalby MJ, Gadegaard N, Riehle MO, Wilkinson CD, Curtis AS. Investigating filopodia sensing using arrays of defined nano-pits down to 35 nm diameter in size. *Int. J. Biochem. Cell Biol.* 2004;36(10):2015–2025.

36. Dalby MJ, Gadegaard N, Tare R, Andar A, Riehle MO, Herzyle P, Wilkinson CDW, Oreffo ROC. The control of human mesenchymal cell differentiation using nanoscale symmetry and disorder. *Nature Mat.* 2007;6(12):997–1003.

37. Biggs MJP, Richards RG, Gadegaard N, Wilkinson CDW, Dalby MJ. Characterisation of S-phase primary osteoblast adhesion complexes on nanopitted materials fabricated by electron-beam lithography. *J. Orthopaedic Res.* 2007;25(2):273–282.

38. Alberts B, Bray D, Lewis J, Raff M, Watson J. *Molecular Biology of the Cell.* New York: Garland; 1994.

39. Cooper GM. *Cell.* Sunderland, MA: Sinauer Associates; 2000.

40. Burridge K, Chrzanowska-Wodnicka M. Focal adhesions, contractility, and signaling. *Annu. Rev. Cell Dev. Biol.* 1996;12:463–518.

41. Amos LA, Amos WB. *Molecules of the Cytoskeleton.* London: MacMillan; 1991.

42. Dalby MJ, Riehle MO, Johnstone H, Affrossman S, Curtis AS. *In vitro* reaction of endothelial cells to polymer demixed nanotopography. *Biomaterials* 2002;23(14):2945–2954.

43. Dalby MJ, Childs S, Riehle MO, Johnstone HJ, Affrossman S, Curtis AS. Fibroblast reaction to island topography: Changes in cytoskeleton and morphology with time. *Biomaterials* 2003;24(6):927–935.

44. Dalby MJ, Riehle MO, Johnstone HJ, Affrossman S, Curtis AS. Polymer-demixed nanotopography: Control of fibroblast spreading and proliferation. *Tissue Eng.* 2002;8(6):1099–1108.

45. Ingber DE. Tensegrity II. How structural networks influence cellular information processing networks. *J. Cell Sci.* 2003;116(Pt 8):1397–1408.

46. Ingber DE. Tensegrity I. Cell structure and hierarchical systems biology. *J. Cell Sci.* 2003;116(Pt 7):1157–1173.

47. Ingber DE. Mechanosensation through integrins: Cells act locally but think globally. *Proc. Natl. Acad. Sci. USA* 2003;100(4):1472–1474.

48. Maniotis AJ, Chen CS, Ingber DE. Demonstration of mechanical connections between integrins, cytoskeletal filaments, and nucleoplasm that stabilize nuclear structure. *Proc. Natl. Acad. Sci. USA* 1997;94(3):849–854.

49. Maniotis AJ, Bojanowski K, Ingber DE. Mechanical continuity and reversible chromosome disassembly within intact genomes removed from living cells. *J. Cell Biochem.* 1997;65(1):114–130.

50. Ingber DE. Cellular tensegrity: Defining new rules of biological design that govern the cytoskeleton. *J. Cell Sci.* 1993;104 (Pt 3):613–627.

51. Charras GT, Horton MA. Single cell mechanotransduction and its modulation analyzed by atomic force microscope indentation. *Biophys. J.* 2002;82(6):2970–2981.

52. Dalby MJ. Topographically induced direct cell mechanotransduction. *Med. Eng. and Phys.* 2005;27(9):730–742.

53. Dalby MJ, Riehle MO, Sutherland DS, Agheli H, Curtis AS. Use of nanotopography to study mechanotransduction in fibroblasts—Methods and perspectives. *Eur. J. Cell Biol.* 2004;83(4):159–169.

54. Heslop-Harrison JS. Comparative genome organization in plants: From sequence and markers to chromatin and chromosomes. *Plant Cell* 2000;12(5):617–636.

55. Heslop-Harrison JS, Leitch AR, Schwarzacher T. The physical organisation of interphase nuclei. In: Heslop-Harrison JS, R.B. F, editors. *The Chromosome.* Oxford: Bios; 1993. pp. 221–232.

56. Mosgoller W, Leitch AR, Brown JK, Heslop-Harrison JS. Chromosome arrangements in human fibroblasts at mitosis. *Hum. Genet.* 1991;88(1):27–33.

57. Foster HA, Bridger JM. The genome and the nucleus: A marriage made by evolution. Genome organisation and nuclear architecture. *Chromosoma* 2005;114(4):212–229.

58. Bustamante C, Bryant Z, Smith SB. Ten years of tension: Single-molecule DNA mechanics. *Nature* 2003;421(6921):423–427.

59. Curtis ASG, Dalby MJ, Gadegaard N. Cell signaling arising from nanotopography: Implications for nanomedical devices. *Nanomedicine* 2006;1(1):67–72.

60. Curtis ASG, Dalby MJ, Gadegaard N. Nanoimprinting onto cells. *J. R. Soc. Interface* 2006;3:393–398.

61. Dalby MJ, Berry CC, Riehle MO, Sutherland DS, Agheli H, Curtis AS. Attempted endocytosis of nano-environment produced by colloidal lithography by human fibroblasts. *Exp. Cell Res.* 2004;295(2):387–394.

62. Balasundaram G, Webster TJ. Nanotechnology and biomaterials for orthopaedic medical applications. *Nanomedicine* 2006;1(2):169–176.

63. Getzenberg RH. Nuclear matrix and the regulation of gene expression: tissue specificity. *J. Cell Biochem.* 1994;55(1):22–31.

64. Engler AJ, Sen S, Sweeney HL, Discher DE. Matrix elasticity directs stem cell lineage specification. *Cell* 2006;126(4):677–689.

4 Biological Applications of Optical Tags Based on Surface-Enhanced Raman Scattering

William E. Doering, Michael Y. Sha,
David Guagliardo, Glenn Davis, Remy Cromer,
Michael J. Natan, and R. Griffith Freeman
Oxonica, Inc., Mountain View, California, USA

CONTENTS

4.1 INTRODUCTION: OPTICAL DETECTION TAGS BASED ON GOLD (AU) NANOPARTICLES

Detection tags are needed in biological assays to identify specific products or events from among the thousands of molecules and events present in a cell or biological extract. Although there are measurement techniques that can directly detect a wide variety of native molecules (including mass spectrometry and nuclear magnetic resonance [NMR]), in complex biological systems, it is often useful to tag a specific molecule of interest to make it "visible" to a detection system. Optical tagging techniques are often favored because they can be applied to virtually any measurement format, from microarrays to cellular assays to *in vivo* applications.

The use of colloidal gold (Au) nanoparticles for biological applications dates back at least 30 years, when they found use as electron-dense markers for transmission electron microscopy (TEM) studies of cells [1]. Since then, a plethora of additional applications have emerged for Au nanoparticles in bioanalysis, from quantitation tags for scanning electron microscopy (SEM) [2] to

45

electrochemical tags [3,4], mass tags in gravimetric devices [5], and tags based on atomic absorption analysis of Au [6]. In this document, we focus only on optical tags involving Au or silver (Ag) nanoparticles and nano- or microparticles with a Raman-based readout. This manuscript does not explain the physical basis of Raman spectroscopy or surface-enhanced Raman scattering (SERS). Numerous informative references are available containing more background information about these techniques [7–10].

The most widespread optical detection tags are organic fluorophores. Fluorescence labels were first used in biology as early as 1941 [11] and were already popular in the 1950s, with a review of fluorescence labeling of tissue appearing in 1961 [12]. Fluorescence detection is both quantitative and very sensitive, as has been demonstrated by the ability to detect single molecules [13]. However, organic fluorophores suffer from saturation and photobleaching effects [14] that limit the rate and the total number of photons that can be acquired from a single tag. In addition, the relatively broad emission spectra and generally strong correlation between positions of the excitation and emission spectra make it difficult to excite and detect multiple fluorophores simultaneously without using multiple excitation wavelengths. This problem has been partially overcome with the introduction of energy transfer dyes for DNA sequencing [15,16]. These dyes allow the use of a single excitation wavelength for all four fluorescent dyes, while the emission maxima remain well separated and report the identity of the attached nucleotide. It seems likely, however, that multiplexed quantitative detection methods, based on organic fluorophores, will be limited to ten or fewer distinct tags. If the analytical method benefits from detection in the near infrared (IR), then only two or three organic fluorophores can be used simultaneously.

Quantum dots (semiconductor nanocrystals) offer substantial advantages over organic dye molecules in that they have brighter emission and significantly narrower emission spectra [17,18]. The average emission spectrum of a quantum dot is typically 30 nm to 50 nm wide and thus allows for higher levels of multiplexing than with organic fluorophores. Other advantages of quantum dots compared to organic fluorophores are their significantly decreased photobleaching and broad excitation profiles, allowing excitation with a single wavelength source. Due to these advantages, and after many years of basic research, quantum dots are being used in a number of real-world biological applications. For example, Wu et al. [19] described a HER2/neu assay for breast cancer; Parak et al. [20] reported on imaging of cell motility; Howarth et al. described the use of a biotin ligase for *Escherichia coli* that allows the use of a streptavidin-conjugated Qdot for surface tagging [21], and Zhang and Johnson have reported a two-color quantum dot method for the detection of DNA [22].

However, quantum dots have their unique set of disadvantages. Because full realization of their multiplexing capabilities requires the excitation source to be in the ultraviolet (UV)/blue region of the spectrum, autofluorescence from a wide range of potentially interfering species is likely. Despite the clear success that quantum dots have achieved as detection tags, the multiplexing promise of this technology has only been partially realized. Today, only approximately nine species are available from vendors, limiting their advantage as multiplexed quantitation tags. If one desires to work in the near infrared (NIR) to avoid problems with background fluorescence, only one or two varieties are available. Finally, there continue to be concerns about the potential toxicity of quantum dots due to their chemical constituents [23].

4.2 OPTICAL TAGS BASED ON SURFACE-ENHANCED RAMAN SCATTERING

There are multiple approaches in the peer-reviewed and patent literature to the use of SERS-based detection tags. The original work was that of Cotton, Rohr, and Tarcha, dating back to 1989 [24,25]. They immobilized a capture antibody on a macroscopic Ag surface and carried out a sandwich immunoassay, in which the detection antibody was labeled with a SERS-active dye. Proof-of-concept was demonstrated, but problems with the assay format, not the least of which were maintaining the biological activity of the capture antibody on the Ag surface and the large distance between the SERS label and the surface, prevented this assay from becoming useful.

The next advance was by Keating and Natan, who demonstrated that binding a biomolecule to a Au colloid improved its detectability by SERS upon adsorption to a SERS-active surface (in this case, aggregated Ag colloid) [26–28]. Although not an assay per se, the work demonstrated unequivocally that placing an analyte between two SERS-active materials yielded the benefit of the heightened electromagnetic fields present there. In order for a SERS-based optical tag to be used for biological measurements, it must generate a specific Raman spectrum; be stable under tagging conditions and optical interrogation; and be readily attached to molecules of interest.

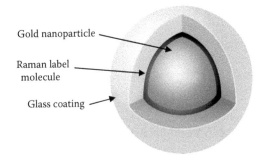

FIGURE 4.1 Cartoon showing architecture of Nanoplex Biotags.

Oxonica's Nanoplex™ Biotags [29–31] are composed of one or more SERS-active metal nanoparticles (typically Au, 50 to 90 nm in diameter), a reporter in very close proximity (ideally adsorbed) to the metal surface, and an encapsulant (preferably silica [glass], typically 20 nm thick). Figure 4.1 shows a cartoon of an idealized tag particle with this geometry. The first reports of SERS tags incorporating Au cores, Raman label molecules, and glass encapsulation occurred in publications by Mulvaney [29] and Doering [32]. The two approaches are similar, with some differences in preparation methods and label molecules used. Irradiation of these tags with monochromatic light yields the SERS spectrum of the reporter. To make a different tag, one simply employs a different reporter molecule. Because SERS features are narrow compared to fluorescence, and as there is a large spectral window in molecular vibrational spectra, it is possible to create many distinct and simultaneously quantifiable tags. Glass encapsulation provides a barrier between the Au surface and external solution components. This prevents desorption of the Raman label molecule, adsorption of other potentially interfering species, and degradation of the Raman-enhancing surface.

Several alternative approaches to nanoparticulate SERS taggants have been reported. Doering described an approach to SERS-based labels in which a reporter molecule and a biomolecule are directly adsorbed to a SERS-active metal nanoparticle surface [33]. Such a system will generate specific SERS signals from the reporter molecule but does not have the protection of the glass coating.

Driskell et al. developed an approach to ultrasensitive assays using thiol-functionalized nanoparticulate SERS-active species [34,35]. In this approach, the Au particle is coated with an organothiol SAM with a covalently attached reporter species (typically a nitroaromatic), and a terminal carboxylate for attachment of detection antibody. Assays are carried out using a capture antibody immobilized on a SAM-coated Au surface, a key point being that the assay apparently depends on optical coupling between the metal nanoparticle and the macroscopic Au surface. Because these particles are not encapsulated, one assumes that the SERS signal could be highly influenced by variation in ambient conditions, including pH and temperature.

Cao, Jin, and Mirkin have demonstrated surface-enhanced resonance Raman scattering (SERRS) as a detection mechanism for DNA [36]. In the approach, a small Au particle is coated with a capture DNA that has been modified to bind to the gold through a dye molecule. After binding to the target, Ag is deposited onto the bound gold nanoparticles to create a SERS-active surface, and the SERRS spectrum of the dye is measured.

Although potentially quite sensitive, the need for *in situ* creation of the SERS surface as well as lack of encapsulation of the dye molecule are likely to lead to practical difficulties with this system.

It is clear that encapsulation has benefits, and Duncan Graham and Ewen Smith of the University of Strathclyde (Glasgow, Scotland) have developed an approach to SERS-based optical labels based on polymer encapsulation of reporter-loaded Ag aggregates [37]. The approach is basically similar to the previously described Oxonica method, except that Ag is used instead of Au, resonant dyes are used (SERRS not SERS), particle aggregates are used rather than particle singlets or dimers, and a thick polymer coating is used (as opposed to silica), the final particle size being in the 300 to 500 nm range. Essentially the same approach has been reported by Intel [38]. In a recent addition to the list of possible SERS tagging solutions, Lee et al. described the use of a system containing tag built on an Au seed particle incorporating rhodamine 6G as the label molecule and both bovine serum albumin and metallic Ag as the shell [39].

4.3 NANOPLEX BIOTAGS

The remainder of this chapter will focus on Nanoplex Biotags that are used by the authors and incorporate predominantly single Au particles encapsulated by glass. Nanoplex Biotags have many useful features that make them ideal for bioassay applications:

- All tags are made using the same core particle, and all have the same shell. Thus, all are the same size and therefore react similarly, including having the same diffusion coefficient.
- Spectral features in the SERS spectrum are independent of particle size and depend only on the reporter molecule. Spectral intensities in the SERS spectra depend only mildly on particle size. Thus, small variations in particle size are far better tolerated than with, say, quantum dots, where different sizes of particles have different emission spectra.
- Nanoplex Biotags have been designed to operate at 785 nm, a near-IR wavelength exceptionally well suited for ultrasensitive detection because of low Raman/fluorescence background in this region. They have also been shown to be active at 980 nm and 1064 nm. In each case, moving to the near-IR lowers the background signal, whether the sample is a polymer-based blood collection tube, nitrocellulose on plastic, a silicon chip, a leaf, or human tissue.
- The signal in Nanoplex Biotags comes from inside the particles (i.e., from the reporter molecules adsorbed to the metal nanoparticle surface). The silica coat prevents other molecules from getting in and prevents the reporter molecules from leaching out. Moreover, this "internal signal generation" makes the particles exceptionally invariant to environmental variation, especially in comparison to other nanoparticles.
- The number of Nanoplex Biotags that can be simultaneously quantified depends on the level of precision required, and also on the cost of the instrument, but should exceed 10 for the lowest-cost handheld spectrometers. With a higher-resolution instrument (and with appropriate synthetic effort), it should be possible to simultaneously quantify 20 or more different tags.
- The silica coat is a perfect surface for biomolecule attachment, either by direct, covalent attachment, or by indirect means. Nanoplex Biotags have been prepared with –SH, –COOH, –NH$_2$, and other functional groups on the surface, enabling biomolecular attachment via several different strategies. A second benefit of silica is the lowered stickiness compared to polymeric coatings. This leads to reduced nonspecific binding (NSB), both of biomolecules to the nanoparticles, and of the nanoparticles to surfaces.
- Because excitation of the biotags does not require the creation of an electronically excited state, the particles can withstand large photon fluxes for extended periods with no loss in signal. To test this theory, we deposited trans(1,4) bis(pyridyl)ethylene (BPE)-labeled tags on a quartz slide and illuminated with 60 mW of 647.1 nm radiation through a 50×, 0.9 NA microscope objective. Raman spectra were collected every hour for 6 hours resulting in an approximately 20% decrease in Raman scattering after 6 hours of exposure. By comparison, the signal from a sample of quantum dots, known to bleach much more slowly than organic fluorophores, exposed to the same amount of power at 488 nm on the same system, increases 19% in the first 40 minutes followed by a drop to 70% of the initial signal over 4

hours (30% loss). Because Raman spectra can be readily collected in a few seconds or less, the tags are more than stable enough for most applications.

4.3.1 SYNTHESIS AND SURFACE FUNCTIONALIZATION

The synthesis of the Nanoplex Biotags has been described in detail elsewhere [29,31]. In brief, we use spherical Au colloid, generally between 50 nm and 90 nm in diameter, label molecules such as 4,4′-dipyridyl (dipy), d^8-4,4′dipyridyl (d^8), BPE, and quinolinethiol (QSH), and typically coat the biotags with between 10 nm and 20 nm of silica. Figure 4.2A shows the Raman spectra obtained from tags using these four different label molecules. The tags show clearly distinguishable spectra, making multiplexed measurements feasible. Figure 4.2B shows a typical TEM image obtained from a set of tags, showing both the Au core and the glass coating of individual tags.

The silanol surface can be readily functionalized with bioreactive moieties such as thiols or amines using methods similar to those reported by Schiestel et al. [40]. Standard cross-linking reagents can then be used to attach proteins or oligonucleotides. For example, one of the preferred routes for attachment of antibodies is to use the heterobifunctional cross-linker sulfosuccinimidyl-4-(N-maleimidomethyl)cyclohexane-1-carboxylate (sulfo-SMCC) to link native amines present on the antibodies to thiol-functionalized tags.

In addition, it is straightforward to manipulate the glass surface using similar methods to those commonly used in production of oligonucleotide or protein microarrays [41]. Typical functionalities that can be readily incorporated using standard silane reagents include the aforementioned amines and thiols, as well as carboxylates, epoxides, and aldehydes. This broad repertoire of chemical functionalities can be an important tool for production of high-quality bioconjugates, as the resulting activity of conjugated biomolecules is dependent upon many factors, including orientation after conjugation, reaction and storage buffer composition, charge of both biomolecule and nanotag, and cross-linker length.

The most gentle reaction conditions are generally achieved by coating the glass surface with epoxide or aldehyde groups. Both of these are desirable in that reaction of biomolecules requires no additional cross-linking reagents. At appropriate pHs, the epoxide groups will react readily with

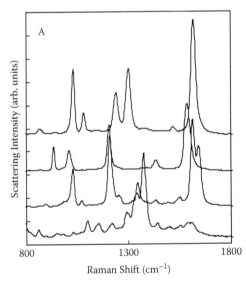
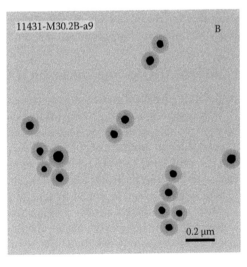

FIGURE 4.2 (A) Raman spectra obtained from Nanoplex Biotags prepared with four different label molecules (from top to bottom): 4,4′dipyridyl, d^8-4,4′-dipyridyl, trans-1,2-bis(4-pyridyl)ethylene, and quinolinethiol. (B) Transmission electron microscopy (TEM) image of surface-enhanced Raman scattering (SERS) tags showing metal core and glass coating.

amines, thiols, and even hydroxyls. This broad range of reactivity is often viewed as beneficial, because proteins can bind in many orientations, thus increasing the chance of some orientations resulting in limited loss of activity. However, if the pH is more carefully controlled, the reaction can be tuned to preferentially link to amine groups. Likewise, the aldehyde-modified tags will react with amines, forming a Schiff base complex. Use of a mild reducing agent, such as sodium cyano-borohydride, will reduce it to a stable secondary amine linkage.

Some of the more interesting possibilities lie in gentle oxidation of antibodies using sodium meta-periodate. Careful control of reaction conditions results in conversion of carbohydrate moieties on the Fc portion of an antibody into reactive aldehydes. These aldehydes can then be linked directly to amine-functionalized tags, or can be coupled to a cross-linker and coupled to alternate reactive groups. The resulting linkage should avoid deleterious conjugation of an antibody through an antigen-binding site.

In addition to direct attachment schemes, there are multistep conjugations that may have particular importance when dealing with the biotags. The silica surface can be chemically functionalized and then conjugated to a spacer molecule. The spacer molecule can be carefully chosen to exhibit low, nonspecific binding properties (for example, a PEG or dextran derivative, or a wide variety of other polymers [42–48]), lessening interactions between biotags as well as controlling unwanted interactions between the biotags and other components in the assay system. The spacer also removes the biologically relevant molecule from the particle's surface. Thus, steric hindrance issues that are known to occur as the tag:biomolecule ratio becomes larger can be minimized. With some linkers, such as large, branched polymers or copolymers, the available surface area can be dramatically increased. This can further increase the number of bound biomolecules by further reducing the potential for steric crowding. More sophisticated surfaces may be created by using multiple layers of polymers: created by either electrostatic interaction between polymers of opposite charge or through covalent bonds.

To summarize, it is clear that there are many ways in which the silica surface of Nanoplex Biotags can be modified to present reactive functional groups. Bifunctional cross-linkers can be used to perform carefully controlled two-step conjugation reactions (directly or through spacers), or they can be incorporated into "one-pot" reactions when reaction conditions are not as critical. Considering the ease with which silanes can be used to modify the biotags to present nearly any chemical functionality and the availability of cross-linkers and conjugation schemes, there are countless possibilities for linkage of biomolecules to the biotags [49,50].

4.4 APPLICATION: NANOPLEX BIOTAGS IN MICROARRAYS

To demonstrate that Nanoplex Biotags can be used in a multiplexed assay, antibodies to interleukin-4 (IL4) and interleukin-7 (IL7) were printed on a glass slide (Telechem International, Sunnyvale, California) to our specifications. Secondary antibodies to IL4 and IL7 were attached to Nanoplex Biotags coded with different label molecules. A hydrophobic pen was used to provide a liquid barrier on the glass slide, to isolate the array spots. After the ink dried, the slide was blocked for 60 minutes with 5% BSA in 10 mM PBS. All incubation and wash steps were done on an orbital rocker, and spent solutions were aspirated by pipette prior to addition of new solutions. After blocking, slides were washed three times for 5 minutes each with 0.5% BSA in PBS. IL4 and IL7 solutions were prepared in the same buffer and incubated on the arrays for 45 minutes to 2 hours, after which arrays were once again washed three times for 5 minutes each in the same buffer. The antibody-modified biotags were then incubated on the arrays for 90 minutes, followed by washing with the BSA–PBS solution and quick rinses in PBS and water. Slides were blown dry by a jet of nitrogen immediately after the water rinse. All spots were interrogated using 785 nm excitation with the scattered radiation measured through an Ocean Optics USB2000 spectrometer. Figure 4.3a shows spectra from two spots, one with the capture antibody for IL4 one with the antibody for IL7, each of which had been exposed to a mixture of IL4 and IL7, followed by development with anti-IL4-

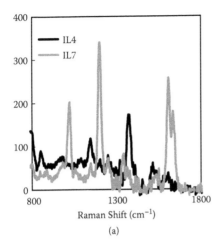
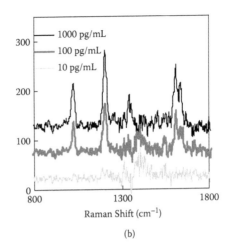

(a)   (b)

FIGURE 4.3 Detection of cytokines with Nanoplex Biotags. (a) The slide was exposed to a mixture of inter-leukin-4 (IL4) and IL7 antigen. After washing, the slide was exposed to a mixture of anti-IL4-quinolinethiol (QSH) labeled biotags and anti-IL7- trans(1,4) bis(pyridyl)ethylene (BPE) labeled biotags. The Raman spectra from the spots were collected. (b) Spectra from experiments in which the slide was exposed to decreasing concentrations of IL7 antigen.

QSH-labeled and anti-IL7-BPE-labeled tags. The spectra demonstrate specific binding of the tags to the correct target spot. Spectra from three spots designed to detect IL7 are shown in Figure 4.3b. The spots were exposed to decreasing amounts of antigen before rinsing and analysis. The signal intensity clearly corresponds with concentration, and the 100-pg/mL spot is readily detected above the background noise in the measurement.

These data show that attachment of antibodies to the glass shell of the biotags does not prevent them from binding specifically to their target antigens.

4.5 RAMAN MICROSCOPY: DETECTION OF INDIVIDUAL BIOTAGS

The relative detectability or "brightness" of the biotags is an important parameter to understand before attempting to detect them on cell surfaces. To better understand this parameter, we prepared samples by scattering biotags across the surface of a quartz slide. A mixture containing 10 µL biotags labeled with three different molecules (4,4′-dipyridyl (dipy), d^8-4,4′-dipyridyl (d^8dipy), and trans-1,2-bis(4-pyridyl)ethylene [BPE]) was deposited on a polylysine treated slide and allowed to sit for 30 minutes. A mixture was used rather than a single type of tag because two biotags in the same observation volume can be independently counted if they have different spectra, making it possible to obtain a higher sample particle density without overly high counting errors. After rinsing gently with water and drying, the slide was ready to view. Raman maps were generated using a Renishaw (Gloucestershire, United Kingdom) inVia Raman microscope equipped with 785 nm excitation. A 120 µm × 90 µm region was examined in 5 µm steps, resulting in a 24 × 18 pixel image. The Renishaw data analysis program (WIRE2.0) was used to create Raman maps of the surface. A grayscale map was created for each biotag after which each map was color coded such that red = BPE, blue = dipy, and green = d^8dipy. The color-coded images were then combined to create the false-color map seen in Figure 4.4. To generate quantitative data from the 432 acquired spectra, they were analyzed using a program written in house (described below). The results of this analysis were exported to Excel, and each pixel was examined to determine whether a signal from one or more of the components was present.

For comparison, the same sample was coated with a thin layer of Au to improve conductivity. SEM images were then collected using a Hitachi S-800 SEM. Nineteen images were collected at

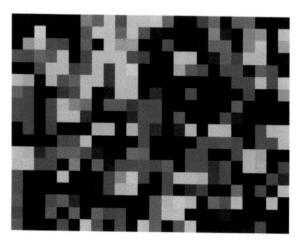

FIGURE 4.4 120 μm × 90 μm false-color map of Nanoplex Biotags spread on a quartz slide. Raman spectra were collected at 5 μm intervals resulting in 24 × 18 = 432 pixels total. Spectra were collected using a 50× (0.75 NA) objective and 1 second integration on a Renishaw inVia Raman microscope. Red pixels correspond to strong signal from trans(1,4) bis(pyridyl)ethylene (BPE), green corresponds to d^8-4,4′-dipyridyl, blue to 4,4′-dipyridyl. **(See color insert following page 112.)**

20k× magnification to determine the surface particle density. Particle counting was carried out by examining each image and tabulating single particles, doublets, and triplets. The area of each SEM image was calculated to be 29.3 μm². By comparing the results from the Raman measurements and the SEM measurements, we can calculate the fraction of particles that appear to be Raman active, or at least can be detected by this system.

A typical scanning electron micrograph of the surface is shown in Figure 4.5. As can be seen, the sample is primarily composed of individual particles, although doublets and triplets are also present. For the complete sample, the particle density was determined to be 0.90 particles/μm² with a 95% confidence interval of ±0.13 particles/μm² based on the standard deviation of the nineteen acquired images. The most likely source of uncompensated error in this measurement would be from additional heterogeneity in the surface coverage that was not accounted for by our attempt to randomly sample the surface when collecting the SEM images. The average number of particles per SEM image was 26.4 with a standard deviation of 6.8, slightly larger than the 5.1 that would be predicted by pure Poisson statistics. The SEM analysis resulted in 66.5% single particles, 21.6% of the particles in doublets, and 12% in triplets. For this analysis, doublets and triplets were defined as particles being within half a particle radius of each other. These doublets may or may not include two Au particles in direct contact. With an average coverage of 0.90 particles/μm² and a Raman image spot size of 1.2 μm², we expect to measure an average of 1.08 particles/Raman map pixel. Assuming a Poisson distribution of particles on

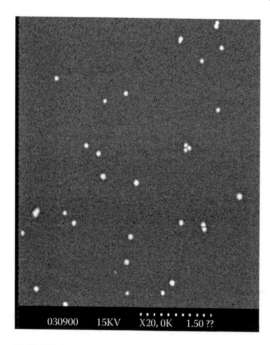

FIGURE 4.5 Typical scanning electron microscopy (SEM) image of surface-enhanced Raman scattering (SERS) nanotags distributed across a quartz slide.

the surface, 34% of the pixels should be empty, 37% should contain a single particle, 20% should contain two particles, and 7% should contain three.

Figure 4.5 is the Raman map collected from the same surface used for the SEM experiment. The map clearly shows many pixels where only background is detected (black), pixels where the Raman signal appears to be from an individual species of tag is detected (red, green, or blue), and numerous pixels where the Raman signal appears to be a combination of two or more spectra.

The number of SERS particles detected in the map depends strongly on the S/N threshold set for determining whether a measured signal is or is not actually a tag. If a S/N = 3 threshold is adopted and the noise is assumed to be the standard deviation of the blank, and the blank measurement is calculated by rejecting outliers from a $\pm 3\sigma$ range around 0 in an iterative fashion, a particle density of 0.72 particles/μm^2 is found, with 41% empty, 38% singles, 17% doubles found. Although the particle density is lower than we predict from the SEM data, it is considerably larger than would be predicted if only dimers or higher aggregates were generating detectable signal. If pixels with extremely large signals from single tag molecules are counted as two particles, the calculated particle density becomes 0.94/μm^2. This is strong, if not irrefutable, evidence that the Renishaw inVia microscope, when operated under these conditions is able to detect Raman scattering from individual Nanoplex Biotags.

4.6 APPLICATION: HER2 LABELING OF SKBR3 CELLS WITH NANOPLEX BIOTAGS

As a model system for cell surface labeling, Nanoplex Biotags were used to label the human epidermal growth factor receptor 2 (HER2) that is overexpressed on the surface of SKBR3 cells. The cells were purchased from ATCC (Manassas, Virginia) and grown on McCoy's 5A medium with 1.5 mM L-glutamine, 10% fetal bovine serum, 5% CO_2 at 37°C. The biotags were prepared by attaching Neutravidin via sulfo-SMCC coupling. For experiments with a single flavor of biotag, the cultured SKBR3 cells were first grown on a chambered slide, then washed and fixed at room temperature, followed by washing and blocking with BSA. The cells were incubated with mouse-derived HER2 antibodies, then washed and incubated with biotinylated antimouse IgG. After washing with PBS, the cells were incubated with Neutravidin-conjugated Nanoplex Biotags (at OD ~1.2) at room temperature for 1 hour. Finally, the cells were washed with PBS and coverslipped using 90% glycerol in PBS. After mounting, the edges were sealed using nail polish. For the multiplex experiment, three separate samples of the cells were treated as above except that each sample was mixed with a different type of biotag (BPE, QSH, and d⁸dipyridyl labels). Raman spectra were obtained using a Renishaw inVia microscope with 785 nm excitation.

Figure 4.6 shows two of the SKBR3 cells after reaction with BPE-labeled biotags. Due to the large scattering coefficients of the tags, they often appear as bright spots in the bright field microscope image. When the Raman microscope was positioned above the cell, the spectra shown were obtained. The BPE spectrum can be clearly identified regardless of the position on the cell as the HER2 receptor is expressed across the cell surface. When the microscope is directed to part of the slide where cells are absent, only background spectra (primarily contributions from the glass coverslip) are obtained.

To demonstrate the potential for multiplexed measurements, three samples of SKBR3 cells were independently labeled with BPE, d⁸dipyridyl, and QSH biotags, respectively. The labeled cells were then mixed and data collected as before. Figure 4.7 shows three cells from a larger population and representative spectra collected from each cell. The spectra of d⁸dipyridyl, BPE, and QSH can be clearly distinguished.

Recently, Kim et al. showed similar detection results using SERS nanotags of their design to label HER2 and CD10 [51]. Their tags were created by depositing Ag nanoparticles on a silica core, adding a label molecule, and encapsulating with glass. The most significant difference between

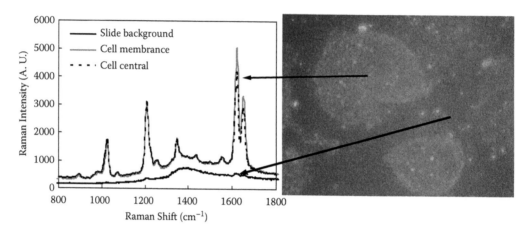

FIGURE 4.6 Microscope image showing cells labeled with Nanoplex Biotags and corresponding surface-enhanced Raman scattering (SERS) spectra. Spectra collected from top and edges of the cell show the distinctive SERS spectrum from the tag, and the background has the broad background typical of glass.

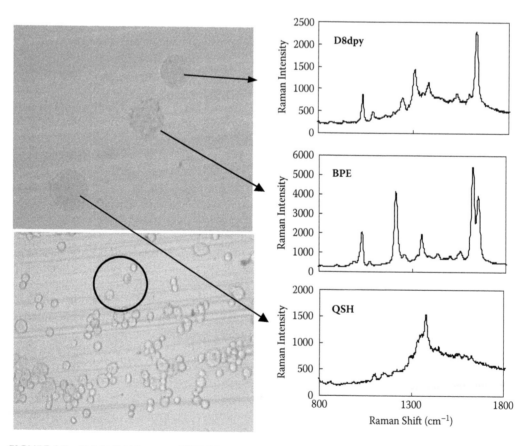

FIGURE 4.7 Bright field images of SKBR3 cells dispersed on a slide. Raman spectra obtained from three adjacent cells in point mode. Each cell originated as separately tagged populations (HER2 targeted) and clearly shows the spectrum from individual tags.

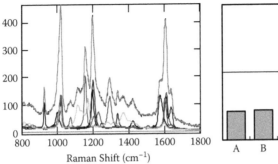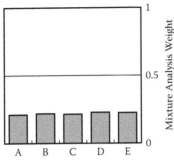

FIGURE 4.8 Graph of the analysis of a mixture containing five Nanoplex Biotags. The tags are mixed in equal parts in a 1 mL glass tube. The spectrum of the mixture is recorded with an Ocean Optics QE65000 spectrometer. The spectrum is then analyzed using in-house software. The graph on the left shows the spectrum of a mixture and the individual components in the mixture. The bar graph on the right shows that the instrument detects nearly identical amounts of each tag in the captured spectrum.

these structures and the Nanoplex Biotags are the somewhat larger size (they start with a ~150 nm silica core), and the use of green excitation (514.5 nm) with the Ag particles.

4.6.1 Spectral Identification of Mixtures and Raman Mapping

Quantitative spectral mixture analysis using linear least squares is readily applied to mixtures of Raman spectra and has been implemented in our labs. This custom software can be used to determine percent composition of components in a spectral mixture. A dialog box is used to load the files containing the spectral data. The program then uses classical least squares regression to determine the amount of each component in an unknown spectrum. Numerical results are displayed in bar graph form and can be exported (either the bar graph or the numerical values) to the Windows Clipboard for use in other programs. The user of the program can choose the wave-number range over which the analysis is to be performed and add a variable component (line, quadratic, etc.) as one element of the least squares fit, making it possible to account for broad changes in the background. Finally, the user can visually inspect the results of the analysis in a window that shows the collected spectrum and each of the components used to create the best fit.

Figure 4.8 illustrates the results obtained from the analysis of a mixture containing equal amounts of Nanoplex Biotags with five different label molecules. The large panel shows the mixture spectrum as well as the contributions to the mixture from the five individual tags. In the smaller inset is a bar graph displaying the relative amounts of each biotag. In this case, they are all nearly the same. In Figure 4.9 we see the case where one component is in great excess, and very small amounts of four other components are present. The very small contributions of the four trace constituents can be seen in the baseline of the plots, and the bar graph shows almost equal but, in this example, very small amounts of the other four tags.

To illustrate the potential for multiplexed Raman mapping, the inVia microscope was used to create a Raman map of the mixed population of cells similar to that described earlier. The only difference was that we did not use QSH labeled tags but instead added 4,4′-dipyridyl as the label on one of the Nanoplex Biotags. After scanning the sample, the software supplied by Renishaw (Wire2.0) was used to create three independent grayscale intensity maps corresponding to each of the three Nanoplex Biotags used. Each grayscale image was then converted to a color (red, blue, or green) and recombined into a single image using ImageJ (free from NIH). Figure 4.10 shows the bright field image of six cells that happened to be aligned on the slide surface. The Raman map shows that three of these cells came from the population labeled with dipy, two from the d^8dipy population, and a single cell labeled with BPE tags. As can be seen, there is a clear distinction between cells indicating

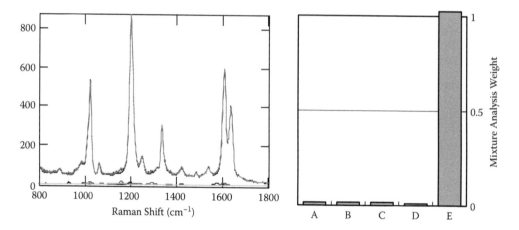

FIGURE 4.9 Graph of the analysis of a mixture containing five Nanoplex Biotags. A single tag is present in large excess, and the other four are present in equal but small amounts. Data are collected and analyzed as above. As expected, one tag is found in large amounts, and the others are found in small and nearly equal amounts.

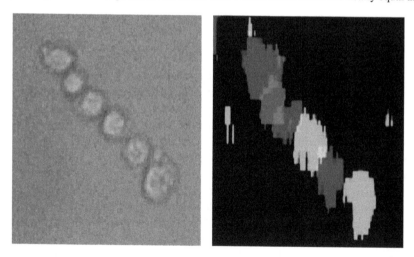

FIGURE 4.10 Bright field image and Raman map of six cells labeled with Nanoplex Biotags. Spectra analysis reveals that three of the cells were labeled with dipyridyl tags (blue), two with d^8-dipyridyl tags (green), and a single cell was labeled with biotags carrying the trans(1,4) bis(pyridyl)ethylene (BPE) label molecule (red). **(See color insert.)**

both successful labeling and successful detection of the biotags. In a similar approach, workers at Intel have used their composite organic–inorganic nanoparticles (COINs) prepared with either acridine orange or basic fuchsin to multiply label prostate-specific antigen in tissue samples [52]. These results clearly show the potential for using SERS-based labels in imaging experiments.

4.7 POTENTIAL ADDITIONAL APPLICATIONS

The number of possible applications that could benefit from Oxonica's novel nanoparticles is extensive, and there are multiple facets of biology, biochemistry, and biophysics that have yet to be explored. For example, it should be possible to internally label individual cells and track them *in vivo*. This could be particularly important for developmental biology, and for understanding cell differentiation, including for stem cells. Given recent advances such as spatially offset spectroscopy [53], it may one day be possible to generate high-resolution three-dimensional images *in vivo* using

SERS, perhaps at appreciable depths. It should be possible to detect low-molecular-weight species and ions *in vivo* using permselective SERS-active particles; a basic proof of concept for pH has been demonstrated by two groups [54,55]. On the intracellular side, a variety of experiments made possible with quantum dots can be extended with SERS tags.

Given the success of the Porter group and our efforts in using SERS-active labels to carry out quantitative measurements of proteins using planar arrays, there are a large number of possible applications in related systems, including but not limited to DNA arrays, carbohydrate arrays, peptide arrays, and arrays of low-molecular-weight species. Of course, it should be understood that such work could be extended to nonplanar arrays, including three-dimensional arrays composed of other types of beads and particles (e.g., Nanobarcodes™ particles [Nanoplex Technologies], magnetic particles, Luminex beads, and so forth).

A key feature of fluorophores which has been exploited to create useful assays is energy transfer and quenching, which allows the proximity of two species to be ascertained. Such types of measurements should be possible with SERS-active particles. Thus, it should be possible to develop proximity assays with a SERS-based readout, in which the key step is electronic communication between two different SERS-active species. This should allow, for example, intracellular protein–protein networks to be elaborated in highly multiplexed fashion (compared to yeast-two hybrid and protein complementation assays which are difficult to multiplex).

Finally, there are sure to be "combination assays" involving different types of SERS responses. For example, Boss has described a magnetic capture assay using SERS [56]. It would be simple to combine this type of assay with one involving the detection of a specific biological target using a SERS biotag. This would allow simultaneous detection of, say, a protein and a small molecule. Likewise, SERS biotags can be used in conjunction with conventional SERS substrates, either particulate, mesoscopic, or macroscopic, again allowing the sensitivity and molecular identification capabilities of SERS to be leveraged at the same time that specific analytes are quantified using the reporter approach described herein.

4.8 SUMMARY

Nanoplex Biotags are part of a new class of nanoparticles that exploit SERS to generate unique optical signatures. Because of these unique optical signatures and because of the flexibility of the glass coating of the particles, a wide range of biological measurements can be envisioned. In this chapter, we reviewed results from the biotags employed in a microarray approach for the detection of cytokines IL4 and IL7. In addition, the possibility of using the biotags for cell surface measurements has been demonstrated in single-plex and pseudo-multiplex schemes. Finally, our preliminary data indicate that multiplexed Raman mapping is possible with these tags.

ACKNOWLEDGMENTS

This work was supported in part by the National Cancer Institute Grant #1R43CA108234-01 and Department of Defense (DOD) award W81XWH-04-1-0751.

REFERENCES

1. Hayat, M.A., *Colloidal Gold: Principles, Methods, and Applications*, Academic Press, San Diego, 1989.
2. Levit-Binnun, N. et al., Quantitative detection of protein arrays, *Anal. Chem.*, 75, 1436, 2003.
3. Ozsoz, M. et al., Electrochemical genosensor based on colloidal gold nanoparticles for the detection of Factor V Leiden mutation using disposable pencil graphite electrodes, *Anal. Chem.*, 75, 2181, 2003.
4. Rochelet-Dequaire, M., Limoges, B., Brossier, P., Subfemtomolar electrochemical detection of target DNA by catalytic enlargement of the hybridized gold nanoparticle labels, *Analyst*, 8, 923, 2006.

5. Zhou, X.C., O'Shea, S.J., Li, S.F.Y., Amplified microgravimetric gene sensor using Au nanoparticle modified oligonucleotides, *Chem. Comm.*, 11, 953, 2000.
6. Zhang, C., et al., Application of the biological conjugate between antibody and colloid Au nanoparticles as analyte to inductively coupled plasma mass spectrometry, *Anal. Chem.*, 74, 96, 2002.
7. Smith, E., and Dent, G., *Modern Raman Spectroscopy: A Practical Approach*, Wiley, New York, 2005.
8. McCreery, Richard, *Raman Spectroscopy for Chemical Analysis*, Wiley, New York, 2000.
9. Cohen, L.F., Editor, Surface Enhanced Raman Spectroscopy, Faraday Discussions, Vol. 132, Royal Society of Chemistry, 2006.
10. Kneipp, Katrin, Moskovits, Martin, Kneipp, Harald (Eds.), *Surface-Enhanced Raman Scattering: Physics and Applications, Topics in Applied Physics*, Vol. 103, Springer, New York, 2006.
11. Creech, H.J., and Jones, R.N., Conjugates synthesized from various proteins and the isocyanates of certain aromatic polynuclear hydrocarbons, *J. Am. Chem. Soc.*, 63, 1670, 1941.
12. Coons, A.H., The beginnings of immunofluorescence, *J. Immunol.*, 87, 499, 1961.
13. Ha, T., Single-molecule fluorescence methods for the study of nucleic acids, *Curr. Opin. Struct. Biol.*, 11, 287, 2001.
14. Xu, H., et al., Multiplexed SNP genotyping using the Qbead™ system: A quantum dot-encoded microsphere-based assay, *Nucleic Acids Res.*, 31, e43, 2003.
15. Smith, L.M., et al., Fluorescence detection in automated DNA sequence analysis, *Nature*, 321, 674, 1986.
16. Ju, J., et al., Fluorescence energy transfer dye-labeled primers for DNA sequencing and analysis, *Proc. Natl. Acad. Sci. USA*, 92, 4347, 1995.
17. Steigerwald, M.L., et al., Surface derivatization and isolation of semiconductor cluster molecules, *J. Am. Chem. Soc.*, 110, 3046, 1988.
18. Rosetti, R., and Brus, L., Electron-hole recombination emission as a probe of surface chemistry in aqueous CdS colloids, *J. Phys. Chem.*, 86, 4470, 1982.
19. Wu, X., et al., Immunofluorescent labeling of cancer marker Her2 and other cellular targets with semiconductor quantum dots, *Nature Biotech.*, 21, 41, 2003.
20. Parak, W.J., et al., Cell motility and metastatic potential studies based on quantum dot imaging of phagokinetic tracks, *Adv. Mater.*, 14, 882, 2002.
21. Howarth, M., et al., Targeting quantum dots to surface proteins in living cells with biotin ligase, *Proc. Nat. Acad. Sci.*, 102, 7583, 2005.
22. Zhang, C.-Y., and Johnson, L.W., Homogenous rapid detection of nucleic acids using two-color quantum dots, *Analyst*, 131, 484, 2006.
23. Hardman, R., A toxicologic review of quantum dots: Toxicity depends on physicochemical and environmental factors, *Env. Health Perspec.*, 114, 165, 2006.
24. Rohr, T.E., et al., Immunoassay employing surface-enhanced Raman spectroscopy, *Anal. Biochem.*, 182, 388, 1989.
25. U.S. Patent Nos. 5,567,628; 5,445,972; 5,376,556; 5,266,498.
26. Keating, C.D., Kovaleski, K.M., Natan, M.J., Heightened electromagnetic fields between metal nanoparticles: Surface enhanced Raman scattering from metal-cytochrome c-metal sandwiches, *J. Phys. Chem. B*, 102, 9414, 1998.
27. Keating, C.D., Kovaleski, K.M., Natan, M.J., Protein:colloid conjugates for surface-enhanced Raman scattering: Stability and control of protein orientation, *J. Phys. Chem. B*, 102, 9404, 1998.
28. US Patent No. 6,149,868.
29. Mulvaney, S., et al., Glass-coated, analyte tagged nanoparticles: A new tagging system based on detection with surface-enhanced Raman scattering, *Langmuir*, 19, 4784, 2003.
30. Freeman, R.G., et al., Detection of biomolecules using nanoparticle, surface enhanced Raman scattering tags, in *Nanobiophotonics and Biomedical Applications II*, Cartwright, Alexander N., and Osinski, Marek, Eds., *Proceedings of SPIE* Vol. 5705, SPIE, Bellingham, WA, 2005, pp. 114–121.
31. US Patent Nos. 6,514,767 and 6,861,263.
32. Doering, W.E., and Nie, S., Spectroscopic tags using dye-embedded nanoparticles and surface-enhanced Raman scattering, *Anal. Chem.*, 75, 6171, 2003.
33. Doering, William E., Mechanisms and Applications of Single-Nanoparticle Surface-Enhanced Raman Scattering, Ph.D. Thesis, University of Indiana, 2003.
34. Driskell, J.D., et al., Low level detection of viral pathogens by a surface-enhanced Raman scattering based immunoassay, *Anal. Chem.*, 19, 6147, 2005.

35. Park, H.-Y., et al., Ultrasensitive immunoassays based on surface-enhanced Raman scattering by immunogold labels, in *Surface-Enhanced Raman Scattering: Physics and Applications*, Kneipp, K., Moskovits, M., Kneipp, H., Eds., Springer-Verlag, Berlin, 2006, p. 427.

36. Cao, C.Y.W., Jin, R., Mirkin, C.A., Nanoparticles with Raman spectroscopic fingerprints for DNA and RNA detection, *Science*, 297, 1536, 2002.

37. Stokes, R.J., et al., SERRS-active nanoparticle-polymer beads for ultra-sensitive biodiagnostic applications, *Micro & Nano Lett.*, 1, 57, 2006.

38. Xing, S., et al., Composite organic-inorganic nanoparticles (COINs) with chemically encoded optical signatures, *Nano. Lett.*, 5, 49, 2005.

39. Lee, S., Kim, S., Choo, J., Shin, S.Y., Lee, Y.H., Choi, H.Y., Ha, S., Kang, K., Oh, C.H., Biological imaging of HEK293 cells expressing PLCγl using surface-enhanced Raman microscopy, *Anal. Chem.*, web publication date 01/06/2007.

40. Schiestel, T., Brunner, H., Tovar, G.E.M., Controlled surface functionalization of silica nanospheres by covalent conjugation reactions and preparation of high density streptavidin nanoparticles, *J. Nanosci. Nanotech.*, 4, 504, 2004.

41. Aslam, M., and Dent, A., *Bioconjugation: Protein Coupling Techniques for the Biomedical Sciences*, Macmillan Reference, London, 1998.

42. Vandevondele, S., et al., RGD-modified PLL-g-PEG copolymers, *Biotechnol. Bioeng.*, 82, 784, 2003.

43. Amanda, A., and Mallapragada, S.K., Comparison of protein fouling on heat-treated poly(vinyl alcohol), poly(ethersulfone) and regenerated cellulose membranes using diffuse reflectance infrared Fourier transform spectroscopy, *Biotechnol. Prog.*, 17, 917, 2001.

44. Park, S., et al., Surface modification of poly(ethylene terephthalate) angioplasty balloons with a hydrophilic poly(acrylamide-co-ethylene glycol) interpenetrating polymer network coating, *J. Biomed. Mater. Res.*, 53, 568, 2000.

45. Matsuda, T., et al., Photoinduced prevention of tissue adhesion, *ASAIO J*, 38, M154, 1992.

46. Holland, N.B., et al., Biomimetic engineering of non-adhesive glycocalyx-like surfaces using oligosaccharide surfactant polymers, *Nature*, 392, 799, 1998.

47. Scott, M.D., and Murad, K.L., Cellular camouflage: Fooling the immune system with polymers, *Curr. Pharm. Des.*, 4, 423, 1998.

48. Lee, P.J., et al., An efficient binding chemistry for glass polynucleotide microarrays, *Bioconj. Chem.*, 13, 97, 2002.

49. Kusnezow, W., Jacob, A., Walijew, A., Diehl, F., and Hoheisel, J.D., Antibody microarrays: An evaluation of production parameters, *Proteomics*, 3, 254–264, 2003.

50. Hermanson, G.T., *Bioconjugate Techniques*, Academic Press, San Diego, 1996.

51. Kim, J.-H., et al., Nanoparticle probes with surface enhanced Raman spectroscopic tags for cellular cancer targeting, *Anal. Chem.*, 78, 6967, 2006.

52. Sun, L., et al., Composite organic-inorganic nanoparticles as Raman labels for tissue analysis, *Nano Lett.*, web release 1/17/2007.

53. Eliasson, C., and Matousek, P., Noninvasive authentication of pharmaceutical products through packaging using spatially offset Raman spectroscopy, *Anal. Chem.*, web release 1/18/2007.

54. Bishnoi, S.W., et al., All-optical nanoscale pH meter, *Nano Lett.*, 6, 1687, 2006.

55. Talley, C.E. et al., Intracellular pH sensors based on surface-enhanced Raman scattering, *Anal. Chem.*, 76, 7064, 2004.

56. Mosier-Boss, P.A., and Lieberman, S.H., Surface-enhanced Raman spectroscopy substrate composed of chemically modified gold colloid particles immobilized on magnetic microparticles, *Anal. Chem.*, 77, 1031, 2005.

5 Nanostructured Titanium Alloys for Implant Applications

Yulin Hao, Shujun Li, and Rui Yang
Institute of Metal Research, Chinese Academy of
Sciences, Shenyang, Liaoning, China

CONTENTS

5.1 INTRODUCTION

Titanium and its alloys have become one of the most attractive classes of biomedical implant materials owing to their light weight, superior biocompatibility and biocorrosion resistance, good mechanical properties, and low elastic modulus. Commercial pure (CP) titanium, Ti-6Al-4V, and Ti-6Al-7Nb alloys have been widely used in biomedical applications to replace stainless steel and cobalt-based alloys. Pure titanium has a hexagonal close-packed (HCP) crystal structure at low temperature, but this changed to body-centered cubic (BCC) above 882°C. Alloy phases based on these two structures are denoted as α and β, respectively. Broadly speaking, simple metals (Al, Ga) and interstitials (C, N, O) are α-stabilizers, and most transition metals, such as Nb, Mo, and V, act as β-stabilizers. Depending on composition, titanium alloys can generally be classified as near-α, $\alpha+\beta$, and β type. For biomedical applications, the primary requirements of an alloy include high strength, good corrosion and wear resistance, low elastic modulus, and no toxicity. Although $\alpha+\beta$ alloys such as Ti-6Al-4V and Ti-6Al-7Nb have adequate strength, they are far from ideal in satisfying other stringent requirements for implant applications. In order to overcome the long-term health problem caused by the release of Al and V ions from these alloys [1,2] as well as stress shielding related to elastic modulus which is still inadequately large as compared to that of the surrounding human bone [3–5], novel β-type titanium alloys with greater biocompatibility, lower elastic modulus, and better processability have been developed in the past decade [6,7].

    In addition to alloy development, several kinds of nanotechnologies have been applied to fabricate nanostructure (NS) titanium and its alloys in recent years. This provides new opportunities to further improve the biochemical and biomechanical properties of these materials [8–10]. Nanostructuring is known to be beneficial biochemically, and positive cell responses have been reported in cell cultures grown on the surfaces of nanophase ceramics [11–14], polymer demixed nanoto-pography [15,16], nanofiber alumina [17], and nanofiber carbon [18]. There have been only a few investigations on NS metallic alloys, possibly because of difficulty in preparing high-quality NS materials. Most research on cell response has focused on NS roughness produced by grinding [19] and acid etching [20] of coarse-grained titanium and its alloys as well as pressed ultrafine powders [52]. These experiments show clearly the positive effect of NS roughness and particle boundaries on cell response [19–21].

In this chapter, we first summarize the commonly used processing technologies to prepare NS titanium and its alloys, then outline the mechanical properties, bioactive surface treatment, and cell response of this class of materials, and follow with a brief discussion of future research directions.

5.2 PROCESSING TECHNOLOGY

Several kinds of processing technologies are currently available to produce NS metallic materials [22–25]. Because NS specimens obtained from mechanical alloying, gas-phase condensation, and electrodeposition technologies have shortcomings of size limitation, residual porosities, or con-taminations, here we focus only on technologies that are capable of producing large-dimension, fully-dense NS titanium and its alloys which have potential for practical and hopefully immediate biomedical application.

5.2.1 SEVERE PLASTIC DEFORMATION

For metallic materials, heavy plastic deformation at low temperature, for example, cold rolling, swaging, or drawing, results in grain refinement by the formation of cellular-type substructure with low-angle misoriented boundaries. The mechanisms involved are usually dislocation interaction, deformation twinning, and stress-induced martensitic (SIM) transformation [24,25]. The evolution of grain refinement processes through the above mechanisms is sluggish in nature, and severe plas-tic deformation (SPD) is usually necessary to fabricate ultrafine structure and even NS grains with high-angle grain boundaries.

Equal channel angular pressing (ECAP) is a specific SPD technology that introduces severe plastic strain into the billet under processing to refine coarse-grained materials by repeated extru-sions through a special die [8]. The two segments of the die channel have equal cross section and an intersection angle of usually 90°, so large-dimension NS billets with identical cross section can be produced after 8 to 12 extrusion passes. Because titanium and its alloys are generally not sufficiently ductile at room temperature, high temperatures are needed for ECAP processing, for example, 450°C for CP-Ti and 650°C for Ti-6Al-4V [26–29]. At high temperatures, recovery and even recrystallization may occur which partially release the stored energy of deformation, com-promising the efficiency of grain refinement. For both CP-Ti and Ti-6Al-4V, it is difficult to refine grains to less than 100 nm even if the ECAP process is followed by conventional cold deformation. They are generally classified as ultrafine-grained (UFG) materials with grain size in the range of hundreds of nanometers.

Another SPD technology is multistep plastic deformation that is capable of producing large-dimension NS materials from ($\alpha+\beta$) titanium alloys. For Ti-6Al-4V, multistep isothermal forging can refine grains to size about 400 nm [30,31]. Because hydrogen is a strong β stabilizer in tita-nium alloys, hydrogenation results in the increase of volume fraction of β phase and the decrease of lamellar thickness of α phase. This contributes to the improvement of processability and grain refinement during plastic deformation. As a result, multistep isothermal forging together with pre-hydrogenation and postdehydrogenation can refine coarse grains of ($\alpha+\beta$) type titanium alloys to 20 nm to approximately 40 nm [32,33].

Surface mechanical attrition treatment (SMAT) is a variant of SPD technology to produce a NS surface layer with thickness up to tens of micrometers on metallic materials [34]. The impact

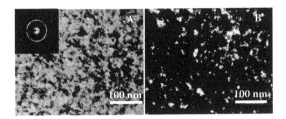

FIGURE 5.1 Transmission electron microscopy (TEM) microstructure of 1.5 mm cold-rolled sheet of Ti-24Nb-4Zr-7.9Sn (TNZS) alloy. (A) and (B) are a pair of bright- and dark-field images showing grain size less than 50 nm. Inset in (A) shows corresponding selected-area electron diffraction pattern with continual diffraction rings.

by high-speed balls with velocity about 1 to 20 meters per second is capable of producing localized plastic deformation on the surface of materials. Because these shooting balls are stimulated by a high-power ultrasonic system with frequency of tens of kilohertz, the entire surface of a component can be peened with a very large number of impacts in a short period of time. As to CP-Ti, SMAT technology can produce equiaxed NS grains with sizes of 50 to approximately 250 nm in a surface layer of 15 to approximately 30 μm thickness or ultrafine grains (100 to approximately 300 nm in size) in a surface layer about 60 μm thick [35].

5.2.2 CONVENTIONAL PLASTIC DEFORMATION

In the past few years, extensive investigations have focused on the development of industrial-scale SPD technologies using CP-Ti as a model material. Although large-dimension UFG CP-Ti billets with strength even higher than Ti-6Al-4V and good room-temperature ductility have been achieved, the low efficiency of SPD technologies and their limited capacity in grain refinement due to dynamic effects such as recovery and recrystallization are yet to be improved.

Recently, significant grain refinement during a conventional plastic deformation process was found in a β-type Ti-Nb-Zr-Sn biomedical titanium alloy with high strength and low elastic modulus [36,37]. The composition of the alloy is such that the BCC phase exhibits nonlinear elastic behavior as well as significant localized plastic deformation. During compression tests, narrow shear bands with width about 1 μm formed inside grains of 100 μm in size, and these bands intersected each other. Within the bands, a single crystal was broken into nanocrystallites as revealed by transmission electron microscopy (TEM) observation [38,39]. Such a peculiar mechanism of plastic deformation enabled 90% thickness reduction during conventional cold rolling from an original thickness of 15 mm, although the ductility of the alloy is only half that of CP-Ti. Bright- and dark-field TEM observations found that grain size of the cold-rolled sheet is less than 50 nm with most between 20 and 30 nm (Figure 5.1). Like NS samples obtained by SPD [40], grain boundaries of the cold-rolled sheet exhibit wavy, curved, or corrugated characteristics. X-ray diffraction analysis yielded crystallite size of about 10 nm, probably due to the presence of substructure in NS grains introduced by severe distortion.

5.2.3 RAPID SOLIDIFICATION

Rapid solidification from melt can also produce NS titanium alloys [41], depending on the glass-forming ability of the alloy system and cooling rate [9]. Multicomponent Ti-Cu-Ni-Sn-Nb nano-structure-dendrite composite was fabricated by copper mold casting [42,43]. For example, the as-cast cylinders of 2 mm and 10 mm in diameter have NS grains about 30 to 50 nm and ultrafine grains about 300 nm, respectively. The dendrite β phase with BCC crystal structure contributed to plastic deformation and low elastic modulus of the alloy, and the NS matrix helped to maintain high strength. Because both Cu and Ni belong to toxic and allergic elements in the human body [6,7], a suitable substitute must be found before biomedical application of the alloys can be considered.

TABLE 5.1

Mechanical Properties of NS CP-Ti at Room Temperature Processed by Severe Plastic Deformation (SPD)

Grain Size (nm)	UTS (MPa)	YS (MPa)	Elongation (%)	SPD	Ref.
100	1150	1020	8	ECAP + CR	[26]
200	730	625	25	ECAP + HPT	[44]
120	950	790	14	HPT	[45]
—	1000	920	14	ECAP + CR + Annealing	[27]
—	1300	1000	20	HPT + Annealing	[46]

Notes: UTS, ultimate tensile strength; YS, yield strength; SPD, severe plastic deformation; ECAP, equal channel angular pressing; CR, cold rolling; HPT, high-pressure torsion.

5.3 MECHANICAL PROPERTIES

Mechanical properties and deformation behavior of NS metallic materials have been extensively investigated, in particular those with face-centered cubic crystal structure [23–25].

Several novel approaches have been developed to achieve both high strength and ductility, for example, bimodal distribution of grain sizes, low temperature/high strain rate deformation, and introduction of twins [25].

Only a few investigations have been performed on the mechanical properties of NS titanium and its alloys with HCP or BCC crystal structure. Most available data for CP-Ti are summarized in Table 5.1. It is clear that NS CP-Ti has high strength comparable to coarse-grained Ti-6Al-4V under the condition of similar room-temperature ductility. Annealing treatment after SPD significantly improved both strength and ductility [46]. Tensile properties of UFG Ti-6Al-4V alloy with grain size of 400 nm may reach that of most high-strength β-type titanium alloys, for example, ultimate tensile strength of 1360 MPa and ductility of 7% [30,31].

Elastic modulus is an important mechanical property for load-bearing implant materials. The balance of high strength and low elastic modulus is one of the key targets in developing novel β-type biomedical titanium alloys [6,7]. Figure 5.2 shows that coarse-grained Ti-Nb-Zr-Sn alloy has strength comparable to that of NS CP-Ti but possesses much lower elastic modulus, and the strength/modulus combination can be varied depending on heat treatment schemes. Precipitation of the α-phase from the NS β-phase matrix contributes to high strength under the condition of similar elastic modulus [47]. Ti-Cu-Ni-Sn-Nb nanostructure dendrite composite also exhibits better balanced properties than NS CP-Ti [9]. The ductility of both NS alloys, however, is lower than that of NS CP-Ti. Due to the barrier effect of nanosized α precipitates on grain growth of NS Ti-Nb-Zr-Sn alloy, an annealing at 600°C for 10 minutes results in average grain size less than 100 nm. It is therefore possible to achieve high strength and reasonable ductility by controlling annealing treatment similar to that for NS CP-Ti [46].

The wear resistance of titanium and its alloys is inferior to Co-Cr-Mo alloys, and this prevented their application in wear-intensive biomedical components, such as artificial joint replacement prostheses [6]. Achieving NS provides a way to enhance wear resistance due to increased hardness over coarse-grained materials, as demonstrated by experimental results for low carbon steel with NS surface layer produced by SMAT [49]. For titanium and its alloys, the formation of oxides such as TiO and Ti_2O during tribo-oxidation interaction makes the situation more complicated [6]. These oxides led to unexpected variation in friction coefficient of UFG CP-Ti produced by ECAP as compared to coarse-grained CP-Ti [50,51]. The effect of grain size on wear resistance of titanium and its alloys therefore needs further investigation.

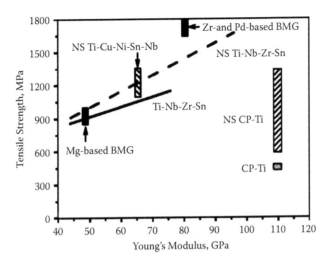

FIGURE 5.2 Relationship between ultimate tensile strength and Young's modulus of coarse-grained and nanostructured (NS) titanium and its alloys, in which data of bulk metallic glass (BMG) based on Mg, Zr, and Pd [48] are plotted for comparison.

With reduced grain size of NS metallic materials, their superplasticity is generally enhanced. As a result, fabrication of structural components by superplastic forming or isothermal forging can be conducted at much lower temperatures. For example, UFG Ti-6Al-4V exhibits superplastic properties at temperatures about 150 to 250°C lower than that for coarse-grained alloy [30].

5.4 CELLULAR RESPONSE

Cell behavior on biomaterial surfaces depends upon implant-cell interactions that play a crucial role in determining biocompatibility and integration between biomaterials and tissues of the human body. A long-standing hypothesis asserts that four material-related surface factors influencing bone-implant interfaces are implant surface composition, surface energy, surface roughness, and surface topography [52]. Different technologies, such as acid and alkaline etching, glow-discharge treatment, and hydroxyapatite (HA) coating, have been applied to improve biomaterial surfaces. Because surface modification technologies generally involve chemical reactions, the high free energy of the NS surface would render these processes easier and cell response more favorable. For example, substantial reduction of nitriding temperature was achieved by means of surface nano-crystallization [53], and NS roughness of oxidation was prepared on the surfaces of NS CP-Ti and Ti-6Al-4V alloy [54].

Recent progress made in SPD technologies and alloy design makes it possible to fabricate large-dimension NS titanium and its alloys for biomedical applications. Their practical applications as implant devices will require detailed understanding of the contribution of NS grains on biocompatibility. Cell responses of bulk NS Ti-24Nb-4Zr-7.9Sn (in weight percent, abbreviated as TNZS) alloy were investigated recently [55]. Polished samples with coarse grains (100 μm), ultrafine grains (400 nm), and NS grains (<50 nm) were studied. Figure 5.3 clearly shows the enhanced effect of grain refinement on adhesion and proliferation of osteoblast cells. Because the polished surfaces have similar roughness, oxidation products, and surface energy, the increasingly positive cell response with reducing grain size as observed from Figure 5.3 can only be attributed to the presence of an increasingly large volume fraction of grain boundaries from coarse-grained to NS materials.

Annealing treatment of NS TNZS alloy in its (α+β) phase field results in a dual-phase microstructure with a grain size of β matrix about hundreds of nanometers. Because the α phase has inferior corrosion resistance as compared to the β matrix, preferential corrosion of the α phase

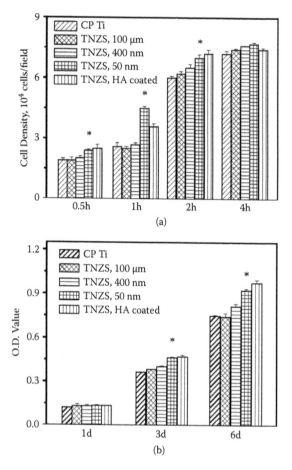

FIGURE 5.3 Osteoblast adhesion (a) and proliferation (b) on polished CP-Ti and Ti-24Nb-4Zr-7.9Sn (TNZS) alloy with different grain size.

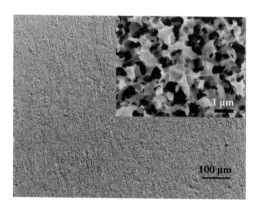

FIGURE 5.4 Optical image of a submicron porous surface produced by acid etching of nanostructured TNZS alloy after solution-treated at 650°C for 0.5 hour. The inset is an enlarged scanning electron microscopy (SEM) image showing details of the porous structure.

within the matrix grains and along grain boundaries that precipitated during high-temperature annealing will produce a porous microstructure. Figure 5.4 shows a submicron porous layer with pore size about 0.5 μm, which was prepared by acid etching in a solution of 40 volume percent of HCl in boiling water cold-rolled sheet of NS TNZS alloy after annealing at 650°C for 0.5 hour. For biomedical applications, such a porous surface layer has the following advantages over coarse-grained materials: First, it enhances the formation of bioactive materials, such as HA coating with bone-like microstructure. Second, it contributes to the adhesion and proliferation of fibroblast cells in the early stage of *in vitro* tests (Figure 5.5). Last, it helps to avoid the "stress shielding" problem by improving elastic matching of implanted materials with adjacent bones.

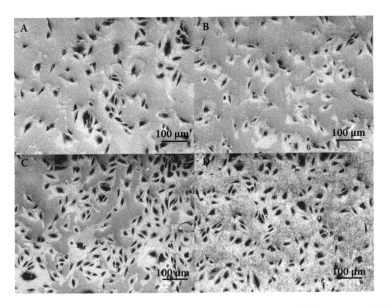

FIGURE 5.5 Scanning electron microscopy (SEM) images of fibroblasts at 24 hours on polished CP-Ti (A), polished Ti-24Nb-4Zr-7.9Sn (TNZS) alloy with grain size of 100 μm (B) and 400 nm (C), and HCl etched TNZS alloy with porous surface (D) as shown in Figure 5.4.

5.5 FUTURE DIRECTIONS

The extensive research on SPD processing of CP-Ti led to the fabrication of large-dimension UFG billets with high strength and ductility (Table 5.1) from which several kinds of implant items were produced [10,56]. However, practical industrial processing of such materials, in particular NS CP-Ti, is still a complicated scientific and technological task. The viability of CP-Ti as implant materials is also challenged by recent development of β-type biomedical titanium alloys [6,7] that generally have higher strength and wear resistance and lower elastic modulus than UFG CP-Ti. Recently developed multifunctional titanium alloys are capable of achieving a better balance of high strength and extra-low elastic modulus [38,57]. Because these materials are usually prepared by conventional technologies, they have great advantages over NS CP-Ti and Ti-6Al-4V alloy in terms of processing efficiency.

Compared to grain refinement mechanisms of dislocation evolution and deformation twinning available for CP-Ti [35], stress-induced martensitic (SIM) transformation is a more powerful mechanism [34]. For example, SPD may result in almost complete amorphization of NiTi owing to a mechanism based on martensitic transformation [58]. Because the reversible SIM transformation from β to α″ martensite exists in a wide range of chemical compositions of β-type titanium alloys [59], this phase transformation may facilitate more significant grain refinement during plastic deformation. Three additional deformation mechanisms in newly developed β-type titanium alloys have been reported recently: reversible SIM transformation between β and a new orthorhombic phase α‴ (similar to α″ martensite) [60], giant fault [57], and a deformation behavior similar to shear banding [38]. These mechanisms all contribute to grain refinement during plastic deformation and, as such, even conventional cold rolling may result in NS sheet with grain size less than 50 nm (Figure 5.1).

It can be concluded from the above discussion that the manyfold grain refinement mechanisms combined with SPD technologies will be able to produce large-dimension forms of NS β-type titanium alloys for biomedical use.

ACKNOWLEDGMENTS

This work is partially supported by the National Science Foundation of China (Grants 50471074 and 30471754) and the Chinese Ministry of Science and Technology (Grant TG2000067105).

REFERENCES

1. Laing, P.G., Fergosum, A.B. Jr., and Hodge, E.S., Tissue reaction in rabbit muscle exposed to metallic implants, *J. Biomed. Mater. Res.,* 1, 135–149, 1967.
2. Steinemann, S.G., Corrosion of surgical implants *in vivo* and *in vitro* test, in *Evaluation of Biomaterials,* Winter, G.D., Leray, J.L., and de Groot, K., Eds., Wiley, New York, 1980, p. 1.
3. Huiskes, R., Weinans, H., and van Rietbergen, B., The relationship between stress shielding and bone resorption around total hip stems and the effects of flexible materials, *Clin. Orthop. Relat. Res.,* 274, 124–134, 1992.
4. Sumner, D.R., and Galante, J.O., Determinants of stress shielding: Design versus materials versus interface, *Clin. Orthop. Relat. Res.,* 274, 202–212, 1992.
5. Dujovne, A.R., Robyn, J.D., Frygier, J.J., Miller, J.E., Brook, C.E., Mechanical compatibility of noncemented hip prostheses with human femur, *J. Arthroplasty,* 8, 7–22, 1993.
6. Long, M., and Rack, H.J., Titanium alloys in total joint replacement—A materials science perspective, *Biomaterials,* 19, 1621–1639, 1998.
7. Niinomi, M., Recent metallic materials for biomedical applications, *Metall. Mater. Trans. A,* 33, 477–486, 2002.
8. Valiev, R.Z., Islamgaliev, P.K., and Alexandrov, I.V., Bulk nanostructured materials from severe plastic deformation, *Prog. Mater. Sci.,* 45, 103–189, 2000.
9. He, G., and Hagiwara, M., Ti alloy design strategy for biomedical applications, *Mater. Sci. Eng. C,* 26, 14–19, 2006.
10. Latysh, V., Krallics, G., Alexandrov, I., and Fodor, A., Application of bulk nanostructured materials in medicine, *Current Appl. Phys.,* 6, 262–266, 2006.
11. Webster, T.J., Siegel, R.W., and Bizios, R., Osteoblast adhesion on nanophase ceramics, *Biomaterials,* 20, 1221–1227, 1999.
12. Webster, T.J., Ergun, C., Doremus, R.H., Siegel, R.W., and Bizios, R., Specific proteins mediate enhanced osteoblast adhesion on nanophase ceramics, *J. Biomed. Mater. Res.,* 51, 475–483, 2000.
13. Webster, T.J., Ergun, C., Doremus, R.H., Siegel, R.W., and Bizios, R., Enhanced functions of osteoblasts on nanophase ceramics, *Biomaterials,* 21, 1803–1810, 2000.
14. Webster, T.J., Schadler, L.S., Siegel, R.W., and Bizios, R., Mechanisms of enhanced osteoblast adhesion on nanophase alumina involve vitronectin, *Tissue Eng.,* 7, 291–301, 2001.
15. Dalby, M.J., Riehle, M.O., Johnstone, H., Affrossman, S., and Curtis, A.S.G., *In vitro* reaction of endothelial cells to polymer demixed nanotopography, *Biomaterials,* 23, 2945–2954, 2002.
16. Dalby, M.J., Yarwood, S.J., Riehle, M.O., Johnston, H.J., Affrossman, S., and Curtis, A.S., Increasing fibroblast response to materials using nanotopography: Morphological and genetic measurements of cell response to 13-nm-high polymer demixed islands, *Exp. Cell. Res.,* 276, 1–9, 2002.
17. Webster, T.J., Hellenmeyer, E.L., and Price, R.L., Increased osteoblast functions on theta + delta nanofiber alumina, *Biomaterials,* 26, 953–960, 2005.
18. Elias, K.L., Price, R.L., and Webster, T.J., Enhanced functions of osteoblasts on nanometer diameter carbon fibers, *Biomaterials,* 23, 3279–3287, 2002.
19. Deligianni, D.D., Katsala, N., Ladas, S., Sotiropoulou, D., Amedee, J., and Missirlis, Y.F., Effect of surface roughness of titanium alloy Ti-6Al-4V on human bone marrow cell response and on protein adsorption, *Biomaterials,* 22, 1241–1251, 2001.
20. De Oliveira, P.T., and Nanci, A., Nanotexturing of titanium-based surface upregulates expression of bone sialoprotein and osteopoutin by cultured osteogenic cells, *Biomaterials,* 25, 403–413, 2004.
21. Webster, T.J., and Ejiofor, J.U., Increased osteoblast adhesion on nanophase metals: Ti, Ti6Al4V, and CoCrMo, *Biomaterials,* 25, 4731–4739, 2004.
22. Gleiter, H., Nanostructured materials: Basic concepts and microstructure, *Acta Mater.,* 48, 1–29, 2000.
23. Suryanarayana, C., Mechanical alloying and milling, *Prog. Mater. Sci.,* 46, 1–184, 2001.
24. Kumar, K.S., van Swygenhoven, H., and Suresh, S., Mechanical behavior of nanocystalline materials and alloys, *Acta Mater.,* 51, 5743–5774, 2003.
25. Suryanarayana, C., Recent developments in nanostructured materials, *Adv. Eng. Mater.,* 7, 983–992, 2005.
26. Stolyarov, V.V., Zhu, Y.T., Lowe, T.C., and Valiev, R.Z., Microstructure and pure Ti processed by ECAP and cold extrusion, *Mater. Sci. Eng. A,* 303, 82–89, 2001.
27. Stolyarov, V.V., Zhu, Y.T., Alexandro, I.V., Lowe, T.C., and Valiev, R.Z., Grain refinement and properties of pure Ti processed by ECAP and cold rolling, *Mater. Sci. Eng. A,* 343, 43–50, 2003.

28. Stolyarov, V.V., Zhu, Y.T., Raab, G.I., Zharikov, A.I., and Valiev, R.Z., Effect of initial microstructure on the microstructural evolution and mechanical properties of Ti during cold rolling, *Mater. Sci. Eng. A*, 385, 309–313, 2004.

29. Kim, S.M., Kim, J., Shin, D.H., Yo, Y.G., Lee, C.S., and Semiatin, S.L., Microstructure development and segment formation during ECA pressing of Ti-6Al-4V alloy, *Scripta Mater.*, 50, 927–930, 2004.

30. Salishchev, G.A., Galeyev, R.M., Valiakhmetov, O.R., Safiullin, R.V., Lutfullin, R.Y., Senkov, O.N., Froes, F.H., and Kaibyshev, O.A., Development of Ti-6Al-4V sheet with low temperature superplastic properties, *J. Mater. Proc. Tech.*, 116, 265–268, 2001.

31. Zherebtsov, S.V., Salishchev, G.A., Galeyev, R.M., Valiakhmetov, O.R., Mironov, S.Y., and Semiatin, S.L., Production of submicrocrystalline structure in large-scale Ti-6Al-4V billet by warm severe deformation processing, *Scripta Mater.*, 51, 1147–1151, 2004.

32. Murzinova, M.A., Mazurski, M.I., Salishchev, G.A., and Afonichev, D.D., Application of reversible hydrogen alloying for formation of submicrocrystalline structure in ($\alpha+\beta$) titanium alloys, *Int. J. Hydrogen Energy*, 22, 201–204, 1997.

33. Murzinova, M.A., Salishchev, G.A., and Afonichev, D.D., Formation of nanocrystalline structure in two-phase titanium alloy by combination of thermohydrogen processing with hot working, *Int. J. Hydrogen Energy*, 27, 775–782, 2002.

34. Lu, K., and Lu, J., Nanostructured surface layer on metallic materials induced by surface mechanical attrition treatment, *Mater. Sci. Eng. A*, 375–377, 38–45, 2004.

35. Zhu, K.Y., Vassel, A., Brissct, F., Lu, K., and Lu, J., Nanostructure formation mechanism of α-titanium using SMAT, *Acta Mater.*, 52, 4101–4110, 2004.

36. Hao, Y.L., Li, S.J., and Yang, R., Chinese Patent Application 1011998.5, 2003; PCT Patent Appl. PCT/CN2004/001352, 2004.

37. Yang, R., Hao, Y.L., and Li, S.J., Titanium with bone-matching elastic modulus and super-elasticity, in *Materials and Processes for Medical Devices 2005*, Venugopalan, K. and Wu, M.H., Eds., ASM, Materials Park, OH, 2006.

38. Hao, Y.L., Li, S.J., Sun, S.Y., Zheng, C.Y., Hu, Q.M., and Yang, R., Super-elastic titanium alloy with unstable plastic deformation, *Appl. Phys. Lett.*, 87, 091906, 2005.

39. Hao, Y.L., Li, S.J., and Yang, R., Nanostructured β-type titanium alloy for biomedical application, in *Materials and Processes for Medical Devices 2005*, Venugopalan, K. and Wu, M.H., Eds., ASM, Materials Park, OH, 2006.

40. Valiev, R.Z., Nanostructuring of metals by severe plastic deformation for advanced properties, *Nature Mater.*, 3, 511–516, 2004.

41. He, G., Eckert, J., Löser, W., and Schultz, L., Novel Ti-based nanostructure-dendrite composite with enhanced plasticity, *Nature Mater.*, 2, 33–37, 2003.

42. He, G., Eckert, J., Löser, W., and Hagiwara, M., Composition dependence of the microstructure and mechanical properties of nano/ultrafine-structured Ti-Cu-Ni-Sn-Ni alloys, *Acta Mater.*, 52, 3035–3046, 2004.

43. Sun, B.B., Sui, M.L., Wang, Y.M., He, G., Eckert, J., and Ma, E., Ultrafine composite microstructure in a bulk Ti alloy for high strength, strain hardening and tensile ductility, *Acta Mater.*, 54, 1349–1357, 2006.

44. Stolyarov, V.V., Zhu, Y.T., Lowe, T.C., Islamgaliev, R.K., and Valiev, R.Z., A two step SPD process of ultrafine-grained titanium, *Nanostrut. Mater.*, 11, 947–954, 1999.

45. Sergueeva, A.V., Stolyarov, V.V., Valiev, R.Z., and Mukherjee, A.K., Advanced mechanical properties of pure titanium with ultrafine grained structure, *Scripta Mater.*, 45, 747–752, 2001.

46. Valiev, R.Z., Sergueeva, A.V., and Mukherjee, A.K., The effect of annealing on tensile deformation behavior of nanostructured SPD titanium, *Scripta Mater.*, 49, 669–674, 2003.

47. Hao, Y.L., and Yang, R., High strength nano-structured Ti-Nb-Zr-Sn alloy, *Acta Metall. Sinica*, 41, 1183–1189, 2005.

48. Inoue, A., and Takeuchi, A., Recent progress of bulk glassy, nanoquasicrystalline and nanocrystalline alloys, *Mater. Sci. Eng. A*, 375–377, 16–30, 2004.

49. Wang, Z.B., Tao, N.R., Li, S., Wang, W., Liu, G., Lu, J., and Lu, K., Effect of surface nanocrystalline on friction and wear properties in low carbon steel, *Mater. Sci. Eng. A*, 352, 144–149, 2003.

50. Stolyarov, V.V., Shuster, L.S., Migranov, M.S., Valiev, R.Z., and Zhu, Y.T., Reduction of friction coefficient of ultrafine-grained CP titanium, *Mater. Sci. Eng. A*, 371, 313–317, 2004.

51. La, P.Q., Ma, J.Q., Zhu, Y.T., Yang, J., Liu, W.M., Xue, Q.J., and Valiev, R.Z., Dry-sliding tribological properties of ultrafine-grained Ti prepared by severe plastic deformation, *Acta Mater.*, 53, 5156–5173, 2005.

52. Schwartz, Z., and Boyan, B.D., Underlying mechanisms at the bone-biomaterial interface. *J. Cell Biochem.*, 56, 340–347, 1994.

53. Tong, W.P., Tao, N.R., Wang, Z.B., Lu, J., and Lu, K., Nitriding iron at lower temperatures, *Science*, 299, 686–688, 2003.

54. Nagy, P.M., Ferencz, B., Kálmán, K., Djuričić, B., and Sonnleitner, R., Morphological evolution of Ti surface during oxidation treatment, *Mater. and Manuf. Processes*, 20, 105–114, 2005.

55. Li, S.J., Zheng, C.Y., Tao, X.J., Hao, Y.L., Yang, R., and Zhao, Y.K., Effect of grain refinement on fibroblast adhesion and proliferation to Ti-24Nb-4Zr-7.9Sn alloy, in *Materials and Processes for Medical Devices 2005*, Venugopalan, K. and Wu, M.H., Eds., ASM, Materials Park, OH, 2006.

56. Zhu, Y.T., Lowe, T.C., and Langdon, T.G., Performance and applications of nanostructured materials by severe plastic deformation, *Scripta Mater.*, 51, 825–830, 2004.

57. Saito, T., Furuta, T., Hwang, J.H., Kuraoto, S., Nishino, K., Suzuki, N., Chen, R., Yamada, A., Ito, K., Seno, Y., Nonaka, T., Ikehata, H., Nagasako, N., Iwamoto, C., Ikuhara, Y., and Sakuma, T., Multifunctional alloys obtained via a dislocation-free plastic deformation mechanism, *Science*, 300, 464–467, 2003.

58. Sergueeva, A.V., Song, C., Valiev, R.Z., and Mukherjee, A.K., Structure and properties of amorphous and nanocrystalline NiTi prepared by severe plastic deformation and annealing, *Mater. Sci. Eng. A*, 339, 159–165, 2003.

59. Collings, E.W., *The Physical Metallurgy of Titanium Alloys*, ASM, Materials Park, OH, 1984.

60. Zhang, L.C., Zhou, T., Alpay, S.P., Aindow, M., and Wu, M.H., Origin of pseudoelastic behavior in Ti-Mo-based alloys, *Appl. Phys. Lett.*, 87, 241909, 2005.

6 Commercializing Bionanotechnology: From the Academic Lab to Products

Michael N. Helmus, Peter Gammel,
Fred Allen, and Magnus Gittins
Advance Nanotech, Inc., New York, New York, USA

CONTENTS

6.1 INTRODUCTION

Bionanotechnology, nanostructured materials/biomaterials (monolithic and particulate), and electronics for enhanced functionality of medical devices, pharmaceutics, as well as hybrid combinations for diagnostics and therapeutics have the potential to enable disruptive medical procedures that are nanoenabled (that is, nanomedicine). The ability to move these technologies from bench to bedside—to bring these products to market—entails unique approaches to guide them through the "Valley of Death."

6.2 NANOTECHNOLOGY: NEW ENABLEMENTS FOR BIOPHARMA

Nanotechnology has received a great deal of attention in both the academic and lay press. As new technologies come forward, the salient question now becomes how does an emerging technology enable improved safety and outcomes? In the parlance of the biopharma world that means the technologies in question need to build upon existing levels of safety and efficacy in medical devices, diagnostics, and pharmaceuticals.

Over approximately the past 30 years, three distinct but interdependent technology disciplines have emerged: biotechnology, information technology, and nanotechnology. Today, we are in the

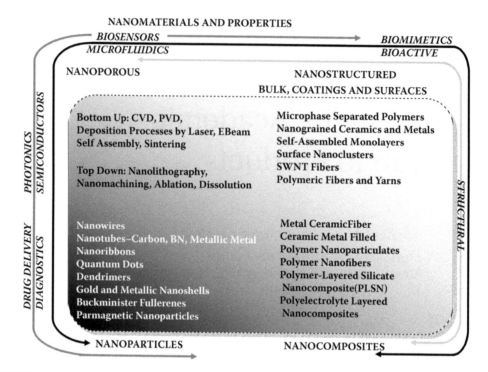

FIGURE 6.1 Nanomaterials and their biopharma applications.

midst of a convergence of all three disciplines, and it is still not clear what this will bring. Nanotechnology is the newest and least understood of these technologies. Not limited to any one discipline of science, nanotechnology is simply defined as the design, characterization, production, and application of structures, devices, and systems measuring between 1 and 100 nanometers. A nanometer is a billionth of a meter, approximately 80,000 times smaller than the width of a human hair. This is the scale at which matter, its properties, and its behavior may be described in terms of atoms and molecules. Nanotechnology is technology that has been developed to take advantage of these properties.

With respect to a nanomaterial, "size" can refer to different things. It can be the length of a single nanocrystal or the diameter of a polycrystalline nanoparticle composing a powder or dispersed in a solid or liquid; the height of a nanofiber grown on a substrate; the geometry of a nanopore opening; the depth of a nanocavity; the thickness of a nanocoating; or the separation distance between nanophases in a multilayer or nanocomposite structure. Figure 6.1 lists some categories of nanomaterials that have utility in the biopharma space.

Unmet needs within the medical sector provide fertile ground for the enablements offered by nanotechnology. Bionanotechnology can provide the enhancements needed for new enablements in devices, diagnostics, and pharmaceuticals. These enhanced enablements include smart biomaterials and devices: bioresponsive, bioactive, biomimetic, tissue engineered, drug/agent/cell delivery for improved diagnostic and therapeutic effectiveness. Examples include noninvasive diagnostics, molecular imaging, and real-time imaging for device placement. Bionanotechnology has the potential to enhance utilization of proteomics, genomics, and gene screening, resulting in improved patient outcomes and reduced health care costs [1].

6.2.1 The Valley of Death

The challenge faced by those working in nanotechnology is to translate these emerging technologies from the bench to bedside—that is, from the laboratory through to product development for clinical

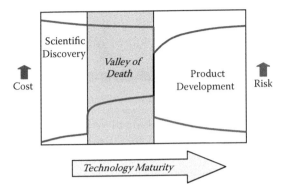

FIGURE 6.2 Crossing the "Valley of Death" from scientific discovery into product development.

use. This journey leads through the Valley of Death [2a],* depicted in Figure 6.2. The Valley of Death, first illuminated by Murphy and Edwards in 2003, is where many research projects end up stalling or dying before having the chance to become viable candidates for commercial products. Most of the technologies developed in academic and research labs rarely make this translation. Out of 1200 or so nanotechnology start-up companies that exist today worldwide, we estimate that only 10% are actually selling products in excess of $1 million/year in the form of nanomaterials, devices, or systems [3]. To our knowledge, none of the companies are profitable as yet. When it comes to nanotechnology, the problem is that neither the scientific researchers nor the product developers have all of the resources in one organization to accomplish the above tasks in a timely manner. There needs to be a reliable, experienced guide who leads the scientific researchers across the Valley of Death and either introduces them to the product developers or educates them on how to run a business venture themselves. This chapter will first explore the options open to disruptive technologies entering the medical field and will then look at how the gap between bench and bedside can be bridged by strategic assistance from development partners.

6.3 BIONANOTECHNOLOGY

Bionanotechnology is the further convergence of biology with nanotechnology and is the utilization of nanostructured materials/biomaterials (monolithic and particulate) and nanostructured electronics for enhanced functionality of medical devices, pharmaceutics, and hybrid combinations (device and pharma combinations) for diagnostics and therapeutics. The application of bionanotechnology into the clinical setting can be described as nanomedicine. Bionanotechnology will drive the thrust for the evolution of personalized medicine and on-demand therapy to mitigate adverse events as they happen.

When taking a view on the benefits of nanotechnology in a certain field, it is important not to fixate on the issue of size. Size in itself does not necessarily impact performance. Product performance should encompass a plurality of unique properties (biocompatibility, electronic, magnetic, optical, mechanical, thermal, catalytic, etc.) that are often regarded as contradictory in the "non-nano" world. An example of a desirable nanomaterial with contradictory properties might be a polymer nanocomposite that is mechanically strong and tough (will not bend or break), hard (resists scratching), electrically conductive, optically transparent, and biocompatible. It would be nice to have all of these properties rolled into one material, but in a typical instance, trade-offs are necessary. Knowing which properties can and cannot be sacrificed depends on the product application. Obtaining a balance of properties is generally the goal, and this is a key part of success in this area, because nanotechnology can result in properties (e.g., high strength, stiffness, and toughness) that

---

* An alternative way of structuring issues that occur in this area—the Darwinian Sea—is offered in [2b].

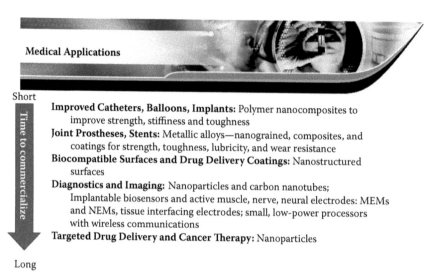

Medical Applications

Short

Time to commercialize

Improved Catheters, Balloons, Implants: Polymer nanocomposites to
   improve strength, stiffness and toughness
Joint Prostheses, Stents: Metallic alloys—nanograined, composites, and
   coatings for strength, toughness, lubricity, and wear resistance
Biocompatible Surfaces and Drug Delivery Coatings: Nanostructured
   surfaces
Diagnostics and Imaging: Nanoparticles and carbon nanotubes;
   Implantable biosensors and active muscle, nerve, neural electrodes: MEMs
   and NEMs, tissue interfacing electrodes; small, low-power processors
   with wireless communications
Targeted Drug Delivery and Cancer Therapy: Nanoparticles

Long

FIGURE 6.3 Potential medical applications for nanotechnology devices.

would be difficult to obtain in the normal world. Examples of how nanotechnology can impact
medical devices are shown in Figure 6.3.

A "technology" is defined here by the ability to deterministically simulate, design, engineer,
and reproducibly manufacture a product to obtain the best balance of properties for a given application. Ultimately, nanotechnology must exist within the constraints of being able to integrate these
unique performance characteristics into controlled systems that operate at the macroscale and can
be made at relatively low cost. This will take some time to achieve, but we are only at the beginning
of realizing the potential of fabricating products by moving and assembling atoms and molecules,
building larger structures, and controlling properties and behavior at the nanoscale.

One of the key areas in which the movement of nanotechnology from the laboratory to the
marketplace is occurring is in the medical sector. Here we are seeing the convergence of disciplines to form a new type of bionanotechnology for diagnostics and therapeutics. The possibilities
that nanotechnology opens up in this area will allow the emergence of a new era of personalized
medicine, offering every patient greater accuracy in his or her diagnosis and treatment, as well
as this being able to occur at a far greater speed. Due to these possibilities, it can be argued that
within the medical field, bionanotechnology can be characterized as a "disruptive technology"
[4], a term coined by the Harvard Business School academic, Professor Clayton M. Christensen.
Professor Christensen defines disruptive technologies as technologies that introduce a different
performance package into a market than that offered by incumbent technologies. Technology disruption occurs when, despite the technology's initially inferior performance according to existing benchmarks, the new technology displaces the mainstream technology from the mainstream
market. The disruptive technology presents a radically different alternative to what is offered by
the status quo. Examples from the past include the introduction of digital photography, the invention of the automobile, and the revolution produced by the introduction of the personal computer
[4]. Drug-eluting stents have many of the same characteristics of a disruptive technology and have
continued the trend of moving patients from coronary bypass surgery to treatments by interven-
tional cardiology.

Disruptive technologies are not disruptive to customers, who actively benefit from their introduction. They are also often not disruptive to incumbent market players in their early stages. Therefore, they often go unrecognized until they have become a significant force within their market.
Bionanotechnology fulfills these characteristics within a number of fields. For the purpose of this
chapter, we will concentrate on its applications within a medical context.

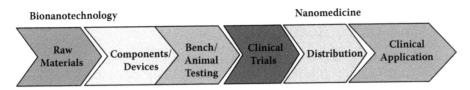

FIGURE 6.4 The supply chain of moving from the laboratory to the marketplace.

Enhanced nano-enabled biomaterials are a disruptive medical technology. Nanobiomaterials will have a huge impact on the future of conventional medicine. Their importance in this field is based on their ability to facilitate the evolution of personalized medicine and on-demand therapy to mitigate adverse events as they happen. They also have the potential to rapidly change the available therapeutic options. One example of this would be implantable sensors for the detection of thrombosis which would then deploy the necessary therapeutics, such as a stent that can allow local release of a drug immediately upon detection of a myocardial infarction. The promise and the challenge of nano-enabled technologies for medical applications are clear. They offer enhanced functionality and biocompatibility for improved healing and neogenesis of functional tissue to replace damaged or diseased tissue and organs.

However, potential new paradigms are required for biocompatibility evaluations of nanostructures and particles. The U.S. Food and Drug Administration (FDA) will be closely watching the testing of nanoenabled devices and pharmaceutics in order to determine if the current test methodologies are adequate or whether new testing will be required [5]. Concerns as to the ultimate deposition of nanoparticles in the body exist because their small size and chemical properties may enable them to easily pass the normal barriers in the body (for example, the blood–brain barrier) and to easily enter cells. The results from ADME (absorption, distribution, metabolism, excretion) in short-term and long-term toxicity studies in rodent and nonrodent species will be carefully monitored. Other important evaluations will focus on pharmacology, safety pharmacology, genotoxicity, developmental toxicity, irritation studies, immunotoxicology, and carcinogenicity studies. The development of efficacious therapeutic and diagnostic procedures based on nanotechnology will require the early collaboration of clinicians and an understanding of the clinical environment.

In order for bionanotechnology to become more than an academic curiosity, it needs to overcome the challenges outlined above. Each individual technology also needs to fight its way from basic laboratory research into the hands of consumers. Figure 6.4 outlines the process that these technologies need to undertake in order to emerge into the marketplace.

The first stage in the supply chain of bionanotechnology along its path to enable nanomedicine is the development of the components and devices (Figure 6.4). The process of developing these components and devices can be divided into the following five stages [2b]:

1. *Basic research*—Funding from National Science Foundation (NSF), National Institutes of Health (NIH) grants, corporate research, and Small Business Innovation Research (SBIR) phase I
2. *Proof of concept/invention*—Previous sources, angel investors, corporations, technology laboratories, and SBIR phase II
3. *Early-stage technology development*—Previous sources and early-stage venture capital funds
4. *Product development*—Traditional venture capital
5. *Production/marketing*—Corporate venture funds, equity, and commercial debt

In moving through these five phases, the product moves from the bench into product development and the commercialization pathway.

6.4 MANAGEMENT OF EARLY-STAGE RESEARCH— CREATION OF A COMMERCIALIZATION PATH

So what is needed to move nanotechnology forward from intellectually stimulating research projects to money-making commercialized products, enabling those involved—from individuals to nations—to prosper? To succeed from an economic standpoint, numerous nanoproducts must be sold that ultimately generate billions of dollars in revenue and yield attractive profits to satisfy investors; practical problems must be solved that benefit customers and positively impact society; new jobs must be created in both large and small companies to sustain existing and emerging businesses; academic institutions must overflow with motivated professors, postdocs, staff, and students; and government officials must propose and pass legislation to secure funding and provide financial assistance to entrepreneurs in support of their ventures. That is a lot to expect, and much can get in the way to slow or prevent the above steps from occurring.

The key question is how to bridge the gap between the realm of scientific discovery and the world of product development. The authors propose four essential ingredients to facilitate rapid product commercialization once a nanotechnology opportunity has been identified and defined and a venture created to move things forward:

1. Investment dollars to resource the venture (people, equipment, space, etc.).
2. Marketing knowledge to orient the venture (develop strategy; point and move people in the right direction).
3. Intellectual property expertise to position the venture (differentiate from and defend against competition).
4. Project management skills to steer the venture (maintain or change the strategic focus and tactical course; schedule milestones; set priorities; assign resources; track and measure progress, etc.).

Advance Nanotech's approach to provide these services is one example of bridging the gap over the Valley of Death. Advance Nanotech provides a toolbox in an effort to ensure the technologies into which we invest reach maximum market potential. This toolbox includes financing and support services, such as commercialization guidance, project and infrastructure management, leadership assets, access to corporate and scientific advisory board networks, and counsel on intellectual property, licensing and regulatory issues, and business and marketing plans. This toolbox significantly increases the probability of enabling nanotechnology discoveries to reach maximum market potential through successful commercialization.

One of the keys to being successful is to structure a research agreement that meets the needs of commercialization and the academic environment. Key items in this research agreement are as follows:

- Right to nonexclusive license of existing intellectual property (IP).
- Right to exclusive license of arising IP (or foreground IP).
- Frequent research reports.
- Right to review publications prior to submission.
- Right to request action on patentable invention.
- Right to participate/initiate patent prosecution.
- Right to redirect research.

The proper management system in conjunction with research agreements focused on key milestones and deliverables enhances the ability to manage and communicate in a secure environment with our collaborators both domestically and internationally. This is a significant enabling tool in the commercialization, monitoring, and development of emerging nanotechnologies.

In this model, emerging technologies that form the potential basis of new companies are provided with technical project management assistance, access to corporate and scientific advisory board networks, competitive and IP landscape analysis, and business and marketing plans. In the case of Advance Nanotech's model, each emerging technology that we invest into must be developing a product for a market with a minimum market value of $200 million, must have a credible path to commercialization which must not exceed 5 years, and must have a protection strategy for IP. Once a subsidiary company's product pipeline is sufficiently matured, it enters the near-to-market technology section of the portfolio and is then at the stage where further funding from capital markets can be sourced.

The financial community is predominantly interested in later-stage products. When the uncertainties are primarily technical, investors are ill equipped to quantify them—trusted experts are needed to do this for them. Research has shown the inefficiencies within the present market in allocating venture capital to early-stage technology ventures despite their innovative research and strong prospects. Mary Good, former Undersecretary of Commerce for Technology, has described this scarcity as an innovation gap [2b]. We utilize our infrastructure and expertise to accelerate the development of multiple early-stage research programs to precommercialization/ready for market status. This process results in substantial value being created at each stage of development. Extensive relationships with academic institutions and industry provide multiple opportunities. At any time, we are assessing more than 30 individual research programs for investment consideration.

Additionally, the portfolio is supplemented by minority interests in businesses that are near-to-market and offer strategic value, and broader development partnerships with business or academia.

6.4.1 EXAMPLE OF A COMMERCIALIZATION MODEL

Will this strategy work? Yes, it will. In fact, there is already clear evidence of its efficacy. Owlstone Nanotech Inc., an Advance Nanotech subsidiary, is a spin-off from the University of Cambridge. Owlstone's sensor technology offers a revolutionary dime-sized device that can be programmed to detect a wide range of chemical agents that may be present in extremely small quantities. Using leading-edge micro- and nanofabrication techniques, Owlstone created a complete chemical detection system that is one hundred times smaller and one thousand times cheaper than existing technology. This makes use of their proprietary Field Asymmetric Ion Mobility Spectrometry (FAIMS) technology that overcomes many of the key problems that exist within the incumbent IMS-based detection technology. This detector has a range of applications across a number of sectors, including homeland security, defense, industrial process control, consumer, medical, and environmental. Within the medical arena, the sensor has a number of applications. For example, at present, diabetes patients have to monitor glucose levels by using a painful, inconvenient pinprick blood test. Owlstone's detector technology offers a compact, painless alternative, enabling rapid, noninvasive measurement of acetone levels in exhaled breath that are directly related to blood sugar levels. The presence of signature chemicals also has the potential to aid in the detection of a wide range of other conditions, from asthma and allergies to organ failures and certain types of cancer. Owlstone has already entered development partnerships with firms from across a number of fields. We believe that the developments at Owlstone are indicative of our strategy of commercializing innovation and are clear representations of our value proposition. It is these types of developments that we expect to foster with our other technologies.

Owlstone offers a clear example of how a carefully chosen technology, a quality team, and the right assistance and support can produce the desired outcome for all involved, benefiting both consumer and investor with the products that are brought to market. Advance Nanotech is taking the small firm environment in which innovation prospers and is partnering it with support and assistance normally available only to a larger firm.

ACKNOWLEDGMENT

The authors would like to acknowledge the research and editing of Tom Ebbutt.

REFERENCES

1. www.nibib.nih.gov/nibib/File/News%20and%20Events/NIBIB_IndustryWorkshop_FinalReport.pdf. (Accessed December 17, 2003.)
2a. Murphy, L.M., and Edwards, P.L., Bridging the Valley of Death—Transitioning from Public to Private Sector Financing, National Renewable Energy Laboratory, Golden, Colorado, NREL/MP-720-34036, May 2003.
2b. Branscomb, L.M., and Auerswald, P.E., Between Invention and Innovation: An Analysis of Funding for Early-Stage Technology Development, Economic Assessment Office Advanced Technology Program, National Institute of Standards and Technology, Gaithersburg, MD, 2003.
3. Lane, Neal, and Kalil, Thomas, The National Nanotechnology Initiative: Present at the Creation, Issues in *Science and Technology*, Summer 2005 (http://www.issues.org/21.4/lane.html).
4. Christensen, C.M., *The Innovator's Dilemma: When New Technologies Cause Great Firms to Fail*, Harvard Business School Press, Boston, 1997.
5. Sadrieh, N., Office of Pharmaceutical Science/CDER/FDA, Nanotechnology: Regulatory Considerations for Drug Development, presentation to ChBSA (www.fda.gov/nanotechnology/ChBSA-nanotech-presentation06-04.ppt).

7 Opportunities for Bionanotechnology in Food and the Food Industry

Frans W.H. Kampers
BioNT, Wageningen Bionanotechnology Centre,
Wageningen, The Netherlands

CONTENTS

7.1 INTRODUCTION

Western societies face large challenges with respect to health care and the well-being of their citizens. Food-related diseases like obesity and diabetes will create serious problems; the aging population will require special nutrition to maintain quality of life; food quality and safety is an increasingly important issue; lifestyle requirements will have to be addressed; and sustainability of food production and processing requires continuous attention. Bionanotechnology can make important contributions to the fulfillment of these challenges. It will allow detection systems with improved sensitivity and specificity with which food quality can be assured and processes can be controlled more accurately. Application of bionanotechnology will make separation and certain processing steps more efficient and will open up possibilities for new products. Derived from drug delivery systems, envelopes for the delivery of nutrients can be designed that will improve uptake by the consumer. Nanoscale control of structures will allow tailoring the texture of food products to the requirements of consumer groups. Results of nanotechnology will have a large impact on packaging and logistics of food.

Although these results are attainable within the foreseeable future, the combined effort of academia, research institutions, and industry is required to develop them into economically viable concepts. Some of the concepts presented here are close to real-world applications, but others are mere ideas that require much more fundamental research before they can actually be developed. They are included to give an idea of the variety of applications and to trigger the imagination of the reader.

7.2 SENSORS AND DIAGNOSTICS

Measuring specific parameters in the complex processes that are associated with food is essential to control these processes and also to enhance and maintain the quality of the products. Sensitivity and specificity are usually limiting factors that need to be improved. Nanotechnology now offers concepts that allow development of sensing principles at the molecular and supramolecular levels. Not only is nanotechnology used for evolutionary improvement of already available sensors, it is also used for devices based on revolutionary new principles for sensing and detection.

7.2.1 QUALITY ASSURANCE

Food production chains become longer, more complex, and more international. To be able to maintain the high level of food quality and safety in the European Union, quality assurance along the full length of the chains is required. A prerequisite to monitor the quality of products at various locations in or at the chain is the availability of low-cost, fast, and easy-to-operate instrumentation that can determine parameters that correlate to the quality, safety, or specific properties of the product. Important parameters in this respect are microbial activity and concentration of certain metabolic products. An example of this approach is the antioxidant (e.g., vitamin C) content of fresh fruit. The concentration of this important compound can vary by several orders of magnitude and is largely influenced by postharvest conditions. To be able to determine these parameters with a very high level of accuracy and specificity, molecular methods are required. Principles that can be used include, for example, detection of characteristic strands of DNA or RNA, detection of specific carbohydrate fragments, and binding of ligands to receptors. However, most of these principles require the extensive use of laboratory equipment, take too much time, or require experienced personnel. Moreover, deriving a physical and measurable signal from these principles usually is difficult. The application of nanotechnology in combination with microtechnology will allow the development of instruments that meet the application demands of the industry. There are two distinct applications of nanotechnology in quality assurance: sensors for monitoring production processes or the presences of unwanted substances (residues, allergens, etc.) and diagnostics for determining microbial activity.

Biosensors are devices that use or mimic the very sensitive and selective detection principles encountered in nature to measure the amount of specific substances in their vicinity. To interface with the control hardware, they must be designed in such a way that they result in a signal that can be converted into an electrical current or voltage. Although there are many principles that fulfill this requirement, sensors based on the principle of a field effect transistor (FET) have the advantage that they do not require a conversion from one physical property to an electrical signal. Chemical changes at the gate electrode directly result in a change in the electric current through the device (see Figure 7.1). A good overview of these types of devices is given by Bergveld [1]. In the case of a biosensor, the principle of a FET is used in combination with a biological receptor. By functionalizing the surface of the FET device [2] with receptor molecules, a binding event of the receptor with a ligand will result in a change of the charge distribution in the receptor molecule. This charge shift is sensed in the silicon substrate and influences the current that flows from the source to the drain. The sensitivity of these devices is enhanced if the cross section of the channel that connects the source and the drain is small compared to the surface of it. It is therefore advantageous to use

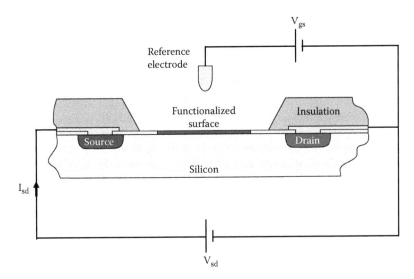

FIGURE 7.1 Schematic diagram of an FET-based sensor. The source drain current I_{sd} that flows through the silicon is influenced by interactions of specific substances in the medium and the functionalized surface.

functionalized nanowires [3]. These nanowires can be of various materials including silicon and carbon nanotubes.

The ultimate goal is to make devices that use many different receptors and to measure the relative strength of the signals of these receptors. The human nose contains approximately 300 different receptors, and our brain has learned to interpret the patterns that are generated to recognize specific smells. A device containing the same receptors could mimic the human nose. Smells contain very much information on the quality of food. An electronic nose could therefore be used to monitor the quality of foodstuff or in food processing.

There is a strong demand for fresher food products. Microbial activity is an important cause for spoilage of these products. After the harvest, the defense mechanisms of products break down and microorganisms can attack the tissue which results in quality deterioration. Some of these microorganisms are pathogenic and threaten the safety of the food product, and others cause organoleptic spoilage. *Diagnostic methods* able to determine the microbial activity in a food product are therefore widely used to assess the quality. Unfortunately, there are no fast and easy-to-use methods to do this. At the moment, the prevailing method requires a well-equipped and staffed laboratory, taking a sample and incubating it on a petri dish. This can take, depending on the type of organism, from 24 hours to several days. Of course this is a problem in the fresh food chains. By combining microtechnology with nanotechnology, new methods for the detection of pathogens and the determination of the amount of spoilage organisms in foodstuff are being developed. They are usually based on the detection of specific DNA or RNA strands in the sample. This requires extraction of the DNA or RNA from the sample, often multiplying it (polymerase chain reaction [PCR]) and detection of characteristic sequences. In other application fields like medicine or homeland security, nanosensing technology for multiplying and detection have already been developed. The difficult part in the application in food quality assessment is the sample pretreatment necessary to extract the DNA or RNA from the food sample. Current regulation prescribes sample volumes of 25 g. It is clear that it will be difficult to process 25 g of sample in a microfluidics device. A change of regulation may therefore be required.

7.2.2 Nutritional Profiling

Our knowledge of nutrigenomics and nutritional processes is developing rapidly [4]. One of the results will be that it will be possible to determine what the nutritional requirements of a certain individual at a certain point in time are from the expression of genes, the metabolism of nutrients, and the excretion of certain chemicals. To exploit this knowledge, it is necessary to monitor gene expression over time and to sense the concentration of relevant molecules in different excretion products. Although still far from realization, the principles used to detect microorganisms will enable the development of new technology to determine in real time the nutritional profile of the individual and to use this profile to optimize nutritional intake. The convergence of nanotechnology, bio(techno)logy, information technology, and cognitive science (NBIC) which is strongly promoted in some parts of the world can result in the systems necessary for such an application.

The concept basically can be divided into two parts: determining the genetic and nutrition-related profile of an individual and measuring the nutritional content of food. The idea can best be demonstrated through the example of allergy. Suppose that in the future it will be feasible to derive sensitivities of an individual for certain substances from the genetic profile. Although the genome is not changing, it is already known that circumstances determine which genes are switched on at a specific point in time. This means that it probably will not suffice to generate the profile and determine the sensitivities once; it will need to be done regularly. This is why fast and easy-to-use devices are necessary for this application. Nanotechnology can provide such devices.

Now that it is known which substances may trigger an allergic reaction, it is necessary to avoid the intake of these substances. Of course, this can be derived from the composition of food products, but very often allergens are present in natural products of which the exact composition is not known. Moreover, processing strongly reduces the allergenicity of substances. Fresh apples can trigger an allergic reaction, whereas the same apples after cooking do not. It is therefore necessary to test the food prior to eating. This also requires an easy-to-use and fast device that is sufficiently sensitive and specific to determine the presence of the substances to be avoided. Nanotechnology will be required to fulfill these demands.

It is important to stress that this is an application of nanotechnology which is still very far from being realized. Much fundamental research, not only in the field of micro- and nanotechnology, is still necessary to make the concepts described here feasible.

7.3 PROCESS INNOVATION

Food production is almost synonymous with food processing. Much of the food produced nowadays is in some way or another processed, either to prepare it, to improve its sensational or nutritional value, or to preserve it. The food industry is therefore constantly searching for ways to improve certain processes or to find new processes that provide specific advantages. With the advent of nanotechnology and earlier microtechnology, new possibilities have become available for process innovation. Here some examples are given.

7.3.1 SEPARATION AND FRACTIONATION

The result of a food process may contain substances that need to be removed. An example is yeast cells in beer. Sometimes the product contains contaminants of microorganisms that are unwanted. Bacteria in milk are an example of this. The food industry has different ways to separate the wanted from the unwanted. In the case of microorganisms, it often relies on heat treatment to kill the organisms. In the case of beer, the product is filtered. Microtechnology as a spin-off of microelectronics has resulted in the ability to produce very accurate and increasingly small three-dimensional structures. Microsieves are a product of microtechnology (see Figure 7.2). Photolithography is used to create a membrane with very many, very small but extremely well-defined holes that can be used for separation [5]. It not only offers very high log reduction ratios, it also improves the sustainability of processes because lower pressures are required or heat treatment can be avoided. Moreover, because of the milder conditions, the original product will maintain its quality. The application of microsieves both in beer and in milk are in the pilot-plant phase.

FIGURE 7.2 Electron microscope photo of a microsieve membrane. (Courtesy of Aquamarijn Microfiltration B.V.)

Food usually is a complex mixture of components with very different properties. By extracting different fractions from the basic food substance, high-value components can be used for different purposes. An example of this principle is milk. By skimming the milk, it is fractionated into two components: cream and low-fat milk. The cream fraction is used for different purposes (butter, cream, etc.). Fractionation in general is a very important process step in the food industry.

The experience of the food for the consumer is strongly related to the relative concentrations of the components and their interaction. If complex mixtures of basic food products can be fractionated in their constituents in such a way that these concentrations and interactions can be restored at a later point in time, the experience of the consumer can be individualized. It can be envisioned that basic ingredients are combined by the retailer, or even in the consumer's home, to tailor the product to the specific tastes or nutritional demands of the consumer.

Fractionation is more complex than separation because it usually cannot be based on size alone. Fortunately, microtechnology can provide complex shapes of the pores in a microsieve; with nanotechnology, the surface of the sieve can be modified to improve fractionation properties. The structures can even be combined with antibodies bound to the surface to capture certain components with a very high degree of specificity (affinity chromatography).

7.3.2 EMULSIFICATION

Emulsions are very important in the food industry. Many products are based on or contain emulsions of some form [6]. New ways of creating improved emulsions are of interest to the sector. Nanotechnology, but to a larger extent microtechnology, can be used to improve emulsification processes.

For low-throughput emulsification processes, microfluidic systems can be used to create small droplets of one phase in the continuous other phase. The discrete phase is entered into the continuous phase via a separate microchannel and forms small droplets. The advantage is that the control of the process is very high, and complex emulsions are feasible. The problem is that it is difficult to scale it up to volumes that are of interest to the food industry. Membrane emulsification has this disadvantage to a much lesser extent.

In cross-flow membrane emulsification, a membrane like the ones used in microsieves (see Figure 7.2) separates one phase (e.g., water) from the other (e.g., oil) [5,7]. The oil is pressed through the pores in the membrane into the water phase that flows across the membrane, and well-defined droplets of oil in water are formed (see Figure 7.3). Because of the uniform pore size, the oil droplets

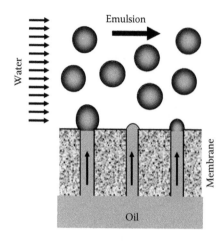

Water

Emulsion

Oil

Membrane

FIGURE 7.3 Cross-flow emulsification used to make an oil in water emulsion. Oil is pressed through the pores and forms equally sized droplets.

are also of equal size, and a monodisperse oil-in-water emulsion is formed.

A monodisperse emulsion is more stable than a polydisperse emulsion, because in the latter it is thermodynamically favorable for small particles to coagulate and form bigger particles. With a monodisperse emulsion, this effect is absent, which makes the emulsion more stable. Another advantage of monodisperse emulsions is that a two-stage process can be devised to create interesting new properties. If you first make a dispersion of water droplets in oil and then an emulsion of water-in-oil droplets in water double emulsions are created that contain less oil (calories) than the original emulsion.

The membrane can be functionalized with nanotechnology (e.g., to make it hydrophilic or hydrophobic). A functionalized membrane can have very interesting and surprising behavior in the emulsification process. If a suitable emulsion of oil droplets in water (or vice versa) is used in the cross-flow membrane emulsification with a hydrophobic membrane, a nice but reversed emulsion of water droplets in oil can be formed. Effects like these show that the combination of micro- and nanotechnology opens up new opportunities for important processes like emulsification.

7.3.3 Micro- and Nanoreactors

Reactors with extremely low volumes have the disadvantage that they produce small amounts of product. However, because of their small volume, they exhibit very specific advantages [8]. An obvious example is derived from the fact that the surface-to-volume ratio is very high. Not only does this offer possibilities for reactions that require close interaction of a solid phase (the wall) with the reactants, it also means that thermodynamic conditions (e.g., speed of temperature change, uniformity of temperature over the reaction volume, etc.) are superior and more controllable. These conditions are very important for the quality of the product, because, for instance, temperature differences over reaction volumes give rise to additional reactions that result in unwanted side products. These side products require additional cleaning steps or, if they resemble the product too closely and cannot be separated effectively, can give rise to the need for additional research into their safety for the user.

In small volumes, potentially dangerous reactions can safely be conducted because of the small amounts of reactants. If the reaction goes out of control, the small amounts of reactants intrinsically limit the extent of the damage.

If larger amounts of product are required for commercial application, parallelization (using more microreactors in parallel) is an elegant solution. The possibility to make as much product as necessary by using more reactors also means that scaling reactions from laboratory scale to pilot scale and ultimately to plant scale is very easy. If the reaction works in the lab in one microreactor, it is possible to just combine the necessary number of microreactors for the required production volume. This means that upscaling reactions to production levels, which in traditional industry often leads to negative results and is therefore an important cost factor in innovation, is linear, easy, and without risk.

The concept of nanoreactors mimics the structures that biology uses to produce specific molecules. Nanostructures like vesicles and micelles can be used to confine reactants and to intensify the contact between reactants and catalysts or enzymes. They can also protect intermediate reaction products from unwanted reactions with other components.

In the first instance, micro- and nanoreactors will be used for products that are produced in small quantities (pharmaceuticals, nutritional components, flavors, and fragrances, etc.). However, because of the linear scaling of production levels in a more distant future it will also be applied to more bulky production processes.

7.4 NANOENGINEERING OF FOOD INGREDIENTS

Nutritional and sensory properties usually originate at the molecular or supramolecular level. Certain food components interact at the molecular level with specific receptors in the consumer's body and trigger a response. This can result in a sensory experience but can also result in biochemical processes that are necessary for the health and well-being of the individual. These interaction events are directly related to the structure of the food components in combination with the receptors. With the knowledge of these processes and the possibilities that nanotechnology offer at the molecular and supramolecular levels, new functionality of food components can be engineered that can overcome food-related health problems or give rise to new sensory experiences.

7.4.1 ENCAPSULATION AND DELIVERY

A substantial number of nutrients can offer potential health benefits to consumers, but the uptake by the body is limited. This may be due to the fact that these substances taste very bad and are therefore not consumed in a sufficient amount or they are broken down during processing or cooking or at an early stage in the digestive tract. In some cases, the substances cannot traverse specific physiological barriers so that they do not reach the sites where they can exercise their beneficial effect. Nanotechnology can provide molecular containers that overcome these delivery problems and make the substances available to the organism. Although the general principles of these envelope-like structures are already known from pharmaceutical applications, to make them applicable in food products requires more research into specific food-related issues.

Specific nutrients are more effective when delivered at specific locations in the body, or when they become available in periods when the organism has most use for them. Nanotechnology can provide systems with which the delivery of nutrients can be controlled both in time and in place. The triggers for the release can be generated externally (e.g., through temperature, light, and magnetic fields) or can originate from the organism itself (e.g., presence of certain molecules, pH, concentration levels of substances, etc.).

During the processing or cooking of food, many flavors and fragrances deteriorate because of their volatile character. By encapsulating them in nanostructured particles, these substances maintain their sensory functions, and the same consumer experience can be obtained with less of the (often expensive) substance, or new taste experiences can be conceived. The contents of the particles need to be released at a very specific moment during consumption. This poses a real challenge to the nanotechnology involved.

There are different strategies to create the nanostructured containers. However, from a cost perspective it is clear that the production of the containers must be simple. Self-assembly is therefore used to form the structures from individual molecules. These structures are well known in nature. Membranes in cells are products of the self-assembly of phospholipid molecules into bilayers. The phospholipids have a hydrophilic head and two hydrophobic tails. The tails find each other in an attempt to get out of the watery environment; the heads protrude into the watery environment. The result is a bilayer of tails with heads on the outside in which the tails form an oily environment. These bilayers can form spheres, called vesicles, with watery contents separated from the watery environment outside. The same principle but with a monolayer can also be used to contain oily contents. These are called micelles. The advantage of vesicles and micelles is that they are based on the biologically safe and well-known phospholipids. A disadvantage is that they are not very good in confining the contents.

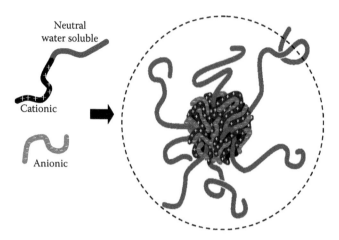

FIGURE 7.4 A solution of block copolymers with positively charged and neutral parts and negatively charged polymers self-assemble into micelles.

Organic chemistry has provided a very elaborate toolbox of different molecules, structures, and principles with which other types of containers can be built. Polymers with different properties can be linked together to form a copolymer. One of the parts can be electrically neutral, whereas the other part is charged. If the solution also contains polymers of opposite charge, they self-assemble into micelles with the charged polymers coiled in the core and the neutral polymer tails sticking into the watery environment (see Figure 7.4). Pluronics are tri-block copolymers that contain hydrophilic tails and a hydrophobic middle part. Depending on the sizes of the tails and the middle part, they self-assemble into micelles or vesicles. The advantage of these structures is that the complete organic chemistry toolbox is available to optimize the properties or to functionalize the containers. A disadvantage is that the polymer building blocks may not be food grade, and expensive procedures must be followed to get them approved for food use.

7.4.2 TEXTURE MODIFICATION

Apart from taste and smell, the structure of food is important in the appreciation of food by the consumer. Texture is the experience of the structure of food and finds its origin in the mind of the consumer. Texture perception is a complex process that involves most of the senses of the consumer [9]. It can be interesting to modify the texture of a food product to make it meet the preferences of certain target consumer groups. It may even be necessary to make a novel food product acceptable to consumers. Meat replacements, for instance, suffer from the fact that they do not have a structure that sufficiently resembles real meat products.

Food is a complex mixture of components, and the structure a consumer experiences is made up from a hierarchy of structures from the nanolevel via the micro- and mesolevels to the macroscopic properties of the food. Modifying the texture therefore not only requires the ability to modify the structure at the nano- and microlevels, but also an understanding of the interaction of the different hierarchy levels. Nanotechnology therefore can only contribute to texture modification of complex food products.

The nanostructure of certain food products often is an important component of the texture. The degree of cross-linking of fibrils in dairy products determines the firmness of the product. A gel is made up of networks of polymers (e.g., protein fibrils) in water (see Figure 7.5). The firmness of the gel is determined by the amount of fibrils and the degree of cross-linking. Modifying the texture of these types of food products or lowering the amount of gelling agent therefore requires control of the nanoscopic structure of the fibrils.

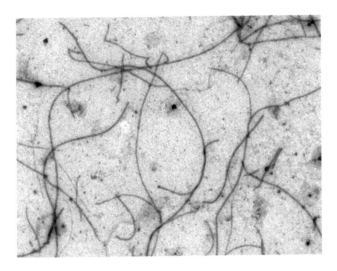

FIGURE 7.5 Nanothick fibrils of nonmeat proteins, allowing tailoring of food texture. (Courtesy of Food Physics Group of Wangeningen University, The Netherlands.)

Traditional meat production with animals is very inefficient in terms of plant protein and water use. It requires 3 to 10 times the amount of plant protein to produce a certain amount of beef. With a growing world population and an even stronger increase of meat consumption it will be interesting to be able to convert the plant protein directly in a meat-like product. Meat predominantly is muscle tissue, so the basis of the structural hierarchy of meat is the structure of muscle. At the nanolevel this starts with fibrils that are connected to each other with linear motor proteins. The fibrils are neatly aligned, and the motor proteins pull one set of fibrils into a second which results in the contraction of the muscle. Bundles of fibrils form muscle cells that make up strands that ultimately form the muscles. Creating a convincing meat-like product from plant protein therefore starts with creating a neatly aligned and interconnected nanostructure from the protein fibrils and building the rest of the hierarchy from this.

7.5 PACKAGING AND LOGISTICS

Packaging of food products plays an essential role in maintaining the product quality over longer periods of time, an important benefit in terms of sustainability and consumer benefit. Many of the problems that result in destruction of food products (in Europe about one-third of all food products are not consumed but are discarded as waste) can be overcome with improved and enhanced packaging. Apart from new materials with improved mechanical properties, nanotechnology will allow concepts that will add functionality to packaging materials.

By adding exfoliated clay platelets to the polymer matrix of packaging material, improved barrier properties can be achieved. The platelets effectively increase the diffusion distance so less oxygen can leak into a modified atmosphere container or, in other products, water or CO_2 out. In this way, it has recently become feasible to package beer into polyethylene terephthalate (PET) bottles without loss of shelf life. If necessary, these barrier materials can be enhanced with scavenger materials that catch the few molecules that do leak in before they can deteriorate the product. Scavenger molecules can also be used to catch odor molecules that are naturally emitted by certain food products and by consumers inadvertently are associated with quality deterioration. Antimicrobial nanocoatings or antimicrobial nanoparticles in the matrix of the packaging material can lower the microbial pressure in the package which results in enhanced shelf life of products.

Indicators based on nanotechnology can inform the consumer about the quality of the product or (for example) the ripeness of contained fruits. Packaging that includes sensors and possibilities based on (printable) radio frequency identification (RFID) technology to communicate with devices in their vicinity can signal problems with storage conditions that could result in quality deterioration or can warn for potential safety problems. Active packaging could release substances at a specific point in time to optimize the quality (e.g., ripeness) when the product is consumed.

7.6   DISCUSSION

Only some of the concepts presented here are already available for application in practice. Nanoenhanced packaging materials are being used in specific applications or in food containers, and there are indicators on the market that signal deterioration of the modified atmosphere or the ripeness of the product. Applications that will be introduced in the near future include the sensing and diagnostics devices and the innovations in certain food processes. Although there are already products on the market that utilize these concepts, large-scale applications of encapsulation and delivery systems and nanoengineered food are further from the market. This is not only because of the complexity of the technology but also because of the acceptance by the consumer. Consumer perception is very important in an application as emotional as food. With applications of nanotechnology in food, the general public is worried about the hazards associated with nanotechnology.

7.6.1   SAFETY ASPECTS OF NANOTECHNOLOGY IN FOOD

The general public has a good recollection of the health problems caused by asbestos, and although there are more differences than similarities between asbestos and products of nanotechnology, the public draws parallels and perceives nanotechnology as hazardous. And research has shown that there are hazards associated with the use of certain nanoparticles. It is known that material in the form of nanoparticles behaves differently than the same material in macroscopic form. A good example is silver that is relatively inert in macroscopic form but is highly antimicrobial in a nano-dispersion. Apart from the geometrical effect, there are two main reasons for this different behavior. First, the ratio between surface and volume in a nanoparticle is very high. This means that reactivity or catalytic properties of the nanoparticle, which are generated at the surface, are extremely enhanced. Second, effects of quantum mechanics start to be important when particles are in the nano range. This results, for instance, in the well-known fluorescent properties of quantum dots and also affects the surface reactivity of the particles. The extent of these effects and their consequences for certain applications are not clear for most of the nanoparticles.

For carbon nanoparticles, it was shown that they can traverse the blood–brain barrier [10]. The blood–brain barrier is one of the major obstacles for modern medicine. Very few molecules can cross it. It was therefore very surprising, and to many people, concerning, that the carbon nanoparticles were able to end up in the brain. Another concern is nanoparticles in the environment [11]. Most of the nanotechnology described is either bound to other macroscopic structures or is highly biodegradable and therefore not persistent in the environment However, nanoparticles in packaging material applications can end up in the environment after the end of their useful lifetime. There is an obligation to show that these particles, even in the long run, do not pose a threat to organisms or the ecosystem.

From the above, it must have become clear that most of the nanotechnology being developed for food applications will not end up in food products. Moreover, most applications make use of one-dimensional nanotechnology (layers or surfaces) or two-dimensional nanotechnology (fibrils). The generally regarded as more hazardous three-dimensional nanotechnology (nanoparticles) is encountered in packaging materials for antimicrobial effects. Encapsulation and delivery systems usually result in particles that are larger than 100 µm and can therefore best be classified as one-dimensional nanotechnology with the wall of the container being the nanostructure that provides the new functionality. Also, the encapsulation systems are relatively large, so the size-effect is weak, and they must be biodegradable for the organism to be able to benefit from the nutrients they contain. They will therefore not persist in the organism. The nanoparticles in packaging materials are of more inorganic, insoluble, nondegradable nature and must therefore be regarded as persistent both in organisms and in the environment. In their application, they are not free but are included in a matrix which means that the exposure to them is limited. However, at some time these materials will be discarded and will end up in a garbage disposal system where they are released from the matrix material. These hazards have not been characterized extensively, and it is recommended that more research be done to the risk assessment before these particles are applied in a large scale in packaging materials.

7.6.2 CONSUMER BEHAVIOR

From a consumer perspective, there are parallels between the use of genetically modified organisms (GMOs) in food and the application of nanotechnology in food. Both are unobservable for the consumer, have hazards associated with them, can have long-term unwanted effects, and are highly unnatural. The GMO debate has shown that consumers can reject a new technology if it is introduced in the wrong way. It is important to learn from the mistakes that were made with the introduction of GMOs and to avoid the same pitfalls in the application of nanotechnology in food. Food is a very emotional issue. Eating is pleasure. Moreover, food is seen as an important component of health and health care. And finally, it comes very close, even inside of you, and you cannot take it out easily when you experience adverse effects. Consumers prefer fresh and natural food because that is perceived to be healthier. Therefore, there is an aversion against any artificial addition to food. Consumers are very traditional when it comes to food. They will only accept unnatural additions or modifications to food when there are distinct advantages and the safety of the product is not compromised.

To overcome the hesitation with consumers and to avoid the GMO pitfalls, it is extremely important to build trust with the consumers. For the acceptance of the technology by the public, it is important that the consumer can make a risk/benefit evaluation. To do that, risks must be discussed openly and uncertainties disclosed. Risk is the product of hazard and exposure. To assess the risks of an application of nanotechnology the hazards must first be characterized. As has been explained above, this requires more research. When the hazards have been characterized, the exposure to the hazards must be assessed. In most of the applications discussed here, the exposure to nanoparticles will be limited, and therefore the risks associated with that application will be limited. For a risk/benefit evaluation, it is also necessary that the benefits to the consumer, not to the manufacturer, are explicit. Usually the benefits of a new technology are highlighted sufficiently. It is important for the trust in both science and industry that these benefits not be exaggerated.

Finally, it is essential for consumer acceptance that the consumer has a choice whether or not to use the nanoenhanced product or the traditional product. It is therefore important to devise a clear labeling system that informs consumers about products that contain nanotechnology to distinguish them from traditional products and to offer the consumer this choice. Unfortunately, the food industry, after the GMO issue, is hesitant toward labeling.

7.6.3 ECONOMIC ASPECTS

The profit margins in the food industry are very small. This means that there is very little room to introduce an expensive new technology. Moreover, stakeholders want to see fast returns on investments. This is a problem for the introduction of nanotechnology in food. Fortunately, many applications discussed here can be introduced in an evolutionary way. The use of new and improved sensors for process control, for instance, does not require a total redesign of the production process and can be done with limited investments.

Consumers are rarely willing to pay much more for a food product, even if it has distinct advantages. The advantages are usually indirect or have effects at different stages of your life. If you are ill, you are easily prepared to pay substantial amounts of money for a product that may cure you; if you are healthy, you usually are not prepared to pay more for a product that has shown to make you healthier. The margin for introduction of nanotechnology is therefore limited. There is no room for sophisticated drug delivery concepts that are currently in development in the pharmaceutical industry. However, there is room for adopting parts of these concepts to provide distinct benefits to the consumer of the food product, especially when these concepts are relatively cheap. To do this, it can be advantageous to "borrow" concepts from other application areas that have more financial room to do the research. Many applications in the medical and pharmaceutical sectors can be adopted for use in food or food industry, but also areas such as homeland security and the defense industry can be fertile sources of inspiration, especially when confidentiality is lifted and patents have expired.

But the ultimate means to control the costs in a high-tech application like nanotechnology is to make use of self-assembly mechanisms for the production of the nanocomponents. By focusing the research on suitable molecules that self-assemble into functional structures, it is possible to create robust applications that do not need large investments in very specific equipment or clean rooms, often associated with nanotechnology. This substantially lowers the threshold for applications of nanotechnology in food. And in the end, it is the combination of industry willing to produce it and consumers willing to buy it that creates the opportunity.

ACKNOWLEDGMENTS

The applications described here are subjects of different research groups of Wageningen UR participating in BioNT. The author is indebted to the researchers in the following groups: Organic Chemistry; Physical Chemistry and Colloid Science; Food Physics; Process Engineering; Human Food; Toxicology; and Marketing and Consumer Behavior.

REFERENCES

1. Bergveld, P., *Sensors and Actuators B* 88 (2003), 1.
2. Smet, L.C.P.M. de, G.A. Stork, G.H.F. Hurenkamp, H. Topal, P.J.E. Vronen, Q.-Y. Sun, A.B. Sieval, A. Wright, G.M. Visser, H. Zuilhof, and E.J.R. Sudhölter, *J. Amer. Chem. Soc.* 125 (2003), 13916.
3. Lieber, Charles M., *MRS Bulletin* (July 2003), 486.
4. Müller, Michael, and Sander Kersten, *Nat. Rev. Genomics* 4 (2003), 315.
5. Rijn, C.J.M. van, Nano and Micro Engineered Membrane Technology (2004), Elsevier Science.
6. Kumar, P., and K.L. Mittal (Eds.), *Handbook of Microemulsion Science and Technology* (1999), CRC Press, Boca Raton, FL.
7. Abrahamse, A.J., R. van Lierop, R.G.M. van der Sman, A. van der Padt, and R.M. Boom, *J. Membr. Sc.* 5247 (2002), 1.
8. Doku, G.N., W. Verboom, D.N. Reinhoudt, and A. van den Berg, *Tetrahedron* 61 (2005), 2733.
9. Wilkinson, G., G.B. Dijksterhuis, and M. Minekus, *Trends in Food Sci. and Technol.* 11 (2000), 442.
10. Oberdörster, G., Z. Sharp, V. Atudorei, A. Elder, R. Gelein, W. Kreyling, and C. Cox, *Inhalation Toxicol.* 16 (2004), 437.
11. Colvin, Vicky L., *Nat. Biotechnol.* 21 (2003), 1166.

8 Engineering Nanostructured Thermal Spray Coatings for Biomedical Applications

Rogerio S. Lima and Basil R. Marple
National Research Council of Canada, Industrial
Materials Institute, Boucherville, Quebec, Canada

CONTENTS

8.1 THERMAL SPRAY COATINGS

8.1.1 THERMAL SPRAY PROCESS

Thermal spraying includes a group of processes wherein a feedstock material (usually a powder, but also in the form of a wire or rod) is heated and propelled as individual molten or semimolten particles toward a substrate surface. Thermal spray torches are employed to deposit coatings on substrates. These thermal spray torches have a heat source that can be the combustion of a fuel gas (e.g., propylene and oxygen), a plasma gas (e.g., Ar/H_2), or an electric arc. At the heat source of the thermal spray

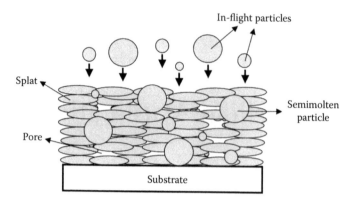

FIGURE 8.1 The lamellar microstructure of a thermal spray coating (cross-section) consisting of splats, semimolten particles, and pores.

torch, the feedstock material is heated and changed to a jet of molten or semimolten particles and is propelled toward the substrate surface via the expansion of the combustion gases, plasma gases, or compressed air. Basically, any material that is stable in its molten state can be deposited by thermal spray, which includes a wide range of metals, ceramics, polymers, and cermets.

At impact with the substrate or previously deposited layers, the molten or semimolten particles flatten and form thin lamellae (splats) that conform and adhere to the irregularities of the substrate surface and to each other. After impact, the splats cool and resolidify very rapidly, generally before the arrival of the next impinging particle. The microstructure of a thermal spray coating is formed by the overlapping and interlocking of splats, thereby creating a lamellar structure (Figure 8.1). The coating is typically nonhomogeneous, anisotropic, and contains a combination of material originating from fully molten and semimolten particles, as well as, pores, cracks (in ceramics), and oxides (in metals) [1].

Thermal spray coatings are normally deposited on metallic substrates; however, ceramic and plastic substrates may also be coated by this technique. Substrate preparation is of high importance. The substrate is generally prepared by grit-blasting the surface with alumina particles, which cleans the surface of contaminants and unwanted species (e.g., oxides) and provides microscopic asperities (roughness) to increase surface area and enhance coating adhesion. The bond between the coating and substrate may be mechanical (anchoring of the splats on the roughness of the substrate), chemical, metallurgical, or a combination of these. The properties of thermal spray coatings are dependent on, among other things, the feedstock material, thermal spray process, thermal spray parameters, and particle temperature and velocity in the thermal spray jet.

The two most widely employed thermal spray processes to spray feedstock powders today are air plasma spray (APS) and high-velocity oxy-fuel (HVOF). The heat source of APS torches is based on a combination of plasma gases (e.g., Ar/H_2). The maximum temperature of a plasma jet is approximately 15,000°C, whereas the particle speed generally varies from 150 to 300 m/s. This process is usually carried out in open air; however, it can be carried out inside a vacuum chamber, and then it is called vacuum plasma spray (VPS). Sometimes metals are sprayed by VPS in order to avoid particle oxidation from air during thermal spraying.

The heat source of HVOF torches is based on the combustion of a fuel gas (e.g., propylene and O_2). The maximum temperature of an HVOF jet is generally below 3000°C, whereas the particle speed generally varies from 600 to 800 m/s.

8.1.2 BIOMEDICAL THERMAL SPRAY COATINGS

There are approximately 435,000 total knee and hip joint replacements per year just in the United States. By the year 2030, a total of 730,000 total knee and hip joint replacement procedures per year is estimated only in the United States [2]. Key components of these implants are generally made of

Ti-6Al-4V alloys, CoCr alloys, or stainless steel. These materials have high mechanical strength, high corrosion resistance, good fatigue life, and are extremely bioinert. Due to their high bioinertness, they do not exhibit good biointeraction with the osteoblast cells once implanted in the human body; therefore, another agent must be used in order to promote the osseointegration between the implant and the bone.

The first hip joint implants, which were developed during the 1960s, employed a cement to provide the fixation of the implant to the bone. These types of implants are called cemented implants. This technique is still employed today, and the cement is based on an acrylic polymer called polymethylmethacrylate (PMMA). Uncemented implants were developed in the 1980s in an attempt to eliminate the possibility of part loosening and the breaking off of cement particles, which occurred more frequently in younger and active patients who had received cemented implants.

The state-of-the-art uncemented implants

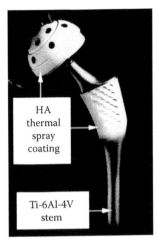

FIGURE 8.2 Typical Ti-6Al-4V hip joint (stem and acetabular cup) coated with an HA thermal spray coating. (From *International Orthopaedics,* 25, 2001, 17–21, C.J.M. Oosterbos, A.I.A. Rahmy, A.J. Tonino, Figure 1. With permission of Springer Science and Business Media.)

used today employ biocompatible thermal spray coatings, such as hydroxyapatite (HA) and titanium (Ti), to promote the osseointegration between the implant and the bone. HA is a calcium-phosphate–based material $(Ca_{10}(PO_4)_6(OH)_2)$, which is the bone mineral found in human bodies. HA is highly biocompatible and bioactive in the human body. It is compatible with various tissue types and can adhere directly to osseous, soft, and muscular tissue without an intermediate layer of modified tissue [3]. Due to this high bioactivity with human cells, synthetic HA powders are thermally sprayed onto the metallic implants. The thermal spray process employed is generally APS. Once the prosthesis is implanted, the osteoblast cells of the bones attach, grow, and proliferate on the surface of these HA coatings, therefore promoting the necessary osseointegration. The HA coatings produced via APS for this application generally have a thickness of 50 to 75 μm, an arithmetic mean roughness (R_a) of 7.5 to 9.5 μm, porosity of 1 to 10%, purity higher than 97%, crystallinity higher than 50%, and bond strength between 20 and 30 MPa [3–6]. Figure 8.2 shows a typical example of a Ti-6Al-4V hip joint (stem and acetabular cup) coated with an HA thermal spray coating [7]. Figure 8.3 shows a typical artificial hip joint implant, like the one of Figure 8.2, implanted in the human body. The stem is implanted in the femur, and the acetabular cup is implanted in the pelvis.

Titanium powders are generally thermally sprayed via VPS in order to avoid the oxidation of the Ti particles in air. Due to the lack of bioactivity between the titanium and the osteoblast cells, the osseointegration mechanism is different from that of HA coatings. The titanium thermal spray coatings are made highly porous. The large pores allow bone in-growth into the microstructure of the coating, filling the porosity of the coating, and thereby promoting mechanical osseointegration. The titanium coatings produced via VPS generally exhibit thickness of 350 to 600 μm, arithmetic mean roughness (R_a) of approximately 30 μm, porosity of 15 to 40%, and bond strength of 25 MPa (minimum) [5,8].

8.1.3 Nanostructured Thermal Spray Coatings

8.1.3.1 Enhanced Mechanical Performance of the Nanostructured Thermal Spray Coatings

It has been demonstrated by different authors that nanostructured thermal spray coatings exhibit enhanced mechanical performance when compared to their conventional counterparts [9–20].

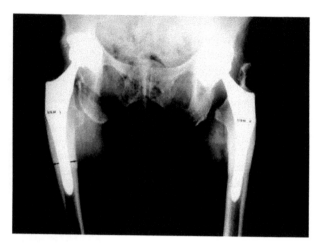

FIGURE 8.3 Typical artificial hip joints implanted in the human body. The stem is implanted in the femur and the acetabular cup is implanted in the pelvis (From *International Orthopaedics*, 25, 2001, 17–21, C.J.M. Oosterbos, A.I.A. Rahmy, A.J. Tonino, Figure 4. With permission of Springer Science and Business Media.)

Different characteristics have been observed, including higher wear resistance, higher bond strength with the substrate, higher resistance to delamination, higher toughness, and higher plasticity. In order to produce these types of coatings and achieve these properties, several important steps must be taken. These are described in the remainder of this section.

8.1.3.2 Nanostructured Powders for Thermal Spray

As previously mentioned, thermal spray coatings are usually made from a powder feedstock. These powder particles typically exhibit a particle size distribution varying from 5 to 100 μm (i.e., the particles are microscopic). Individual nanostructured particles (i.e., smaller than 100 nm) cannot be thermally sprayed using the regular powder feeders employed in thermal spray. These tiny nanoparticles would clog the hoses that transport the powder particles from the powder feeder to the thermal spray torch.

In order to spray nanoparticles using regular powder feeders, the individual nanostructured particles are agglomerated via spray-drying into microscopic particles. This process is usually employed when very fine materials such as nanostructured ceramic or cermet powders are to be thermally sprayed. Figure 8.4 and Figure 8.5 show the morphology of typical conventional and nanostructured titania (TiO_2) powder particles for thermal spray systems [18]. The conventional particle is formed via fusing and crushing of the titania material (Figure 8.4A). When this conventional particle is observed at high magnification, it is not possible to identify any nanostructural character (Figure 8.4B). The nanostructured titania particle produced for thermal spray is shown in Figure 8.5. It exhibits the typical donut shape of spray-dried particles (Figure 8.5A). When analyzed at higher magnifications, it is possible to observe the nanostructure of the feedstock (Figure 8.5B) (i.e., each microscopic titania particle is formed via the agglomeration of individual titania particles smaller than 100 nm).

Nanostructured metallic powders can also be thermally sprayed. In this case, conventional microscopic metallic particles are usually milled in methanol or liquid nitrogen [21]. Due to the excessive plastic deformation of the metallic particles during milling, the submicron grains of the powders are destroyed and transformed into grains with diameters smaller than 100 nm [21].

8.1.3.3 Thermal Spraying Nanostructured Powders

The thermal spray process is intrinsically associated with the melting of particles. Without some particle melting it is extremely difficult to produce thermal spray coatings, particularly with ceramic

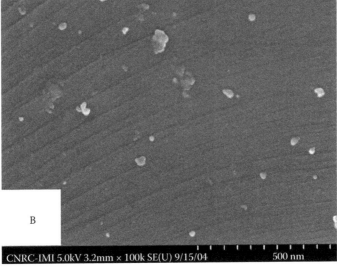

FIGURE 8.4 (A) Conventional titania thermal spray powder particle for thermal spray. (B) Particle of (A) observed at higher magnification; absence of nanostructural character. (Reprinted from *Materials Science and Engineering A*, 395, R.S. Lima and B.R. Marple, Enhanced Ductility in Thermally Sprayed Titania Coating Synthesized Using a Nanostructured Feedstock, 269–280, 2005. With permission from Elsevier.)

FIGURE 8.5   (A) Titania powder particle for thermal spray formed by the agglomeration of individual nano-sized particles of titania. (B) Particle of (A) observed at higher magnification; individual nanosized titania particles smaller than 100 nm. (Reprinted from *Materials Science and Engineering A*, 395, R.S. Lima and B.R. Marple, Enhanced Ductility in Thermally Sprayed Titania Coating Synthesized Using a Nanostructured Feedstock, 269–280, 2005. With permission from Elsevier.)

hESCs   hMSCs

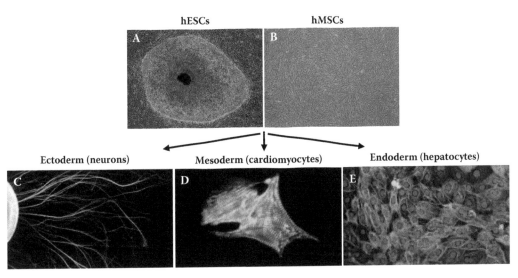

Ectoderm (neurons)   Mesoderm (cardiomyocytes)   Endoderm (hepatocytes)

FIGURE 1.2 Morphology and derivatives of embryonic and adult stem cells. Phase-contrast microscopy of (A) a human embryonic stem cell (hESC) (Royan H5) colony cultured on mouse embryonic fibroblast feeder cells (see Baharvand, H., et al., *Dev. Growth. Differ.* 48, 117–128, 2006), and (B) human bone marrow mesenchymal stem cells (hMSCs). Immunocytochemistry of differentiated ESCs with (C) antineuron-specific tubulin III, (D) antialpha actinin, and (E) anticytokeratin 18.

FIGURE 2.7 Atomic force microscopy (AFM) topographic image (25 × 25 μm²) of a homogeneously covered mica surface with a 1,2-dimyristoyl-*sn*-glycero-3-phosphatitidylcholine (DMPC) layer prepared by vesicle fusion. The measurements were performed at a temperature of 24°C, showing the lipid phase transition from liquid-crystalline to gel.

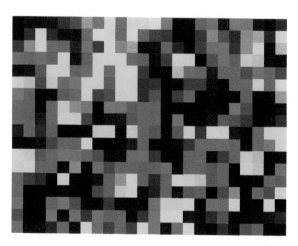

FIGURE 4.4 120 μm × 90 μm false-color map of Nanoplex Biotags spread on a quartz slide. Raman spectra were collected at 5 μm intervals resulting in 24 × 18 = 432 pixels total. Spectra were collected using a 50× (0.75 NA) objective and 1 second integration on a Renishaw inVia Raman microscope. Red pixels correspond to strong signal from trans(1,4) bis(pyridyl)ethylene (BPE), green corresponds to d^8-4,4′-dipyridyl, blue to 4,4′-dipyridyl.

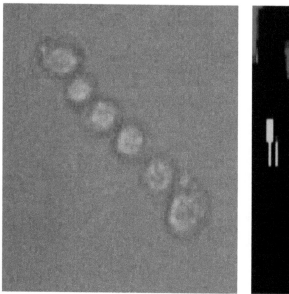 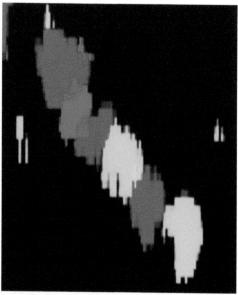

FIGURE 4.10 Bright field image and Raman map of six cells labeled with Nanoplex Biotags. Spectra analysis reveals that three of the cells were labeled with dipyridyl tags (blue), two with d^8-dipyridyl tags (green), and a single cell was labeled with biotags carrying the trans(1,4) bis(pyridyl)ethylene (BPE) label molecule (red).

(a) Abalone Shell [4a]

(b) Freshwater Goby [95]

(c) *Morpho rhetenor* [36]

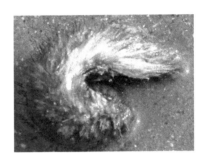

(d) Sea-Mouse [13]

FIGURE 9.1   Sample of structural color in biosystems.

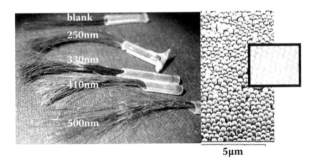

FIGURE 9.6   Photonic cosmetics and "hair jewelry": (a) L'Oreal's announcement of "colorless color" in cosmetics. The inset shows platelets of nanoscale thickness, producing the color (www.wired.com/news/technology/1,68683-0.html). (b) Unilever's patented technology for coloring hair by deposition of nanoparticles as multilayer colloid crystal (R. Djalali, Unilever Research Colworth, *personal communications,* 2007; Patent, Unilever, WO 2006/097332 A2). Inset shows cross-section of the iridescent spine of sea-mouse with similar characteristic lattice dimensions (R.C. McPhedran and N.A. Nicorovici, *Australian Optical Society NEWS* 15(2,3), 7–9, 2001; S. Daily, Marine Life of Channel Islands, 1998; A.R. Parker, R.C. McPhedran, D.R. Mckenzie, L.C. Botten, and N.A. Nicorovici, *Nature* 409, 36–37, 2001).

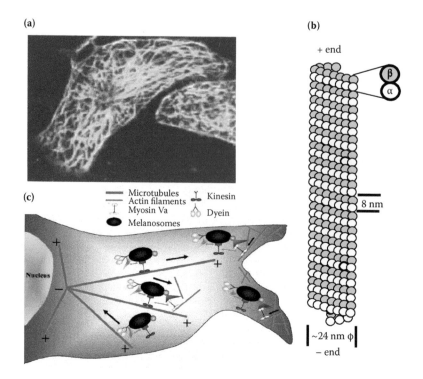

FIGURE 9.9 Multiscale assembly melanosome transport system: (a) microtubule (green) and actin (red) tracks in melanocute; (b) microtubule assembly; and (c) bidirectional transport. (For more information on parts a and c, see D.C. Barral and M.C. Seabra, *Pigment Cell Res.* 17, 111–118, 2004.)

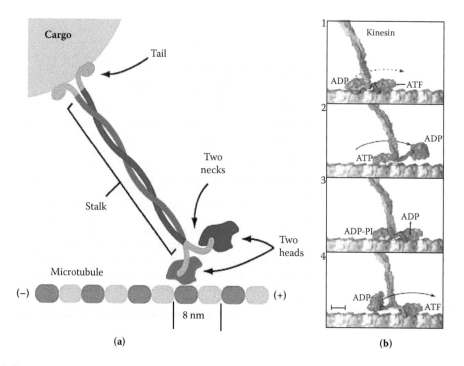

FIGURE 9.10 Cargo transport by "hand-over-head" mechanism of kinesin motor proteins. (See R.D. Vale and R.A. Milligan, *Science* 288, 88–95, 2007; S.M. Block, www.stanford.edu/group/blocklab/ScienceLimping/.)

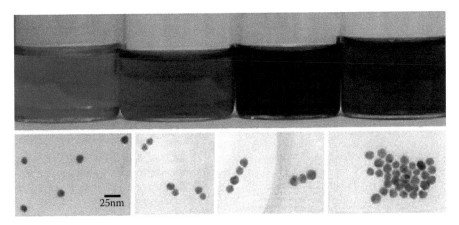

FIGURE 11.5 Chromatic response and evolving gold nanoparticle dimensionality in lysozyme and antilysozyme system. Functional Nanosystems Laboratory RMIT University. (Image courtesy of Professor David Mainwaring, RMIT University, Australia.)

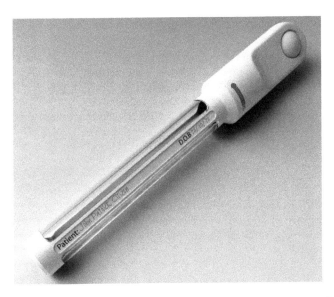

FIGURE 11.6 An overview of plasmon–plasmon colorimetric effects of gold nanoparticle clusters. (Image courtesy of NanoVic, Australia.)

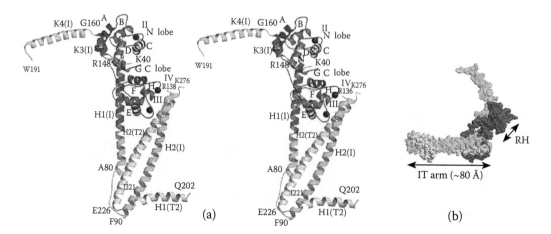

FIGURE 12.3 Stereo view of crystal structure of troponin core domain: (a) TnC and TnT are colored in red and yellow, respectively. TnI is colored in cyan, except for the two stretches of amphiphilic helices (TnC-binding sites), which are dark blue. (b) A space-filling model of the Troponin molecule. RH, regulatory head. (From Takeda S, Yamashita A, et al., *Nature*, 424, 35–41, 2003. With permission.)

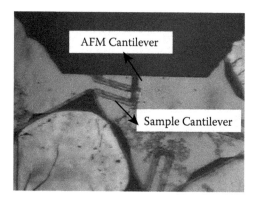

FIGURE 12.17 Atomic force microscopy (AFM) cantilever deflecting sample cantilever taken from the charge-coupled device (CCD) camera.

FIGURE 15.1 Lotus leaf.

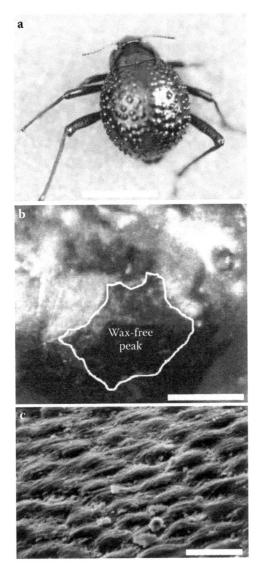

FIGURE 15.3 Details of the water-capturing, hydrophobic surface of the desert beetle *Stenocara* sp. (a) Adult beetle. (b) A close up of a "bump" on the fused overwings. Special staining marks waxy depressed areas as colored, whereas bumps remain unstained (black). (c) SEM of the depressed areas. Scale bars: (a) –1 mm; (b) –0.2 mm; (c) –10 μm. (From Parker, A.R., Lawrence, C.R. 2001. *Nature* 414:33–34. With permission.)

FIGURE 15.5 Nanograss and droplets.

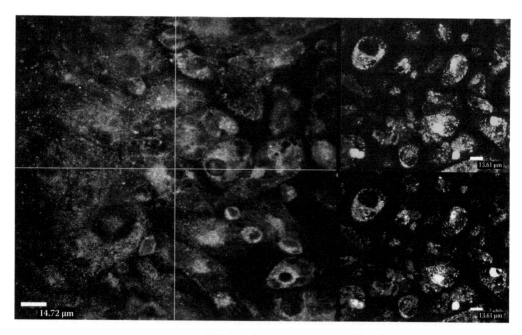

FIGURE 16.4 Nanocapsule transport/uptake by CaCo-2 cells viewed under a confocal microscope. (A) Z-stack of cells treated with nanoparticles over a 5 minute incubation period. (B) Appearance of cells depicting lysotracker stain in cells treated with rhodamine-labeled particles. (C) Same image showing colocalization of nanoparticles with lysosomes as indicated by the orange staining at 60 minutes postadministration. Rhodamine dye was used as a fluorophore.

materials. Some degree of melting is necessary to achieve a sufficient degree of particle adhesion and cohesion. This is a challenge for thermal spraying nanostructured powders; if all powder particles are fully molten in the thermal spray jet, all the nanostructural character of the powder particles will disappear, and therefore the thermal spray coating will not exhibit any nanostructured related property.

In order to overcome this challenge, it is necessary to carefully control the temperature of the particles in the thermal spray jet (i.e., the temperature of the powder particles should be maintained such that it is not much higher than the melting point of the material). The particles must be thermally sprayed in such a way to guarantee that part of the initial nanostructure of the feedstock will be embedded in the coating microstructure.

Such an approach has been employed for titania in which nanostructured titania powders like that of Figure 8.5 (VHP-DCS, 5 to 20 μm; Altair Nanomaterials, Reno, Nevada) were thermally sprayed using an HVOF torch (DJ2700-hybrid, Sulzer Metco, Westbury, New York). During deposition, the temperature and velocities of the sprayed particles were monitored using a diagnostic tool (DPV 2000, Tecnar Automation, Saint Bruno, Quebec, Canada). The diagnostic tool is based on optical pyrometry and time-of-flight measurements to measure the distribution of particle temperature and velocity in the thermal spray jet. The average surface temperature and velocity of the thermally sprayed particles were $1874 \pm 136°C$ and 635 ± 89 m/s [22]. As the melting point of titania is 1855°C [23], it is considered that part of the nanostructure of the titania powder was preserved and embedded in the coating microstructure, as will be seen in the next subsection.

It is important to point out that a new thermal spray process, cold spray, was launched near the end of the 1990s. This process has a particular difference when compared to other thermal spray processes. In cold spray, the powder particles are mixed with a heated gas (He, N_2 or air) at temperatures below 700°C (i.e., there is no particle melting) and accelerated to supersonic velocities on the order of 600 to 1000 m/s through a de Laval nozzle. The particles arrive at the substrate surface at these high speeds, plastically deform, and adhere to the substrate surface [24]. This new process is still in its initial stages of development, but it may become very important in the future for allowing the spraying of nanostructured particles without any degree of melting (i.e., it will be possible to produce Ti coatings consisting of 100% nanostructured material). It is important to point out that cold spray is usually employed to spray metallic powders, due to their deformation capabilities.

8.1.3.4 Bimodal Microstructure of Nanostructured Thermal Spray Coatings

As previously stated, thermally sprayed nanostructured coatings are formed from nanostructured particles that were fully molten and semimolten in the thermal spray jet. The particles that were fully molten in the spray jet lose the nanostructural character of the feedstock, whereas the semimolten particles retain some of their nanostructural features. Due to this characteristic, many authors describe these nanostructured thermal spray coatings as exhibiting a bimodal microstructure [9–13,18–20]. A typical schematic (cross section) of a microstructure of a nanostructured thermal spray coating is shown in Figure 8.6. The semimolten nanostructured particles (nanozones) are spread throughout the coating microstructure. The nanozones are found at the coating/substrate interface, embedded in the coating microstructure, and at the surface of the coating.

Figure 8.7A shows a low-magnification view of the cross-section of the HVOF-sprayed nanostructured titania coating, described in the previous section, which was deposited on a Ti-6Al-4V substrate. The coating is very dense and uniform, not exhibiting the typical lamellar structure of thermal spray coatings. It may be stated that this coating has an isotropic or bulk-like microstructure. When the coating of Figure 8.7A is viewed at higher magnifications (Figure 8.7B), it is possible to observe the nanostructured zones, which were formed by semimolten nanostructured particles that became entrapped in the coating microstructure. It is important to point out that these nanozones (Figure 8.7b) are spread throughout the coating microstructure.

This bimodal microstructure is essential for the enhanced mechanical performance of these nanostructured coatings. The semimolten nanostructured particles (nanozones) located at the coat-

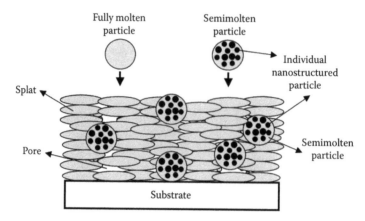

FIGURE 8.6 Cross-section of the bimodal microstructure of nanostructured thermal spray coatings, formed by nanostructured particles that were fully molten and semimolten in the thermal spray jet.

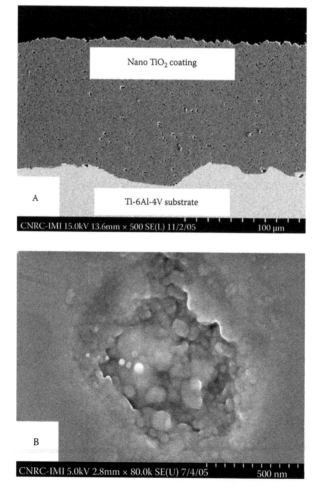

FIGURE 8.7 (A) Low-magnification view of the cross-section of the high-velocity oxy-fuel (HVOF)-sprayed nanostructured titania coating deposited on a Ti-6Al-4V substrate. (B) Coating of (A) observed at higher magnification; semimolten nanostructured TiO_2 agglomerate (nanozone).

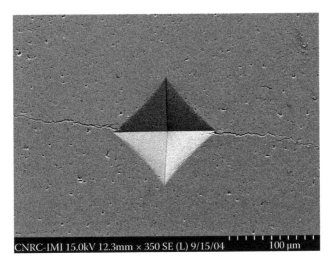

CNRC-IMI 15.0kV 12.3mm × 350 SE (L) 9/15/04   100 μm

FIGURE 8.8 Vickers indentation impression (5 kgf) and crack propagation in the cross-section of a high-velocity oxy-fuel (HVOF)-sprayed conventional TiO$_2$ coating. (Reprinted from *Materials Science and Engineering A*, 395, R.S. Lima and B.R. Marple, Enhanced Ductility in Thermally Sprayed Titania Coating Synthesized Using a Nanostructured Feedstock, 269–280, 2005, with permission from Elsevier.)

ing/substrate interface and embedded throughout the coating microstructure act as crack arresters, thereby increasing the bond strength of the coating, the resistance to delamination, and the coating toughness [9–13,18–20].

A practical example of the mechanism of crack arresting is found in Figure 8.8 and Figure 8.9. Figure 8.8 shows a crack propagation experiment carried out via Vickers indentation (5 kgf) on the cross-section of an HVOF-sprayed conventional TiO$_2$ coating. The Vickers indenter was aligned such that one of its diagonals would be parallel to the substrate surface. It is possible to observe that the cracks propagate parallel to the substrate surface beyond the limits of the picture [18]. Figure 8.9A shows a crack propagation experiment carried out via Vickers indentation (5 kgf) on the cross-section of an HVOF-sprayed nanostructured TiO$_2$ coating described in the previous section. In this case, the Vickers indenter was also aligned such that one of its diagonals would be parallel to the substrate surface. It is also possible to observe that the cracks propagated parallel to the substrate surface but were arrested before the limits of the picture. It is important to point out that both pictures of Figure 8.8 and Figure 8.9A were taken at the same magnification. Therefore, it can be observed that the nanostructured coating exhibits higher crack propagation resistance (toughness) and resistance to delamination when compared to the conventional one. By looking at the tip of the indentation crack of Figure 8.9A at higher magnification, it is possible to notice that the crack is arrested after passing through a nanozone, which was formed by a semimolten nanostructured particle (Figure 8.9B) [18].

It is also important to point out that the nanostructured coating (Figure 8.9) exhibited uniform crack propagation under Vickers indentation (i.e., four cracks with similar length propagating from the corners of the indentation impression), which is a typical characteristic of bulk materials and not thermal spray coatings (e.g., this type of uniform crack propagation was not observed in the conventional coating shown in Figure 8.8). Therefore, it may be stated that this nanostructured coating is so uniform that it is behaving like a bulk material, which is not the regular behavior of a thermal spray coating.

8.2 ENHANCED BIOCOMPATIBILITY OF NANOSTRUCTURED MATERIALS

It has been demonstrated that nanostructured materials, such as alumina (Al$_2$O$_3$), titania (TiO$_2$), HA, Ti, Ti-6Al-4V, and Co28Cr6Mo, exhibit enhanced biocompatibility with osteoblast cells (i.e., bone

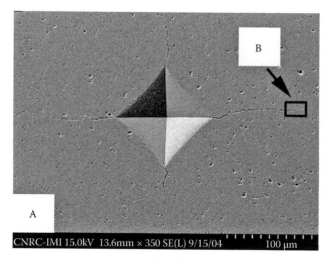

FIGURE 8.9 (A) Vickers indentation impression (5 kgf) and crack propagation in the cross-section of a high-velocity oxy-fuel (HVOF)-sprayed nanostructured TiO_2 coating. (Reprinted from *Materials Science and Engineering A*, 395, R.S. Lima and B.R. Marple, Enhanced Ductility in Thermally Sprayed Titania Coating Synthesized Using a Nanostructured Feedstock, 269–280, 2005. With permission from Elsevier.) (B) Vickers indentation crack being arrested by passing through a nanozone in the coating. (Reprinted from *Journal of Thermal Spray Technology*, R.S. Lima and B.R. Marple, Superior Performance of High-Velocity Oxyfuel-Sprayed Nanostructured TiO_2 in Comparison to Air Plasma-Sprayed Conventional Al_2O_3-13TiO_2, 14(3), 2005, 397–404. Reprinted with permission of ASM International®.)

cells) when compared to their conventional counterparts [25–30]. This enhanced biocompatibility is translated into higher cell reproduction and adhesion on the surface of these materials, which are very important characteristics for making implants with improved bioperformance and longevity. Webster et al. [15] explained this better performance of the nanostructured material as the effect of the nanotexture or nanoroughness of these materials on the adsorption of the adhesion proteins like fibronectin. This phenomenon was experimentally observed by Dalby et al. [31]. Adhesion proteins like fibronectin mediate the adhesion of anchorage-dependent cells (such as osteoblasts) on substrates and coatings [32]. These adhesion proteins are initially adsorbed on the surface of an implant almost immediately upon its implantation in the human body. When the osteoblast cells arrive at the implant surface, they "see" a protein-covered surface that will connect with the transmembrane proteins (integrins) of the osteoblast cells [32]. It is important to point out that these proteins, such as fibronectin, exhibit nanosized lengths and structures [33].

It is interesting to note that the surface of a nanostructured material (nanosized grains) will exhibit predominantly nanoscale features, like nanoroughness, whereas the surface of a conventional material (microsized grains) will tend to exhibit more microsized features [26]. It has been proven that the interaction or the adsorption of a nanosized protein (e.g., fibronectin) to a nanotextured surface will be more effective than that provided by a microtextured one [34,35]. This was shown in an experiment in which proteins were placed on substrate surfaces containing essentially flat regions (no roughness) and nanoprotuberances. It was observed that the proteins tended to attach and anchor on the nanoprotuberances, whereas no significant attachment was noted for the flat regions [34,35]. Therefore, nanostructured materials, containing regions on their surfaces exhibiting nanotexture, have the potential of being the next generation of biomedical materials, with the attributes of exhibiting enhanced cell proliferation and adhesion.

8.3 ENGINEERING THE NEW GENERATION OF BIOMEDICAL THERMAL SPRAY COATINGS FOR UNCEMENTED IMPLANTS

8.3.1 CURRENT PROBLEMS OF HYDROXYAPATITE (HA) THERMAL SPRAY COATINGS

HA coatings thermally sprayed via APS can be considered as one of the state-of-the-art materials used today to promote osseointegration of the implant to the bone in uncemented implants [36]. HA thermal spray coatings have been successfully used since the 1980s in thousands of patients. Despite the success with this coating, there are still drawbacks concerning its application.

HA thermal spray coatings may fail by aseptic loosening or osteolysis. Concerning aseptic loosening, Lai et al. [37] and Reikeras and Gunderson [6] observed that following implantation in humans, the HA coatings dissolved over a period of up to 10 years. They observed that after HA dissolution, the bone did not necessarily interlock with the metallic implant surface. Lai et al. [37] also established a relationship between the amount of residual HA on acetabular cups and the stability of the cup. It was found that when the percentage of residual HA covering the surface was less than 40%, the implant tended to become unstable. Reikeras and Gunderson [6] and Manley et al. [38] observed that some HA coatings did not exhibit dissolution and interlocked very well with the bone; however, the HA coatings were not able to withstand the stresses generated due to the activity of the patients and failed by delamination. Lai et al. [37] and Shen et al. [39] also observed that the initial dissolution of the HA may weaken the structure of the coating, causing it to delaminate.

Osteolysis may also lead to the failure of HA thermal spray coatings. According to Silver et al. [40], the activities of macrophages and osteoclasts, which are present during osteolysis, may lower the pH of the environment surrounding the bone to values equal to or less than 3.6. It is known that HA thermal spray coatings when immersed in simulated body fluid (SBF), which has the pH of the human blood (7.4), exhibit a decrease in the values of hardness, elastic modulus, and bond strength due to the HA dissolution [41]. At a pH of 3.6 or lower it is expected that the impact on the integrity of HA thermal spray coatings would be negative, due to an accelerated dissolution. Lai et al. [37], Reikeras and Gunderson [6], Blacha [42], and Bloebaum et al. [43] observed a correlation between the osteolysis and the failure of HA thermal spray coatings. Bloebaum et al. [43] also observed that the osteolysis can generate particulates of HA via dissolution that may migrate to polyethylene inserts of the hip joints leading to a third-body wear and contributing to an accelerated failure of the implant.

8.3.2 NANOSTRUCTURED THERMAL SPRAY COATINGS FOR BIOMEDICAL APPLICATIONS

Biomedical thermal spray coatings, such as HA and Ti, due to their previous successes, will continue to be employed in the coming years as important agents to promote osseointegration in uncemented implants. However, despite this success, the current implants are not yet optimized. The average longevity of the implants, in general, ranges from 12 to 15 years [2]. The life expectancy in countries like the United States and Canada is close to 80 years [44,45], and this number is increasing. As the longevity of the current implants is about 15 years, it means that many of those who receive an implant at age 65 or younger will require at least one revision surgery. Consequently, the next generation of implants will be required to be more biocompatible and mechanically superior when compared to those of the current generation of implants. Patients implanted with these new generation of prosthesis should have shorter hospitalization times and lower rates of revision surgeries. These improvements will translate into an improvement in the quality of life and reduced medical costs.

It was previously stated that nanostructured thermal spray coatings have been shown to exhibit enhanced mechanical performance, such as higher bond strength, higher toughness, and higher resistance to delamination, when compared to the current conventional thermal spray coatings. It was also previously stated that cell cultures on nanostructured materials, such as alumina (Al_2O_3), titania (TiO_2), HA, Ti, Ti-6Al-4V, and Co28Cr6Mo, have exhibited higher cell reproduction rates and adhesion strength on the surface of these materials when compared to that of conventional ones. Therefore, engineering thermal spray coatings to contain nanostructured features for application in the biomedical field is a new approach and a promising new area that is in its initial stage of development. These coatings may represent the next generation of thermal spray coatings for uncemented implants. Such materials offer the possibility of improved performance by combining the good mechanical characteristics imparted by the nanostructured thermal spray coatings and the enhanced biocompatibility of nanotextured surfaces.

8.3.3　NANOSTRUCTURED TiO_2 THERMAL SPRAY COATINGS FOR BIOMEDICAL APPLICATIONS

Due to the above-mentioned problems, new higher-performance alternatives to HA thermal spray coatings for biomedical applications are required. It is hypothesized that a good coating or material to replace HA thermal spray coatings would have to exhibit three main characteristics: be nontoxic and nonabsorbable by the human body, have excellent mechanical performance, and have good biocompatibility with the osteoblast cells.

Nanostructured titania thermal spray coatings may be an interesting alternative to HA thermal spray coatings. Titania is a nontoxic material and is nonabsorbable by the human body. It has been shown that nanostructured titania thermal spray coatings exhibit excellent mechanical performance [13,18–20]. It also has been demonstrated that nanostructured titania (bulk) has enhanced biocompatibility with osteoblast cells [25–29]. Therefore, HVOF-sprayed nanostructured titania coatings are being considered as alternatives to HA coatings thermally sprayed via APS [22,46].

8.3.3.1　Superior Mechanical Performance of Nanostructured TiO_2 Coatings

HVOF-sprayed nanostructured titania coatings were produced according to the conditions described in Section 8.1.3.3, and their mechanical properties were compared to those of HA. Table 8.1 shows a comparison of Vickers hardness values of HVOF-sprayed nanostructured titania coatings and HA. The Vickers hardness number of the HVOF-sprayed nanostructured titania coating was found to be 61% higher than that of the bulk (sintered) HA and more than three times that of a plasma sprayed HA (Table 8.1) [47,48]. This shows that the nanostructured titania coating exhibits higher cohesive strength, which is an important property for a long-term performance implant.

The bond strength values (ASTM C633 [49]) of the HVOF-sprayed nanostructured titania coating and various HA thermal spray coatings (deposited on Ti-6Al-4V substrates) found in the literature [50–55] are listed in Table 8.2. The mechanical strength of the nanostructured titania coating and its bond with the substrate are higher than the mechanical strength of the epoxy glue used during the bond strength test of the ASTM standard C633. Therefore, during the tensile test for bond strength, the bond between the coating and substrate remains intact when the epoxy glue breaks (fails) before the coating at 77 MPa (i.e., the bond strength value [adhesion to the substrate] of

TABLE 8.1

Vickers Microhardness Values of the HVOF-Sprayed Nanostructured Titania Coating, Plasma Sprayed HA, and Bulk (Sintered) HA

Material	Indentation Load (g)	Vickers Hardness Number
HVOF Nano TiO$_2$	300	324 ± 40 (n = 10)
Bulk (sintered) HA	300	513 ± 52 [47]
HVOF Nano TiO$_2$	100	851 ± 30 (n = 10)
Plasma spray HA	100	275 ± 40 [48]

the nanostructured titania coating is higher than 77 MPa). As can be seen in Table 8.2, the bond strength value of the HVOF-sprayed nanostructured titania coating is at least 2.5 times that of the highest bond strength value shown for an HA thermal spray coating.

These mechanical characteristics are very desirable when engineering an implant for increased longevity. It is important to point out that, unlike HA, titania does not dissolve in the human body. Therefore, these improved mechanical properties of the HVOF-sprayed nanostructured titania coating should remain intact through the years after implantation in the human body.

8.3.3.2 Nanotexture on the Coating Surface

TABLE 8.2

Comparison of Bond Strength Values (ASTM C633) for the Nanostructured Titania Coating and Various HA Thermal Spray Coatings Available in the Literature (Substrate: Ti-6A1-4V)

Material	Powder	Process	Bond Strength (MPa)
TiO$_2$	Nanostructured	HVOF	>77 (n = 5)
HA	Conventional	APS	23 ± 4 [50]
HA	Nanostructured	HVOF	24 ± 8 [51]
HA	Conventional	APS	13 ± 1 [52]
HA	Conventional	APS	14 ± 2 [53]
HA	Conventional	HVOF	31 ± 2 [54]
HA	Conventional	APS	27 ± 2 [55]

It was previously stated that nanostructured thermal spray coatings exhibit semimolten nanostructured particles (nanozones) spread throughout their microstructures. Nanozones found at the coating/substrate interface and embedded in the microstructure help to enhance the mechanical performance of the coatings.

These nanozones can also be found at the coating surface (as shown schematically in Figure 8.6). Figure 8.10 shows a nanozone at the surface of the HVOF-sprayed nanostructured titania coating engineered as described in Section 8.1.3.3. It is hypothesized that these nanozones located at the coating surface may enhance the interaction of the nanosized adhesion proteins (e.g., fibronectin) with the coating surface and, consequently, the proliferation and adhesion of cells, like the osteoblast cells, as described by Webster et al. [28–30] and experimentally observed by Dalby et al. [31].

Therefore, in addition to the superior mechanical performance, the nanostructured thermal spray coatings may exhibit a superior biocompatibility when compared to the conventional ones. The conventional thermal spray coatings would tend to exhibit mainly microirregularities on their surfaces, features that may be less effective for interlocking with the nanosized adhesion proteins. A schematic of the enhanced biocompatibility of nanostructured thermal spray coatings is shown in Figure 8.11. It is important to point out again that this is a hypothesis, not yet experimentally confirmed for nanostructured thermal spray coatings.

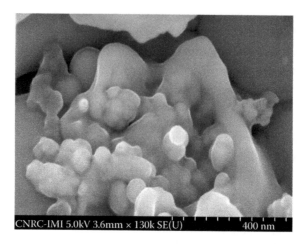

FIGURE 8.10 A nanozone on the surface of the high-velocity oxy-fuel (HVOF)-sprayed nanostructured titania coating formed by a semimolten nanostructured particle. (Reprinted from R.S. Lima, B.R. Marple, H. Li, K.A. Khor, Biocompatible Nanostructured High Velocity Oxy-Fuel (HVOF) Sprayed Titania Coating, *Building on 100 Years of Success: Proceedings of the 2006 International Thermal Spray Conference*, (Eds.) B.R. Marple, M.M. Hyland, Y.-C. Lau, R.S. Lima, and J. Voyer, ASM International, Materials Park, Ohio. Reprinted with permission of ASM International.)

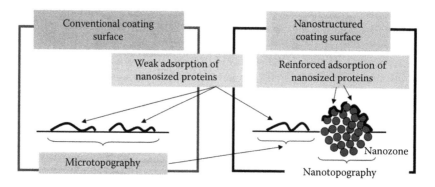

FIGURE 8.11 The surface of a conventional and a nanostructured thermal spray coating. The nanosized adhesion proteins (e.g., fibronectin) would tend to exhibit better interlocking with the nanozones located at the surface of the nanostructured coating.

8.3.3.3 Preliminary Osteoblast Cell Culture

HVOF-sprayed nanostructured titania and APS HA coatings were deposited on Ti-6Al-4V discs. Osteoblast cells, obtained from rat calvaria, were cultured on the surface of these coatings for scanning electron microscopy (SEM) analysis (7-day culture) and alkaline phosphatase activity (15-day culture) in order to compare the degree of cell proliferation and adhesion on these two types of coatings. It is important to point out that the same number of cells was initially seeded on both coatings in order to produce valid statistical results and comparison. The detailed information about the *in vitro* testing is described elsewhere [46].

Figure 8.12 and Figure 8.13 show the SEM analysis of the comparison of an osteoblast cell culture (obtained from rat calvaria) carried out during 7 days on the surface of the HVOF-sprayed nanostructured titania and APS HA coatings (both coatings deposited on Ti-6Al-4V substrates). The osteoblast cells completely covered the surface of the nanostructured titania coating, whereas the surface of the HA coating was partially covered [46].

FIGURE 8.12   Osteoblast cells (obtained from rat calvaria) cultured during 7 days on the surface of the high-velocity oxy-fuel (HVOF)-sprayed nanostructured titania coating (Ti-6Al-4V substrate). The osteoblast cells completely covered the coating surface.

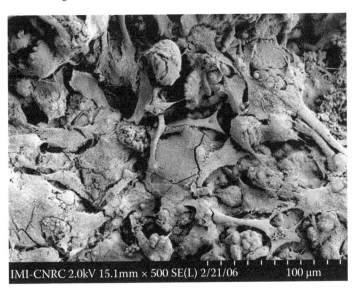

FIGURE 8.13   Osteoblast cells (obtained from rat calvaria) cultured during 7 days on the surface of the air plasma spray hydroxyapatite (APS HA) coating (Ti-6Al-4V substrate). The osteoblast cells partially covered the coating surface.

After a 15-day culture, the cells were stained for alkaline phosphatase activity shown as a red stain over the coatings. The percentage of the coating covered in red is a measure of the osteopro-genitor's ability to adhere, proliferate, and differentiate toward the osteoblast lineage. The results can be found in Figures 8.14, 8.15, and 8.16 [46]. Figure 8.16 quantifies the relative intensity of red staining on the surface of the coatings measured from a threshold.

All the preliminary results of Figure 8.12 through Figure 8.16 indicated the same trend (i.e., the HVOF-sprayed nanostructured titania coatings exhibit a degree of osteoblast cell proliferation and adhesion equivalent or superior to that of HA APS coatings).

FIGURE 8.14 Osteoblast cells (obtained from rat calvaria) stained for alkaline phosphatase activity (shown in red) after 15-day culture on the surface of the high-velocity oxy-fuel (HVOF)-sprayed nanostructured titania coatings. (Reprinted from J.-G. Legoux, F. Chellat, R.S. Lima, B.R. Marple, M.N. Bureau, H. Shen, G.A. Candeliere, *Development of Osteoblast Colonies on New Bioactive Coatings,* (Eds.) B.R. Marple, M.M. Hyland, Y.-C. Lau, R.S. Lima, and J. Voyer, ASM International, Materials Park, Ohio. Reprinted with permission of ASM International.)

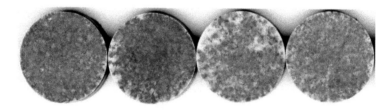

FIGURE 8.15 Osteoblast cells (obtained from rat calvaria) stained for alkaline phosphatase activity (shown in red) after 15-day culture on the surface of the air plasma spray hydroxyapatite (APS HA) coatings. (Reprinted from J.-G. Legoux, F. Chellat, R.S. Lima, B.R. Marple, M.N. Bureau, H. Shen, G.A. Candeliere, *Development of Osteoblast Colonies on New Bioactive Coatings,* (Eds.) B.R. Marple, M.M. Hyland, Y.-C. Lau, R.S. Lima, and J. Voyer, ASM International, Materials Park, Ohio. Reprinted with permission of ASM International.)

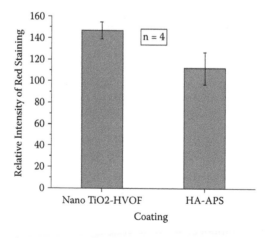

FIGURE 8.16 Relative intensity of red staining for the osteoblast cells on the surface of the high-velocity oxy-fuel (HVOF)-sprayed nanostructured titania and air plasma spray hydroxyapatite (APS HA) coatings after a 15-day cell culture. (Reprinted from *Building on 100 Years of Success: Proceedings of the 2006 International Thermal Spray Conference*, (Eds.) B.R. Marple, M.M. Hyland, Y.-C. Lau, R.S. Lima, and J. Voyer, ASM International, Materials Park, Ohio. Reprinted with permission of ASM International.)

Therefore, the initial mechanical evaluation and preliminary cell culture results indicated that these HVOF-sprayed nanostructured titania coatings may become an interesting alternative to HA thermal spray coatings for uncemented implants. The combination of the good chemical and structural stabilities of titania in the human body, enhanced mechanical performance of the nanostructured titania coatings, and excellent bioperformance of nanostructured titania may help lead toward increased implant life. Of course, results of ongoing *in vivo* testing and future clinical trials will be required to determine if the potential that these coatings hold translates into a successful long-term solution for implants in humans.

8.4   FINAL CONSIDERATIONS

In addition to nanostructured titania, nanostructured HA thermal spray coatings may bring important improvements to the field of biomedical coatings. The good biocompatibility of HA could be enhanced by the presence of the nanostructural character. In addition, an improvement of the mechanical performance of these coatings could be expected due to the effect of the nanozones.

The possibility of cold spraying nanostructured Ti also seems very promising. By using cold spray (i.e., no particle melting), it will be possible to produce Ti coatings consisting of 100% nanostructured material.

One of the challenges of the coming years will probably be the production of biomedical coatings (using thermal spray or other techniques) exhibiting strong nanostructural surface characteristics with the goal of producing biomimetic surfaces.

It is important to point out again that the idea of combining the enhanced mechanical performance of nanostructured thermal spray coatings with the improved biocompatibility of the nanostructured materials is new. Consequently, much work is still required to more fully investigate this potential. In fact, this is an open field with many possibilities and opportunities for development and application.

REFERENCES

1. Kucuk, A., Lima, R.S., and Berndt, C.C., Influence of Plasma Spray Parameters on Formation and Morphology of ZrO_2-8wt% Y_2O_3 Deposits, *Journal of the American Ceramic Society*, 84(4), 693, 2001.
2. www.aaos.org, *American Academic of Orthopaedic Surgeons*, September 29, 2004.
3. Sun, L. et al., Material Fundamentals and Clinical Performance of Plasma-Sprayed Hydroxyapatite Coatings: A Review, *Journal of Biomedical Materials Research*, 58(5), 570, 2001.
4. Rokkum, M. et al., Polyethylene Wear, Osteolysis and Acetabular Loosening with an HA-Coated Hip Prosthesis, *Journal of Bone and Joint Surgery (Br)*, 81-B, 582, 1999.
5. Sharp, R.J. et al., Analysis of the Results of the C-Fit Uncemented Total Hip Arthroplasty in Young Patients with Hydroxyapatite or Porous Coating of Components, *Journal of Arthroplasty*, 15, 627, 2000.
6. Reikeras, O. and Gunderson, R.B., Failure of HA Coating on a Gritblasted Acetabular Cup, *Acta Orthopaedica Scandinavica*, 73(1), 104, 2002.
7. Oosterbos, C.J.M., Rahmy, A.I.A., and Tonino, A.J., Hydroxyapatite Coated Hip Prosthesis Followed Up for 5 Years, *International Orthopaedics*, 25, 17, 2001.
8. www.sulzermetco.com, SUME®PLANT Ti coatings, February 15, 2006.
9. Bansal, P., Padture, N.P., and Vasiliev, A., Improved Interfacial Mechanical Properties of Al_2O_3-13wt% TiO_2 Plasma-Sprayed Coatings Derived from Nanocrystalline Powders, *Acta Materialia*, 51, 2959, 2003.
10. Gell, M. et al., Development and Implementation of Plasma Sprayed Nanostructured Ceramic Coatings, *Surface and Coatings Technology*, 146–147, 48, 2001.
11. Jordan, E.H. et al., Fabrication and Evaluation of Plasma Sprayed Nanostructured Alumina-Titania Coatings with Superior Properties, *Materials Science and Engineering A*, 301, 80, 2001.
12. Luo, H. et al., Indentation Fracture Behavior of Plasma-Sprayed Nanostructured Al_2O_3-13wt% TiO_2 Coatings, *Materials Science and Engineering A*, 346, 237, 2003.

13. Kim, G.E., Walker Jr., J., and Williams Jr., J.B., Nanostructured Titania Coated Titanium, *US Patent 6,835,449 B2*, December 28, 2004.
14. He, J. and Schoenung, J.M., A Review on Nanostructured WC-Co Coatings, *Surface and Coatings Technology*, 157, 72, 2002.
15. Tellkamp, V.L. et al., Thermal Spraying of Nanocrystalline Inconel 718, *NanoStructured Materials*, 9, 489, 1997.
16. Lau, M.G. and Lavernia, E.J., Microstructural Evolution and Oxidation Behavior of Nanocrystalline 316-Stainless Steel Coatings Produced by High-Velocity Oxygen Fuel Spraying, *Materials Science and Engineering A*, 272, 222, 1999.
17. He, J. and Lavernia, E.J., Precipitation Phenomenon in Nanostructured Cr_3C_2-NiCr Coatings, *Materials Science and Engineering A*, 301, 69, 2001.
18. Lima, R.S. and Marple, B.R., Enhanced Ductility in Thermally Sprayed Titania Coating Synthesized Using a Nanostructured Feedstock, *Materials Science and Engineering A*, 395, 269, 2005.
19. Lima, R.S. and Marple, B.R., Superior Performance of High-Velocity Oxyfuel-Sprayed Nanostructured TiO_2 in Comparison to Air Plasma-Sprayed Conventional Al_2O_3-13TiO_2, *Journal of Thermal Spray Technology*, 14(3), 397, 2005.
20. Lima, R.S. and Marple, B.R., From APS to HVOF Spraying of Conventional and Nanostructured Titania Feedstock Powders: A Study on the Enhancement of the Mechanical Properties, *Surface and Coatings Technology*, 200, 3428, 2006.
21. Lau, M.J. et al., Synthesis and Characterization of Nanocrystalline Cu-Al Coatings, *Materials Science and Engineering A*, 347, 231, 2003.
22. Lima, R.S. et al., Biocompatible Nanostructured High Velocity Oxy-Fuel (HVOF) Titania Coating, PDF (CD), in *Building on 100 Years of Success: Proceedings of the 2006 International Thermal Spray Conference*, B.R. Marple, M.M. Hyland, Y.-C. Lau, R.S. Lima, J. Voyer (Eds.) May 15–18 (Seattle, WA), ASM International, Materials Park, Ohio, 2006.
23. Miyayama, M., Koumoto, K., and Yanagida, H., Engineering Properties of Single Oxides, in *Engineered Materials Handbook, Vol. 4, Ceramics and Glasses*, Ed., Schneider, ASM International, Materials Park, OH, 1991, p. 748.
24. Schmidt, T. et al., Development of a Generalized Parameter Window for Cold Spray Deposition, *Acta Materialia*, 54, 729, 2006.
25. Webster, T.J., Siegel, R.W., and Bizios, R., Nanostructured Ceramics and Composite Materials for Orthopaedic-Dental Implants, *US Patent 6,270,347 B1*, August 7, 2001.
26. Webster, T.J., Siegel, R.W., and Bizios, R., Osteoblast Adhesion on Nanophase Ceramics, *Biomaterials*, 20, 1221, 1999.
27. Gutwein, L.G. and Webster, T.J., Increased Viable Osteoblast Density in the Presence of Nanophase Compared to Conventional Alumina and Titania Particles, *Biomaterials*, 25, 4175, 2004.
28. Webster, T.J. et al., Specific Proteins Mediate Enhanced Osteoblast Adhesion on Nanophase Ceramics, *Journal of Biomedical Materials Research*, 51(3), 475, 2000.
29. Webster, T.J. et al., Enhanced Functions of Osteoblasts on Nanophase Ceramics, *Biomaterials*, 21, 1803, 2000.
30. Webster, T.J. and Ejiofor, J.U., Increased Osteoblast Adhesion on Nanophase Metals: Ti, Ti6Al4V, and CoCrMo, *Biomaterials*, 25, 4731, 2004.
31. Dalby, M.J. et al., Investigating the Limits of Filopodial Sensing: A Brief Report Using SEM to Image the Interaction between 10 nm high Nano-Topography and Fibroblast Filopodia, *Cell Biology International*, 28, 229, 2004.
32. Anselme, K., Osteoblast Adhesion on Biomaterials, *Biomaterials*, 21, 667, 2000.
33. Erikson, H.P., Carrell, N., and McDonagh, J., Fibronectin Molecule Visualized in Electron Microscopy: A Long, Thin, Flexible Strand, *Journal of Cell Biology*, 91, 673, 1981.
34. Shi, H. et al., Template-Imprinted Nanostructured Surfaces for Protein Recognition, *Nature*, 398, 593, 1999.
35. Lee, K.B. et al., Protein Nanoarrays Generated by Dip-Pen Nanolithography, *Science*, 295, 1702, 2002.
36. Capello, W.N. et al., Hydroxyapatite in Total Hip Arthroplasty, *Clinical Orthopaedics and Related Research*, 355, 200, 1998.
37. Lai, K.A. et al., Failure of Hydroxyapatite-Coated Acetabular Cups, *Journal of Bone and Joint Surgery (Br)*, 84-B, 641, 2002.

38. Manley, M.T. et al., Fixation of Acetabular Cups without Cement in Total Hip Arthroplasty, *Journal of Bone and Joint Surgery (Am)*, 80-A(8), 1175, 1998.
39. Shen, W.J. et al., Mechanical Failure of Hydroxyapatite- and Polysulfone-Coated Titanium Rods in a Weight-Bearing Canine Model, *Journal of Arthroplasty*, 7(1), 43, 1992.
40. Silver, I.A., Murrills, R.J., and Etherington, D.J., Microelectrode Studies on the Acid Microenvironment Beneath Adherent Macrophages and Osteoclasts, *Experimental Cell Research*, 175, 266, 1988.
41. Kweh, S.W.K., Khor, K.A., and Cheang, P., An *In Vitro* Investigation of Plasma Sprayed Hydroxyapatite (HA) Coatings Produced with Flame-Spheroidized Feedstock, *Biomaterials*, 23, 775, 2002.
42. Blacha, J., High Osteolysis and Revision Rate with the Hydroxyapatite-Coated ABG Hip Prostheses, *Acta Orthop. Scand.*, 75(3), 276, 2004.
43. Bloebaum, R.D., et al., Complications with Hydroxyapatite Particulate Separation in Total Hip Arthroplasty, *Clinical Orthopaedics and Related Research*, 298, 19, 1994.
44. www.who.int/countries/usa/en/, *World Health Organization*, February 16, 2006.
45. www.who.int/countries/can/en/, *World Health Organization*, February 16, 2006.
46. Legoux, J.G. et al., Development of Osteoblast Colonies on New Bioactive Coatings, PDF (CD), in *Building on 100 Years of Success: Proceedings of the 2006 International Thermal Spray Conference*, B.R. Marple, M.M. Hyland, Y.-C. Lau, R.S. Lima, J. Voyer (Eds.) May 15–18 (Seattle, WA), ASM International, Materials Park, Ohio, 2006.
47. Lopes, M.A., Monteiro, F.J., and Santos, J.D., Glass-Reinforced Hydroxyapatite Composites: Fracture Toughness and Hardness Dependence on Microstructural Characteristics, *Biomaterials*, 20, 2085, 1999.
48. Espagnol, M. et al., Effect of Heat Treatment on High Pressure Plasma Sprayed Hydroxyapatite Coatings, *Surface Engineering*, 18(3), 213, 2002.
49. Standard Test Method for Adhesion or Cohesion Strength of Thermal Spray Coatings, ASTM Standard C 633-01. ASTM, West Conshohocken, PA.
50. Gu, Y.W. et al., Microstructure and Mechanical Properties of Plasma Sprayed HA/YSZ/Ti-6Al-4V Composite Coatings, *Biomaterials*, 25, 4009, 2004.
51. Lima, R.S. et al., HVOF Spraying of Nanostructured Hydroxyapatite for Biomedical Applications, *Materials Science and Engineering A*, 396, 181, 2005.
52. Zheng, X., Huang, M., and Ding, C., Bond Strength of Plasma-Sprayed Hydroxyapatite/Ti Composite Coatings, *Biomaterials*, 21, 841, 2000.
53. Tsui, Y.C., Doyle, C., and Clyne, T.W., Plasma Sprayed Hydroxyapatite Coatings on Titanium Substrates Part 1: Mechanical Properties and Residual Stress Levels, *Biomaterials*, 19, 2015, 1998.
54. Li, H., Khor, K.A., and Cheang, P., Effect of the Powders' Melting State on the Properties of HVOF Sprayed Hydroxyapatite Coatings, *Materials Science and Engineering A*, 293, 71, 2000.
55. Yang, Y.C. and Chang, E., The Bonding of Plasma-Sprayed Hydroxyapatite Coatings to Titanium: Effect of Processing, Porosity and Residual Stress, *Thin Solid Films*, 444, 260, 2003.

9 Nanophenomena at Work, for Color Management in Personal Care

Vijay M. Naik
Hindustan Unilever Research Centre, Bangalore, India

CONTENTS

9.1   INTRODUCTION

The mission of the multibillion dollar personal care industry is to raise vitality of life, to help us get more out of life—emotionally and physically. The personal care products such as toothpaste, soap, shampoo, cold cream, antiwrinkle cream, deodorant spray, lipstick, nail polish, hair colorants, and so forth, collectively touch every spot of our body at every moment of our life. They take care of our hygiene, protect us from damage by ultraviolet (UV) light or aggressive climate, and restore or repair damaged body surfaces. They make us look good and feel good. The majority of these personal care products are not pure chemical molecules or their homogeneous solutions. They are multiphase composite systems, the constituents of which are often self-organized into a myriad of multiscale hierarchical structures. The nanoscale features or entities present in these products often uniquely control or contribute to their physical characteristics, the sensorial experience they offer, and the functional interactions they have with the body. The body substrates with which these products interact are also hierarchically assembled, multiorganelle, multicellular tissues, with nanoscale features and entities having unique characteristic structural or functional roles. No wonder, therefore, that understanding the emergence of novel phenomena at nanoscale in products and body substrates, as well as scouting new applications of nanoscience and nanomaterials for exploring novel benefits, is a fascinating topic for research and development (R&D) in the personal care industry. The objective of this short chapter is to introduce this exciting subject with a few illustrations, primarily in the area of color management.

9.2   LESSONS IN PHOTONICS FROM NATURE

Creating products offering novel desirable visual sensory experience, and delivering functional benefits through the product for managing and enhancing the visual appearance of the body, making the body look good and distinctively attractive, are among the most powerful drivers of R&D in

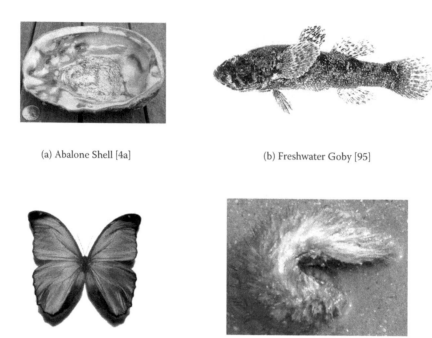

(a) Abalone Shell [4a]   (b) Freshwater Goby [95]

(c) *Morpho rhetenor* [36]   (d) Sea-Mouse [13]

FIGURE 9.1 Sample of structural color in biosystems. (**See color insert following page 112.**)

the personal care world. Therefore, one frontier of nanoscience which is likely to make early forays into the personal care industry is photonics. Biology inspired photonics can offer a number of new ideas and opportunities in this context. If we look around at the diverse ways in which photonic nanophenomena are encountered in the animal kingdom creating a range of structural colors and optical effects, that also with limited use of dyes and pigments, it is truly bewildering [1,2,34]. The metallic luster on the shells of some mollusks, visible light reflectors with 100% efficiency in the tapeta of eyes of many vertebrates for improving their vision, switchable light reflectors in the iridophores or scales of some fish for camouflaging, highly iridescent wing scales of some butterflies which help in identifying their sex or for creating attractive displays visible from a long distance, the gem-stone-like spectacular color of the spines in the hairy felt of the sea-mouse—these are some of the illustrative examples of what can be achieved through structural optics. Figure 9.1 presents a sample of these optical effects.

Although some of these effects have arisen out of accidents in the evolution of structural morphology, as in case of the abalone shell, the others have played an important role in aiding the survival of the species under selection pressures in the evolutionary history. Whatever their origin, they serve as useful models for understanding and creating magnificent structural colors and optical effects.

Figure 9.2 presents electron micrographs of the chromophores responsible for some of the optical effects shown in Figure 9.1.

In the case of the abalone (*Haliotis rufescens*), the brilliant reflectance in the shell is produced by a composite structure that has about 95% calcium carbonate and about 5% organic matter rich in a protein named *conchiolin* [3–5]. The calcium carbonate is present as approximately 500-nm thick and 5-µm wide polygonal platelets of aragonite. The platelets, along with 20-nm thick proteinaceous organic films separating them, are organized in layered lamellar structures.

The freshwater goby (*Odontobutis obscura*) has a mottled appearance. It is capable of changing color by actuating organelles in its scales called iridophores [6]. The iridophores contain a large

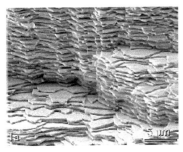
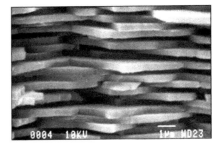

(a) Abalone: Ultrastructure of shell [3,4]

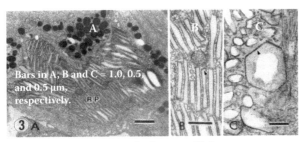
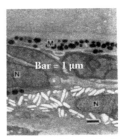

Bright Spots: Guanine platelets assembled as
lamellar multilayers (A, B) and their top view (C)

Dark Spots: Randomized
guanine platelets

Freshwater Goby: Ultrastructure of iridophore [6]

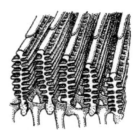

Morpho rhetenor: Ultrastructure of wing scales, and its 3D model [1a]

FIGURE 9.2 Ultrastructure of chromophores responsible for some optical effects in Figure 9.1.

number of polygonal platelet-like crystals of a purine, namely, guanine. The platelets appear to be approximately 150 to 200 nm in thickness and 500 to 1000 nm in width. An interesting fact is that the relative position and orientation of the platelets can be manipulated by the fish. When the platelets are randomly oriented, they produce dark spots. When they are aligned in lamellar stacks with approximately 50 nm thick film of cytoplasm occupying the space between platelets, they produce light reflecting bright spots. This ability and skill helps the goby hide itself in its surroundings.

The butterfly *Morpho rhetenor* has peculiarly structured cuticles on the scales of its wings [1a]. The cuticle on the top surface of the scale is laid in the form of parallel ridges. Each ridge, in turn, carries on its either side parallel stacks of 10 to 12 fins spaced at approximately 170-nm pitch, made up of chitin projections, the intervening space being occupied by air. This morphology selectively reflects low-wavelength visible light, giving a spectacular iridescent blue color to the butterflies that makes them visible to their potential mates, even from a distance of half a mile [1]. Depending upon

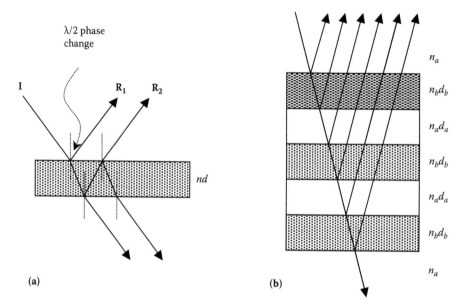

FIGURE 9.3 Refraction and reflection from a multilayer stack of transparent films of two dielectric materials. (a) Thin-film reflection. Constructive interference between R_1 and R_2 occurs when the optical thickness of the film n (refractive index) $\times d$ (actual thickness) $= \lambda/4$. (b) "Ideal" multilayer in which light reflected from every interface interferes constructively. This occurs when $n_a d_a = n_b d_b = \lambda/4$.

the angle of viewing and incidence of light, the hue of the color also changes, producing stunning iridescence.

The biological function of the iridescent hairy spine of sea-mouse (*Aphrodita* sp.) is not fully understood, but its anatomical origin has been discovered [12–14]. The hairy spines of a sea-mouse are made up of chitin. They contain closely packed arrays of hollow cylinders running along the long axis. The geometry can also be visualized as a multilayer stacking of materials with two different averaged refractive indices. The spacing of adjacent layers of cylinders is approximately 510 nm (see inset of Figure 9.6). When illuminated with light, the spines show an array of beautiful colors, with a close to 100% reflectance for red light at a wavelength in the neighborhood of 630 nm.

There are controversies regarding the extent to which different fundamental optical phenomena such as absorption, elastic scattering, reflection, refraction, diffraction, and interference contribute to the observed colors in different biological systems. Different species exploit different physical mechanisms to acquire color. However, a simple model [2,3] of interference among rays of light reflected from interfaces between alternating thin layers of transparent dielectric materials of two different refractive indices provides sufficiently useful insights about the magnificent optical effects observed for the above four cases.

Figure 9.3 presents the situation diagrammatically. When a ray of light is incident on alternatively stacked low- and high-refractive index thin films, there is refraction and partial reflectance at every interface. The split fractional rays of the incident light which are ultimately reflected backward by the multilayer stack of films exhibit mutual interference. Different theoretical expressions have been reported in the literature [8,9] for predicting the intensity of the total reflected light at different wavelengths, with varying degrees of rigor. However, the numerical solution of even a simplified model as summarized by Land [2] provides sufficiently reliable insights. One such solution for the optical behavior of the wings of *Morpho rhetenor* butterfly, employing primary data for the system of interest available in literature [2,7], is presented in Figure 9.4. The results [15] indicate peak reflectance near 430 to 450 nm, as experimentally measured, and an overall brilliant blue display for the wings of the butterfly, clearly demonstrating utility of the simple analysis.

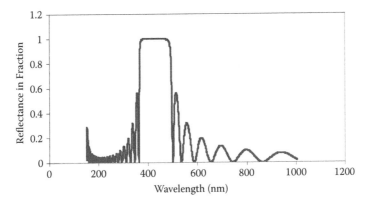

FIGURE 9.4 Theoretically predicted reflectance pattern for the wings of *Morpho rhetenor* (J. Raut, Unilever Research India, personal communications, 2007). Data used (see M.F. Land, *Prog. Biophysics Mol. Biol.* 24, 75–106, 1972; P. Vukusic, J.R. Sambles, C.R. Lawrence, and R.J. Wootton, *Proc. R. Soc. Lond.* B 266, 1403–1411, 1999.) da = 105 nm (airgap), db = 70 nm (chitin projection), na = 1.0 (air), nb = 1.54 (chitin), p = 12 bilayers.

9.3 BIOLOGY-INSPIRED LUSTER AND IRIDESCENCE

The physical insights are used by the personal care industry to produce creams for various applications, having attractive shimmering or lustrous appearance. The formulation base of a typical skin cream may be made up of approximately 16% stearic or palmitic acid and 2% sodium salt of the fatty acid, the rest being water. This formulation when homogenized at elevated temperature results in an emulsion, which upon cooling gives a cream with a matte finish. The same formulation, if subjected to controlled shear with a specific time-temperature cooling program, produces a cream with attractive luster. Figure 9.5 presents images of cream-bases having identical formulations but prepared employing two different processes [16]. The corresponding micrographs are also presented. The cream with the lustrous appearance contains multilayer stacks of crystalline platelets of fatty

(a)   (b)

FIGURE 9.5 Matte and lustrous creams from identical formulations. Micrographs show randomized and layered platelets of fatty acid, for respective cases (S. Zhu, Unilever Research Colworth, personal communications, 2007). Inset shows morphology of abalone shell for comparison [3].

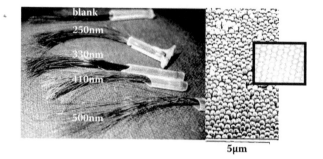

5μm

FIGURE 9.6 Photonic cosmetics and "hair jewelry." Unilever's patented technology for coloring hair by deposition of nanoparticles as multilayer colloid crystal (R. Djalali, Unilever Research Colworth, *personal communications*, 2007; Patent, Unilever, WO 2006/097332 A2). Inset shows cross-section of the iridescent spine of the sea-mouse with similar characteristic lattice dimensions. (A.R. Parker, R.C. McPhedran, D.R. Mckenzie, L.C. Botten, and N.A. Nicorovici, *Nature* 409, 36–37, 2001. With permission.) **(See color insert.)**

acids having characteristic nanoscale thickness. In the absence of orderly multilayer stacking, the cream has a matte finish. The microstructure of "naturally" assembled abalone shell is also shown for comparison, as inset. It is obvious that although the secret of nature has been unveiled, even the best man-made creams have large unrealized potential to improve their structure and thereby raise reflectivity and luster.

The personal care industry is also on its way to deliver novel structured colorants for makeup and hair dressing (Figure 9.6). First, these new-generation structured colorants could potentially present further improved and safer alternatives to conventional molecular dyes or pigments. They can offer unique benefits, such as one claimed by L'Oreal that their lipstick would simply flake off as a colorless dull white powder if transferred from lips or smeared on to some other surface [10]. And of course, they can provide novel surrealistic effects such as butterfly-like iridescence, shimmering vibrant colors that move as we change the point of view, or glittering tresses of hair that change hue as the hair moves [11]. This last-mentioned, novel invention has been patented by Unilever [17,18]. The crux of the technology involves preparation of high-refractive-index particles of narrow size distribution, and their deposition in the form of an ordered colloid crystalline lattice on the surface of hair. Figure 9.6 presents a range of iridescent colors obtained by employing silica particles of different dimensions. Figure 9.6 also shows as an inset, a cross-section of the spine of a sea-mouse that produces gemstone-like spectacular colors that inspired the researchers to invent the technology. The morphological and dimensional similarities are amazing. (Incidentally the sea-mouse spine has been providing lot of food for thought to researchers in other industries as well, such as those interested in creating optical analogues of electrical semiconductors, with structures that do not allow propagation of light in specific wavelength ranges, for all angles of incidence [12].)

9.4 BIOMIMETIC STRUCTURAL SUNSCREENS AND PHOTONIC MODEL FOR TEETH WHITENING

The scientific understanding of structured color effects displayed by periodically arranged multi-layer biological systems can also open up opportunities in other related areas.

It is well known that long exposure to the Sun's rays, especially its UV components, can damage skin. It may lead to faster aging, sunburn, and even malignancy. These effects are likely to be more pronounced if the skin happens to be light in color. Although the natural pigment produced by skin, melanin, is adequate in dark-skinned genotypes to act as an effective barrier to UV, the UV radiation itself upregulates pigmentation in skin, making it look even darker. Thus, many personal care products, whether formulated for light skin or dark skin, contain UV filters to provide protection to skin. A number of molecules both of natural origin as well as man-made, are examined for use as UV filters in such formulations. Unfortunately, most of these molecules are themselves prone to photodegradation. Not only can they provide UV protection for only a short duration of time, but their products of degradation can also have harmful effects on the skin. Therefore, development of UV filters that are safe, photostable, and economical is a big opportunity in the personal care industry (see, for example, [19–23]). It can be readily recognized that the optical behavior of ordered multilayer structures discussed in Section 9.2 is not limited to visible light. The physical principles can be harnessed to achieve completely destructive or constructive reflectance from such nanolayered films in the UV (or for that matter in the infrared [IR] and microwave) frequency bands to deliver the desired benefits of photoprotection using harmless, photostable, and affordable materials. Nature demonstrates this possibility [24]. The crux of the challenge is obtaining high reflectivity for UV light and transparency for visible light so that the filter is "water clear" to the human eye. Figure 9.7a presents one of nature's UV reflecting constructs on the scale of male butterfly *Eurema lisa* [24], as well as theoretically predicted reflectance spectrum for one such periodically nanolayered designer material [15].

As we looked at the virtually specular reflections by orderly multilayered structures, nature has also created perfectly aperiodic, superefficient scatterers to obtain brilliant whiteness.

Whiteness is observed when visible light at all wavelengths is scattered randomly by particles or scattering centers, owing to discontinuity in the refractive index between them and their surroundings. The larger the number of scattering centers in a given volume, the better is the brightness and opacity. However, if the scattering centers are arranged in a periodic manner or are "optically crowded," the effect is diminished. The whiteness in the human tooth is a result of hydroxyapatite

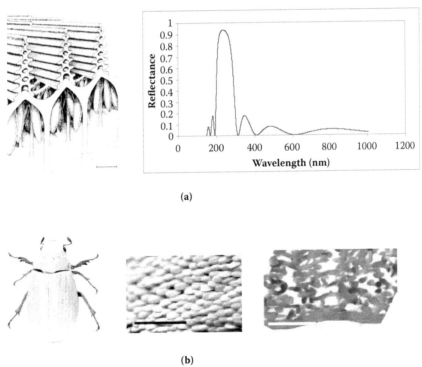

(a)

(b)

FIGURE 9.7 Nanobiomimetic structural sunscreens, and photonic model for tooth whitening. (a) Schematic of UVR reflecting scales of *Eurema lisa* (bar = 5 μm). (From Ghiradella, H. Aneshanseley, D. Eisner, T. Silbergled, R.E., and Hinton, H.E. *Science* 178, 1214–1217, 1972. With permission.) And UV-selective reflectance spectrum of a novel theoretical design of nanolayered film [15]. (b) A *Cyphochilus* sp. Beetle [26]. Its typical white scales (bar = 1 mm), and their intrascale photonic structure (bar = 3.00 mm). (From Vukusic, P., Hallam, B., and Noyes, J. *Science* 315, 348, 2007. With permission.)

crystals acting as such scattering centers in the tooth enamel. The average tooth whiteness displayed by almost 2 mm thick enamel is, however, vastly outperformed by the scales of *Cyphochilus* sp. beetles, which are merely 5 μm thick. These scales seem to accommodate a large density of aperiodically arranged scattering centers without optical crowding. This is achieved by creating an intrascale three-dimensional photonic structure of interconnecting cuticular filaments with diameters of about 250 nm and volume occupancy of 70%. Figure 9.7b presents an image of one species of these beetles and typical views of scales and the intrascale structure [25,26]. It is predicted that the beetle scale architecture would be key to creating futuristic ultrathin coatings, to deliver exceptional whiteness and brightness to teeth [27].

9.5   MOLECULAR MOTORS MANAGING COLOR OF SKIN

Although all *Homo sapiens* on the planet Earth share essentially the same genes, mutations in a tiny number of genes have resulted in a diversity of skin tones, from very dark brown to nearly colorless. The genetic variation in color has been a regrettable basis of prejudice and discrimination, instead of being a cause for celebrating the diversity. One of the objectives of the personal care industry is to provide a safe means to consumers for achieving small changes in their skin tone, to make it darker or lighter or pinkish, as per their personal choice, without loss of their individual identity.

Figure 9.8a schematically presents relevant cellular details of the epidermis of skin to bring out the essential differences between light and dark skin. The color of skin is mainly contributed by melanin pigments formed by enzymatic oxidation of tyrosine in the skin cells called melanocytes residing at the basal layer of the epidermis. The pigment is packaged in the form of organelles called melanosomes and transferred to the upper layers of differentiating cells called keratinocytes. The quantity and state of distribution of the pigment in the keratinocytes determine the depth of intensity of skin color. The higher the quantity and the more uniform the peripheral dispersion are, the deeper is the color. The lesser the pigment level and the more the perinuclear clustering of the pigment, the lighter is the color. Experiments conducted in a model system of tadpole tails have been used in the past to identify bioactive molecules that can cause dispersion or aggregation of the pigment, thereby bringing about changes in the tone of skin color. Figure 9.8b shows some typical observations [28]. In the above context, mechanisms involved in systemic as well as induced transport or dispersion or aggregation of melanin are one of the important areas of study in mammalian skin pigmentation.

It is now well established that the intracellular transport of melanin is carried out by nanosized molecular motors. In fact, melanosome movement by molecular motors in melanocytes is turning out to be an important model for the study of organelle motility in general [29]. A simplified picture of the subject is as follows. A melanocyte, like other animal cells, comprises a sac of cell membrane stretched over a structural network called cytoskeleton (which is a part of the cytoplasm), the cytoplasm itself, and the cell nucleus. One of the components of the cytoskeleton is microtubules. These are approximately 24 nm outer diameter tubes of varying lengths. Figure 9.9a shows a network of microtubules (stained green) within a skin melanocyte [29]. The microtubules themselves are made of α and β tubulin dimers, polymerized end to end into protofilaments. Thirteen protofilaments self-assemble parallel to each other and then coil into a hollow microtubule. The microtubule structure has a polarity. One of its ends displays exposed α subunits, and the other end the β subunits. These two ends are designated (−) and (+), respectively. This layout is also schematically shown in Figure 9.9b. The melanosomal parcels are carried by the molecular motors along these tracks formed by the microtubule network. Two kinds of motors are involved in transport along microtubules, kinesin for transport toward the (+) end, and dynein for transport toward the (−) end. A schematic picture is shown in Figure 9.9c. The figure also shows a third kind of motor called myosin that is involved in the transfer of melanosomes to actin filaments, which are also present in the cell. Let us take the example of kinesin. It is a force-generating homodimer enzyme protein, which specifically binds to and walks along a protofilament of microtubule. It converts chemical energy released by

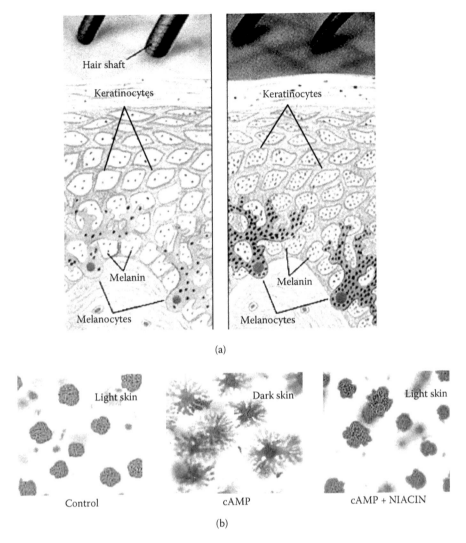

(a)

(b)

FIGURE 9.8 Pigmentation of human skin and pigment distribution studies in model system. (a) Human skin—schematic. (b) Melanin dispersion and clustering studies in *Rana tigrina* tadpole skin cells [28].

hydrolysis of ATP into mechanical energy. Hydrolysis of 1 ATP molecule can provide 40×10^{-21} J of energy and up to 7 pN force, enabling kinesin to move its center of mass in increments of 8 nm length, along the microtubule of several micrometers length, without dissociating, at speeds of 0.1 μm/s to 1μm/s. Figure 9.10 shows this "hand-over-head" mechanism of kinesin walk, schematically. The relevance of studying motor proteins transcends beyond skin pigmentation as they play an essential role in a range of biological operations and cell functions involving active transport of organelles and cell movements. A deeper understanding of their mechanism of motion can also provide novel clues for designing nanoscale machines and devices. The subject, therefore, has received immense attention and is a topic of several experimental as well as theoretical studies [30–33].

9.6   CONCLUSION

The personal care industry is generally very modest about the technology content of its products and the science that goes behind designing these products. Because most of its products qualify for the notorious descriptor "sludge systems," it is not apparent that the rapidly advancing field of

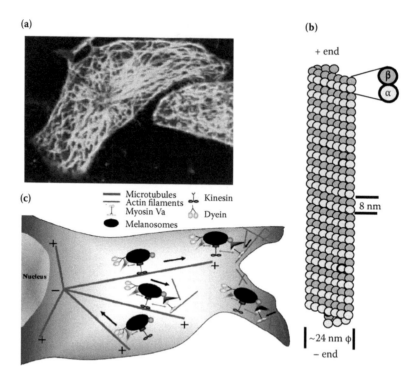

FIGURE 9.9 Multiscale assembly melanosome transport system: (a) microtubule (green) and actin (red) tracks in melanocute; (b) schematic picture of the microtubule assembly; and (c) schematic picture of the bidirectional transport. (Parts a and c provided by D.C. Barral and M.C. Seabra, *Pigment Cell Res.* 17, 111–118, 2004. With permission.) **(See color insert.)**

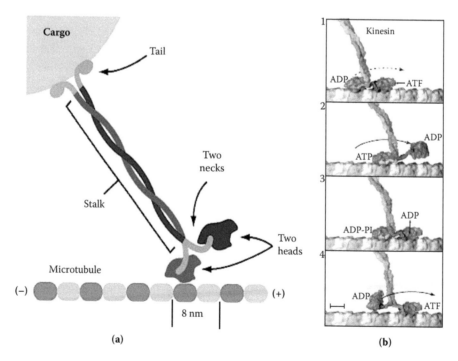

FIGURE 9.10 Schematic image of kinesin with cargo [33a]. And its movement by "hand-over-head" mechanism. (R.D. Vale and R.A. Milligan, *Science* 288, 88–95, 2007; S.M. Block. With permission.) **(See color insert.)**

nanoscience and nanotechnology is likely to intersect with the domain of personal care products. The present chapter is intended to illustrate that the industry is indeed deriving several fascinating and inspirational ideas from the myriad of functional nanostructures encountered in nature in order to enrich its product development and background science, which in turn is likely to contribute toward the development of nanotechnology itself.

ACKNOWLEDGMENTS

The author would like to thank Unilever Research for encouragement to write this chapter, and colleagues Shilpa Vora, Janhavi Raut, Shiping Zhu, and Ramin Djalali for very useful discussions and help in preparing the manuscript.

REFERENCES

1. P. Vukusic and J.R. Sambles (2003) Photonic Structures in Biology. *Nature* 424, 852–855.
1a. Available online at http://www.foresight.gov.uk/Previous_Projects/Exploiting_the_electromagnetic_spectrum/Opticks/RoySamblesR.pdf.
2. M.F. Land (1972) The Physics and Biology of Animal Reflectors. *Prog. Biophysics Mol. Biol.* 24, 75–106; Iridescence, http://en.wikipedia.org/wiki/Iridesence.
3. Biomolecular Materials, Biomineralization and New Materials, www.lifesci.ucsb.edu/mcdb/faculty/morse/research/biomolecular/biomolecular.html.
4. N. Yao, A. Epstein, and A. Akey (2006) Crystal Growth via Spiral Motion in Abalone Shell Nacre. *J. Mater. Res.* 21(8), 1939–1946.
4a. Available online at htpp://en.wikipedia.org/wiki/Abalone.
5. M. Fritz, A.M. Belcher, M. Radmacher, D.A. Walters, P.K. Hansma, G.D. Stucky, D.E. Morse, and S. Mann (1994) Flat Pearls from Biofabrication of Organized Composites on Inorganic Substrates. *Nature* 371, 49–51.
6. A. Matsuno and T. Iga (1989) Ultrastructural Observations of Motile Iridophores from the Freshwater Goby, *Odontobutis obscura*. *Pigment Cell Research* 2, 431–438.
7. P. Vukusic, J.R. Sambles, C.R. Lawrence, and R.J. Wootton (1999) Quantified Interference and Diffraction in Single Morpho Butterfly Scales. *Proc. R. Soc. Lond.* B 266, 1403–1411.
8. A.F. Huxley (1968) A Theoretical Treatment of the Reflexion of Light by Multilayer Structures. *J. Exp. Biol.* 48, 227–245.
9. B.S. Verma, A. Basu, R. Bhattacharyya, and V.V. Shah (1988) General Expression for the Reflectance of an All-Dielectric Multilayer Stack. *Applied Optics* 27(19), 4110–4116.
10. Wired News: Butterfly Wings on Every Eyelid, www.wired.com/news/technology/1,68683-0.html.
11. Iridescent Hair, www.halfbakery.com/idea/Iridescent_20Hair.
12. R.C. McPhedran and N.A. Nicorovici (2001) Learning Optics in Nature's School. *Australian Optical Society NEWS* 15(2,3), 7–9.
13. Available online at http://en.wikipedia.org/wiki/Images:S70000590.JPG.
14. A.R. Parker, R.C. McPhedran, D.R. Mckenzie, L.C. Botten, and N.A. Nicorovici (2001) *Nature* 409, 36–37.
15. J. Raut, Unilever Research India, *personal communications* (2007).
16. S. Zhu, Unilever Research Colworth, *personal communications* (2007).
17. R. Djalali, Unilever Research Colworth, *personal communications* (2007).
18. Patent, Unilever, WO 2006/097332 A2, Colourant compositions and their use.
19. H. Gonzalez, N.T. Wahlberg, B. Stromdahl, A. Juzeniene, J. Moan, O. Larko, A. Rosen, A. Wennberg (2007) Photostability of Commercial Sunscreens upon Sun Exposure and Irradiation by Ultraviolet Lamps. *BMC Dermatology* 7, 1.
20. D. Dondi, A. Albini, and N. Serpone (2006) Interactions between Different Solar UVB/UVA Filters Contained in Commercial Sun Creams and Consequent Loss of UV Protection. *Photochem. Photobiol. Sci.* DOI: 10.1039/b606768a.
21. C.A. Bonda (2004) Photostabilization of a Sunscreen Composition with Low Levels of an Alpha-Cyano-Beta, Beta-Diphenylacrylate Compound. *US Patent*: US2004/0047818 A1 (2004).
22. S. Pattanaargson and P. Limphong (2001) Stability of Octyl Methoxycinnamate and Identification of Its Photo-Degradation Product. *International. Journal of Cosmetic Science* 23, 153–160.

23. W. Schwack and T. Rudolph (1995) Photochemistry of Dibenzoyl Methane UVA Filters Part 1. *Journal of Photochemistry and Photobiology: B Biology* 28, 229–234.

24. H. Ghiradella, D. Aneshansley, T. Eisner, R.E. Silberglied, and H.E. Hinton (1972) Ultraviolet Reflection of a Male Butterfly: Interference Color Caused by Thin-Layer Elaboration of Wing Scales. *Science* 178, 1214–1217.

25. P. Vukusic, B. Hallam, and J. Noyes (2007) Brilliant Whiteness is Ultrathin Beetle Scales. *Science* 315, 348.

26. Cyphochilus crataceus, www.imdap.entomol.ntu.edu.tw/Commoninsectimages.

27. Sparkling White Beetle Dazzles Scientists, www.scenta.co.uk/Home/1443274/sparking-white-beetle-dazzles-scientists.htm.

28. S. Vora, Unilever Research India, *personal communications* (2007).

28a. G.P. Mathur, K.K.G. Menon, S. Varadarajan, Lever Brothers Company, Skin lightening composition and method of using the same. U.S. Patent US3937810 (1976).

29. D.C. Barral and M.C. Seabra (2004) The Melanosome as a Model to Study Organelle Motility in Mammals. *Pigment Cell Res.* 17, 111–118.

30. R.D. Vale and R.A. Milligan (2007) The Way Things Move: Looking under the Hood of Molecular Motor Proteins. *Science* 288, 88–95.

31. A.B. Kolomeisky and M.E. Fisher (2007) Molecular Motors: A Theorist's Perspective. *Ann. Rev. Phys. Chem.* 58, 675–695.

32. M.A. Welte (2004) Bidirectional Transport along Microtubules. *Current Biology* 14, R525–R537.

33. C.L. Asbury, A.N. Fehr, and S.M. Block, Kinesin moves by an asymmetric hands-over-head mechanism, *Science*, 19, 2130–2134 (2003).

33a. S.M. Block, www.stanford.edu/group/blocklab/ScienceLimping/.

34. A.R. Parker, 515 million years of structural colour, *J. Opt. A: Pure Appl Opt.*, 2, R15–R28 (2000). DOIL 10.1088/1464-4258/2/2/201.

35. K.T. Shao, *Odontobutis obscura*. Available at http://www.fishbase.org/Photos/ThumbnailsSummary.php?ID=15879.

36. Iridescence of Blue Morpho butterfly wings. Available at http://en.wikipedia.org/wiki/Iridescence.

10 Proteoliposome as a Nanoparticle for Vaccine Adjuvants

Oliver Pérez, Gustavo Bracho, Miriam Lastre,
Domingo González, Judith del Campo,
Caridad Zayas, Reinaldo Acevedo, Ramón Barberá,
Gustavo Sierra, Alexis Labrada, and Concepción Campa
Finlay Institute, Havana, Cuba

CONTENTS

10.1 INTRODUCTION

Nanotechnology is an emergent area with main applications in diagnosis, preventive, and therapeutic medicine. Nevertheless, its use in vaccines as the best cost-effective intervention to prevent and treat not only infectious diseases but also chronic diseases like cancer has not been extensively explored. New vaccines certainly require new adjuvants, because only a limited number and not potent adjuvants exist. In addition, the use of nanotechnology in the adjuvant field is very limited. Therefore, the present chapter will be focused on highlighting the promissory results of our team in the field of adjuvants.

Adjuvants are substances that nonspecifically (acting on the innate response) increase the specific (adaptative) response to the antigens, enabling a successful vaccine subunit development. Adjuvants can be particularly important in the following:

- Influencing the required antigen dose and permitting the use of lesser quantity and dose number.
- Polarizing the immune response patterns, first described in the 1980s [1,2], inducing functional bactericidal and opsonophagocitic antibodies (Th [T helper] 1 pattern) or passive cutaneous anaphylaxis mediated by IgE antibodies (Th2 pattern). The functional neutralizing antibodies frequently searched do not lead to any polarization because they are

produced by both Th1 and Th2 patterns. The most important functional polarization is the induction of cytotoxic T lymphocyte (CTL) response. This can be induced by both TCD8+ and TCD4+ in the presence or absence of Th functions, but for the induction of a good CTL response resulting in a memory response; it should be mediated by TCD8$^+$ in the presence of Th1 assistance.

- Increasing the speed and memory (duration) of the induced immune response.
- Making effective responses in extreme ages (very young and old people).
- Functioning in immune compromised people.
- Solving the problems of different routes of delivery, particularly the mucosal one.
- Overcoming the competition of the immune-dominant antigens in combination vaccines.

The essential signal for an efficient immune induction is related to signals 1, 2, and 3 (Table 10.1). Signal 1 is sent by the interaction of the antigen in the form of peptide coupled to major histocompatibility molecules to the T-cell receptor. Signal 2 is divided in signal 0 and real signal 2. Signal 0 arrives by the interaction of pathogen-associated molecular patterns (PAMPs) with pathogen recognition receptors (PRRs) with the induction of chemokines and proinflammatory cytokines. Signal 2 is costimulation arriving mainly through CD40, CD80, and CD86 and their ligands in T cells. Signal 3 is related to immune polarization where IL-2/IFNγ and IL-4/IL-5 are mainly involved in Th1 and Th2 polarization, respectively. In addition, there is also evidence that polarization occurs from different subsets of dendritic cells, and it is known that IL-12 is involved in driving to Th1 [2]. The cytokines responsible for Th3 polarization are not totally clear, but TGFβ is involved in the switching from μ to α antibody heavy chain with the subsequent production of IgA by B cells, hallmark of the Th3 pattern. In addition, coordinated actions of Th1 and Th2 cytokines are also involved in some steps of Th3 stimulation [4].

There are several classifications of the adjuvants, but a simple and useful one is that which divides them in immune potentiator and delivery systems [5]. For the importance of the type of immune response induced in a vaccine protection, we add a third group to this classification: the immune polarization agents. This focuses the evaluation of the immune response induced in the determination of Th1 with or without CTL, Th2, or Th3 responses.

TABLE 10.1

Comparison of Essential Signs Induced by Several Adjuvants Including AFPL1

	Essential Signals for Immune Induction				
	1		2		3
			0	2	
	MHCII-Peptide	MHCI-Peptide	Recognition by PRR	Costimulation	Polarization
Adjuvant	Help	Cross Priming/CTL	ChK†: IL8, Rantes, MIP-1α,β	PI Ck†: IL12, IL6 TNFα	CD80, CD86, CD40	IL2/IFNγ, IL4/5, TGFβ?
FCA	+	+	+	+	+	Th1
AFPL1	+	+	+	+	+	Th1
AFCo1	+	+	+	+	+	Th1/3 (IgA)
MPL	+	+	+, rantes?	+, IL6?	+	Th1/Th2 (IL5)
Alum	+	−	?	?	−	Th2

Notes: PRR, pathogen recognition receptor; Chk†, chemokines; PI Ck†, proinflammatory cytokines; AFPL1, proteoliposome; AFCo1, proteoliposome-derived cochleate; FCA, freund; Alum, AL(OH)3; ?, unknown response.

Taking into account the above classification, some adjuvants are immune potentiators (like lipopolysaccharide [LPS] and its derivatives, oligos [unmethylated CpG, ISS, IMO], QS21/QuilA, SMIP, MDP and derivatives), others delivery systems (like aluminum salts, calcium phosphate, liposomes, emulsions [MF59, montanide], virosomes, virus-like particles, iscoms), and others have immune polarization activities. The delivery system works better when the immune potentiator is incorporated inside it. Then, when all these functions are joined together in the same structure, such adjuvants work better. Moving in that way, many pharmaceutical companies are looking for adjuvant systems combining different products to drive the immune response to the desired effectors. In our case, we obtained structures directly from bacteria that include all the necessary components for making a complex adjuvant.

10.2 SEARCHING FOR A NEW ADJUVANT BASED ON PAMP–PRR INTERACTIONS

The understanding of the immune system operation has evolved very fast in the last years. The modern view proposes the immune system as a very regulated and redundant system precisely tuned to maintain or recover the body homeostasis following dangerous insults. The activation of host defense mechanisms is a complex process triggered by danger signals that should be strong enough to disturb the balanced or regulated state switching to the activated state [6–8].

The sense of dangerous situations by the immune system is mediated by a network of evolutionary fixed receptors that have evolved to recognize structural molecular patterns that are well conserved among infectious pathogens [9]. The PRR recognize PAMPs and activate the innate immune system upon ligand binding. Following activation, the innate immune system releases unspecific mechanisms to control infection but simultaneously collects and processes information about the insult from the environment needed to further activate the specific adaptive response and to induce immune memory [7,9,10]. The concept of the need of danger signals to activate effective adaptive immunity and memory has revolved the field of vaccinology, the approaches to explore host immunity, therapies, and especially adjuvant development [11].

The immune system also discriminates between particulate and nonparticulate structures. But because immune system cells better phagocytize particulate antigens, the presence of activating signals from pathogens is decisive for immune system activation. However, the presence of particles containing such danger signals should be a stronger activator for the immune system than particles and danger signals as separated components.

Toll-like receptors (TLRs) constitute a well-defined family that has already been described. Twelve TLRs and their ligands have been identified to date (Table 10.2). TLRs are differently expressed on different cells, most of them present on the cytoplasmic membrane but TLR9, TLR8, TLR7, and TLR3 present only on the endosomal compartment. The ligands of TLR are PAMPs, including LPS, external lipoproteins and lipopeptides, dsRNA, microbial DNA, peptidoglycan, flagella components, and polysaccharides [12]. Upon ligand binding, the TLRs activate adaptor proteins to further activate transduction pathways, resulting in transcription factor activation and selective gene expression. Despite the fact that most TLRs share a common transduction system, the Myd88 molecule to deliver signal to the nucleus, an alternative cascade is used by TLR4, and TLR3 does not use Myd88 [13,14]. However, TLR activation frequently results in cell activation with increasing antigen uptake, the activation of unspecific pathogen elimination mechanisms, and the upregulation of the expression of surface molecules on antigen-presenting cells (APCs) that are involved in antigen presentation and costimulation for T cells. Particularly in dendritic cells, the professional APC, the recognition of PAMPs by TLR becomes a crucial step that decides further maturation and the acquisition of special features (increased expression of antigen-presenting and costimulatory molecules) required for the activation of naïve T cells that further differentiate into helper T cells that orchestrate the adaptive immune response [15].

The strongest PAMPs described because of their capability to activate the innate immune system and to generate fully mature dendritic cells are the LPSs from Gram-negative bacteria. Several strategies of adjuvant development are currently based on LPS derivatives [16].

TABLE 10.2

Known Toll-Like Receptors and Their Ligands—Cellular Localization and Comments Are Also Shown

TLR	Ligands	Localization	Comments
TLR-1	Triacyl lipopeptides	Cytoplasmic membrane	Combines with TLR2
TLR-2	Lipoprotein, PGN, LTA, porins, LPS, lipoarabinomannan, zymosan, trypanosomal phospholipids, HSP60, HSP70, defensins, Pam3Cys	Cytoplasmic membrane	
TLR-3	ds RNA, poly (I:C), endogenous mRNA	Endosomal compartment	Synergy with TLR8
TLR-4	LPS, Pseudomonas exoenzyme S, HSP60, RSV F protein, MMTV envelope protein, trepanosomal lipids, HSP70, HSP90, hyaluronic acid, fibrinogen, fibronectin, β-defensin, heparin sulfate, taxol	Cytoplasmic membrane	Synergy with TLR8
TLR-5	Flagellin	Cytoplasmic membrane	
TLR-6	Diacyl lipopeptides	Cytoplasmic membrane	Combines with TLR2
TLR-7	ssRNA, imiquimod	Endosomal compartment	
TLR-8	ssRNA, resquimod	Endosomal compartment	
TLR-9	Bacterial/viral DNA, CpG DNA	Endosomal compartment	
TLR-10	Unknown		
TLR-11	Urophatogenic bacteria	Cytoplasmic membrane	

Notes: HSP, heat-shock protein; LTA, lipoteichoic acid; MMTV, mouse mammary tumor virus; Pam3CysÑ N-palmitoyl (S)-[2,3-bis(palmitoyloxy)-(2, RS)-propyl]-cysteine; PGN, peptidoglycan; poly (I:C): polyinosinic–polycytidylic acid; RSV, respiratory syncytial virus; and TLR, toll-like receptor.

Other members of the PRR not belonging to the TLR family have been described, including the scavenger receptor, CD1 molecules, mannose, lectins (β-glucan) receptors, and the phosphoglycan receptor proteins (PGRPs) family. The actions of such receptors are similar to the TLR activating the innate immune system upon the presence of molecules representing other danger signals. Particularly, the recognition of the phosphoglycan present on the bacterial wall of either Gram-positive or Gram-negative bacteria through membrane-bound proteins (PGRP) and intracellular proteins (NOD family) have recently been reported to be a key system for detecting intracellular pathogens [17].

Although most of PRR recognize membrane-bound molecules, a soluble recognizing system was also developed for the indirect recognition of PAMPs. The complement system and the complement receptor network as well as the Fc receptors are involved in the recognition of PAMPs by means of soluble molecules like complement cascade components and antibodies [18,19].

These recognition systems act as a sentinel network to sense pathogens and to activate the appropriate host defense mechanisms. Following the activation of the innate immune response, APC migrate to the secondary lymphoid organs to activate naïve T cells that differentiate into Th cells. However, the type of Th cell generated defines the type of mechanisms to be activated. Two main systems of Th cells have been described depending on the profile of cytokine production. Th1 cells secrete IFNγ, and Th2 cells produce IL-4, as major cytokines. The cytokines they secrete function as soluble mediators for the activation of different cell types and further development of defense mechanisms. Because different Th cells activate different mechanisms, the polarization of the immune response to one subset becomes a key step to induce protection against or eliminate pathogens [20]. The relevance of this polarization has been well documented in several diseases. Furthermore, some pathogens have developed efficient mechanisms to divert the immune response to the noneffective type to escape the immune system and perpetuate the infection. Several factors have been identified as modulators of the Th polarization, including cytokines and costimulatory molecules, all produced by APC. The production of IL-12 by APC during antigen presentation induces the Th1 type, and its absence promotes differentiation into Th2 [21]. Thus, APC activation status defines the activation of Th cells and conditions the polarization. Furthermore, the T-cell activation properties of the APC are conditioned by the environment recognized by the APC during danger-signal-dependent activation of the innate immune system [22,23]. Consequently, PAMPs recognition not only activates the innate system to respond against pathogens and activates the adaptative system but also defines the type of the immune response. Recently it was reported that the combinatorial and synergistical stimulation of different PRRs by corresponding PAMPs could be used to drive the polarization process [21]. Therefore, the presence of PAMPs on vaccine adjuvants appears to be highly important, and the PAMPs combination should have a great impact on better activating the immune system and manipulating the immune response and the host defense mechanisms to confer protection.

Although the use of PAMPs as activators of the immune system constitutes a promising strategy to enhance the response to vaccine antigens, its combination with an appropriate delivery system could dramatically impact and improve the adjuvant effect [11,21,24]. Recent advances of nanotechnology bring the possibility to develop such a combination. Despite nanoparticles having found several applications as delivery systems in the pharmaceutical industry, the incorporation of danger signals could extend their use in the adjuvant field. However, the specific delivery of immunologically interesting antigens is desirable in order to target relevant APC, but unspecific delivery is also applicable in order to control or increase the time of contact of the immune system with the antigen, favoring by this way the induction of the strongest immune response.

The application of the above knowledge and the nanotechnologies could be exemplified by the use of bacterial-derived nanoparticles as vaccine strategy and adjuvant development, particularly discussing the evidence obtained from the use of meningococcal B outer membrane proteoliposome as vaccine and adjuvant.

10.3   PROTEOLIPOSOME AS A POTENT ADJUVANT

Proteoliposome is a detergent-extracted outer-membrane vesicle from *Neisseria meningitidis* serogroup B bacterium (B:4:P1,19,15:L3,7,9 strain) with potent adjuvant function [25]. It is called AFPL1 (meaning adjuvant Finlay proteoliposome 1), as the first PL found by us with adjuvant activity. AFPL1 induces a preferential Th1 with CTL responses [26–28].

The immune potentiator capacity of AFPL1 is a consequence of the presence of native inserted LPS, the most potent PAMP described [27] that interacts with TLR4; proteins with porin properties (Por A and Por B) [29] that interact with TLR2; traces of bacterial DNA that interact with TLR9; phospholipids that may interact with TLR1, TLR2, and TLR6; and possibly peptidoglycan that interacts with TLR2 and Nodl1 and Nodl2.

The delivery system capacity of AFPL1 is related to its nanoparticle structure and phospholipid constitution. It has around 70 nm of diameter and its phospholipids permit the interaction with membranes. This delivery system capacity increases when proteoliposome is absorbed onto Alum.

The immune polarization capacity is related to its LPS content and possibly the presence of porins. The LPS permits the earliest and highest production of IL-12, the highest production of nitric oxide, TNFα, IFNγ, specific IgG and IgG2a (mice)/IgG1 and IgG3 (human), and its functional activities.

The potent adjuvant activity of AFPL1 is mainly caused by the combination of the immune potentiator, delivery system, and immune polarization activities in the same natural nanoparticle structure; therefore, it is not necessary to mix several compounds to prepare an adjuvant.

10.4   STATUS OF PROTEOLIPOSOME FOR A VACCINE ADJUVANT

AFPL1 has been the most used PAMP in human vaccination, as more than 60 million doses of VA-MENGOC-BC™ vaccine have been delivered from 1990 to 2007 [30–32] (Table 10.3). In addition,

TABLE 10.3

Clinical Experience with Lipopolysaccharide (LPS)-Derived Adjuvants

Adjuvant	AFPL1	VSSP	AFCo1	MPL	Ribi.529
Doses	55×10^6	3×10^3	4	273×10^3	>138
Approved	1986	1997	2006?	2005	—
Vaccine	1986	1997	2006?	2005	—
	P Neisseria	T (cancer, HIV)	P? Neisseria	P, T (Hep B, malaria, HSV2, S. pneumoniae, allergy, cancer)	P (Hep B)
DS	Alum	O/w	—	Alum, O/w	—
Other IP	—		—	QS21, CWS	
or Tyrosine	—				
License (countries)	15	3	—	NI	NI

Notes: AFPL1, proteoliposome; VSSP, very small AFPL1; AFCo1, AFPL1-derived cochleate; MPL, mono-phosphoril Lip A; Ribi.529, synthetic LPS derivate; P, prophylactic; T, therapeutic; DS, delivery system; Alum, $AL(OH)_3$; o/w, oil in water; IP, immune potentiator;? In process; NI, not information.

other proteoliposomes from different *N. meningitidis* B strains have been obtained and used in humans [33]. Nevertheless, only AFPL1 has been used in infants (first dose at 3 months of age and second 6 to 8 weeks later) after the inclusion of the vaccine in the Cuban National Immunization Schedule since 1991. Another AFPL1-derived compound is the VSSP (very small size proteolipo-some) which is also being used in humans in cancer vaccination [34,35]. The second more used PAMP is MPL™, which is applied in 273×10^3 doses [36].

The adsorption of AFPL1 onto $AL(OH)_3$ (Alum) changes the Th2 pattern induced by Alum to a Th1 pattern [26]. It is characterized by the induction of TNFα, IL-12, IFNγ, delayed-type hypersensitivity, and specific IgG subclasses related to complement fixing and opsonophagocytic activities [26,37,38]. The adsorption of LPS derivatives onto Alum also changes its Th2 pattern [39]. Therefore, AFPL1 is currently used as adjuvant in an allergy vaccine. *Dermatophagoides siboney* mite major allergens were selected and adsorbed onto Alum [40]. This allergy vaccine candidate has concluded its preclinical studies. The change from a Th2 allergen pattern into a Th1 one was demonstrated [40,41].

Last, we are transforming the AFPL1 into an AFCo1 (proteoliposome-derived cochleate) to obtain a microparticle [42]. This maintains the constituents of AFPL1 and its immune response pat-tern, but it is more resistant and useful for mucosal routes. Several candidate vaccines are currently being explored, including a nasal *Neisseria* vaccine [43].

10.5 CONCLUSIONS

The AFPL1 nanoparticle is a potent adjuvant and is used successfully as a platform of adjuvant at Finlay Institute. Its main success is the presence of different synergistically acting PAMPs in its structure that reduce the toxicity of LPS, but maintain its potent activity to induce the innate immu-nity and consequently to bridge the adaptative response with polarization to Th1 response with the induction of CTL activity; but keeping a very low toxicity profile.

ACKNOWLEDGMENT

The authors are indebted to E.M. Fajardo for her assistance.

REFERENCES

1. Mosmann, T.R., et al., Two types of murine helper T cell clone. I. Definition according to profiles of lymphokine activities and secreted proteins. J Immunol, 1986. 136(7): 2348–2357.
2. Mosmann, T.R. and R.L. Coffman, TH1 and TH2 cells: Different patterns of lymphokine secretion lead to different functional properties. Annu Rev Immunol, 1989. 7: 145–173.
3. Moser, M. and K.M. Murphy, Dendritic cell regulation of TH1-TH2 development. Nat Immunol, 2000. 1(3): 199–205.
4. Fujihashi, K., et al., A revisit of mucosal IgA immunity and oral tolerance. Acta Odontol Scand, 2001. 59(5): 301–308.
5. O'Hagan, D.T. and N.M. Valiante, Recent advances in the discovery and delivery of vaccine adjuvants. Nat Rev Drug Discov, 2003. 2(9): 727–735.
6. Matzinger, P., Tolerance, danger, and the extended family. Annu Rev Immunol, 1994. 12: 991–1045.
7. Medzhitov, R., Toll-like receptors and innate immunity. Nat Rev Immunol, 2001. 1(2): 135–145.
8. Matzinger, P., The danger model: A renewed sense of self. Science, 2002. 296(5566): 301–305.
9. Schnare, M., M. Rollinghoff, and S. Qureshi, Toll-like receptors: Sentinels of host defence against bacterial infection. Int Arch Allergy Immunol, 2006. 139(1): 75–85.
10. Majewska, M. and M. Szczepanik, [The role of toll-like receptors (TLR) in innate and adaptive immune responses and their function in immune response regulation]. Postepy Hig Med Dosw (Online), 2006. 60: 52–63.
11. Gallucci, S., M. Lolkema, and P. Matzinger, Natural adjuvants: Endogenous activators of dendritic cells. Nat Med, 1999. 5(11): 1249–1255.
12. Ushio, H., [Toll-like receptor (TLR)]. Nippon Rinsho, 2005. 63 Suppl 5: 15–20.
13. Kawai, T. and S. Akira, TLR signaling. Cell Death Differ, 2006. 13(5): 816–825.
14. O'Neill, L.A., How toll-like receptors signal: What we know and what we don't know. Curr Opin Immunol, 2006. 18(1): 3–9.
15. Matzinger, P., An innate sense of danger. Ann N Y Acad Sci, 2002. 961: 341–342.
16. Hemmi, H. and S. Akira, TLR signalling and the function of dendritic cells. Chem Immunol Allergy, 2005. 86: 120–135.
17. Cario, E., Bacterial interactions with cells of the intestinal mucosa: Toll-like receptors and NOD2. Gut, 2005. 54(8): 1182–1193.
18. Hawlisch, H. and J. Kohl, Complement and toll-like receptors: Key regulators of adaptive immune responses. Mol Immunol, 2006. 43(1–2): 13–21.
19. Kohl, J., The role of complement in danger sensing and transmission. Immunol Res, 2006. 34(2): 157–176.
20. Guerder, S. and P. Matzinger, Activation versus tolerance: A decision made by T helper cells. Cold Spring Harb Symp Quant Biol, 1989. 54 Pt 2: 799–805.
21. Napolitani, G., et al., Selected toll-like receptor agonist combinations synergistically trigger a T helper type 1-polarizing program in dendritic cells. Nat Immunol, 2005. 6(8): 769–776.
22. Schiller, M., et al., Immune response modifiers—Mode of action. Exp Dermatol, 2006. 15(5): 331–341.
23. Corthay, A., A three-cell model for activation of naive T helper cells. Scand J Immunol, 2006. 64(2): 93–96.
24. Pandey, S. and D.K. Agrawal, Immunobiology of toll-like receptors: Emerging trends. Immunol Cell Biol, 2006. 84(4): 333–341.
25. Pérez, O., et al., Método de obtención de estructuras cocleares. Composiciones vacunales y adyuvantes basados en estructuras cocleares y sus intermediarios. 2002: Cuba.
26. Perez, O., et al., Immune response induction and new effector mechanisms possibly involved in protection conferred by the Cuban anti-meningococcal BC vaccine. Infect Immun, 2001. 69(7): 4502–4508.
27. Rodriguez, T., et al., Interactions of proteoliposomes from serogroup B Neisseria meningitidis with bone marrow-derived dendritic cells and macrophages: Adjuvant effects and antigen delivery. Vaccine, 2005. 23(10): 1312–1321.
28. Pérez, O., et al., Proteoliposomas y sus derivados como adyuvantes inductores de respuesta citotóxica y las Formulaciones resultantes. 2003: Cuba.
29. Massari, P., et al., Cutting edge: Immune stimulation by neisserial porins is toll-like receptor 2 and MyD88 dependent. J Immunol, 2002. 168(4): 1533–1537.
30. Sierra VG, C.H.C., Preclinical and clinical studies with the antimeningococcal Vaccine BC: VA-MEN-GOC-BC. Rev. Interferón y Biotecnología., 1990. Special vol.

31. Sierra GV, C.C., L. García, F. Sotolongo, L. Izquierdo, M. Valcárcel, V. Casanueva, M. Baró, F. Leguen, R. Rodríguez, and H. Terry, Efficacy evaluation of the Cuban vaccine VA-MENGOC-BC against disease caused by serogroup B Neisseria menigitidis. Proceedings of the 7th International Pathogenic Neisseria Conference. Walter de Gruyter, Berlin, Germany, 1990: 129–134.

32. Sierra VG, C.C., N.M. Valcárcel, L. García, L. Izquierdo, F. Sotolongo, V. Casanueva, C.O. Rico, C.R. Rodríguez, and H. Terry, Vaccine against group B Neisseria meningitidis: Protection trial and mass vaccination results in Cuba. NIPH Ann., 1991. 14: 195–210.

33. Fredriksen, J.H., et al., Production, characterization and control of MenB-vaccine "Folkehelsa": An outer membrane vesicle vaccine against group B meningococcal disease. NIPH Ann, 1991. 14(2): 67–79; discussion 79–80.

34. Mesa, C., et al., Very small size proteoliposomes derived from Neisseria meningitidis: An effective adjuvant for dendritic cell activation. Vaccine, 2006. 24 Suppl 2: S2–42–43.

35. Mesa, C., J. de Leon, and L.E. Fernandez, Very small size proteoliposomes derived from Neisseria meningitidis: An effective adjuvant for generation of CTL responses to peptide and protein antigens. Vaccine, 2006. 24(14): 2692–2699.

36. Persing, D., et al., Toll-like receptor 4 agonists as vaccine adjuvants, in Immunopotentiators in Modern Vaccines, V. Schijns, E.J.C. and D.T. O'Hagan, Editors. Elsevier Academic, San Diego, 2006: 93–107.

37. Lapinet, J.A., et al., Gene expression and production of tumor necrosis factor alpha, interleukin-1β (IL-1β), IL-8, macrophage inflammatory protein 1α (MIP-1α), MIP-1β, and gamma interferon-inducible protein 10 by human neutrophils stimulated with group B meningococcal outer membrane vesicles. Infect Immun, 2000. 68(12): 6917–6923.

38. Pérez, O., J. Lapinet, A. Pérez, M. Díaz, C. Zayas, A. Batista, Y. Quintero, F. Aguiar, R. Sánchez, and G. Sierra, Long-lasting cellular immune response in babies, children, and pre-teenagers vaccinated with a proteoliposome based anti-meningococcal BC vaccine. Immunología, 2001. 20(4): 177–183.

39. Vernacchio, L., et al., Effect of monophosphoryl lipid A (MPL) on T-helper cells when administered as an adjuvant with pneumocococcal-CRM197 conjugate vaccine in healthy toddlers. Vaccine, 2002. 20(31–32): 3658–3667.

40. Lastre, M., et al., Allergic vaccine composition, production method thereof and use of same in allergic treatment. 2002.

41. Lastre, M., et al., Bacterial derived proteoliposome for allergy vaccines. Vaccine, 2006. 24(Supplement 2): 34–35.

42. Perez, O., et al., Novel adjuvant based on a proteoliposome-derived cochleate structure containing native lipopolysaccharide as a pathogen-associated molecular pattern. Immunol Cell Biol, 2004. 82(6): 603–610.

43. Perez, O., et al., Proteoliposome-derived Cochleate as an immunomodulator for nasal vaccine. Vaccine, 2006. 24 Suppl 2: S2–52–53.

11 Australian BioNanotechnology for Delivery, Diagnostics, and Sensing

*Jeanette Pritchard, Michelle Critchley,
Sarah Morgan, and Bob Irving*
Nanotechnology Victoria Ltd., Victoria, Australia

CONTENTS

11.1 INTRODUCTION

Australia has a history of worldwide successes in the biological and molecular sciences, and is ideally suited to gain a strong position in bionanotechnology applications. The Victorian State Government has declared that it aims for Victoria to be in the top five biotechnology locations in the world by 2010. This has led to increased funding for biotechnology research and development not only in Victoria but on a national basis, and in turn, much attention is being directed toward what bionanotechnology can offer for human benefit.

Within Australian industry, there are several companies working in the area of bionanotechnology. Starpharma, a Melbourne-based biotechnology company, is using novel dendrimer-based platforms for medical, material, chemical, and electronic applications. Perth-based pSivida are applying their BioSilicon™ product platform in the health-care sector, with a focus on drug delivery, and Telesso Technologies Ltd. (formerly Eiffel Technologies) in Sydney uses supercritical fluid technology to reengineer pharmaceuticals and proteins, improving bioavailability and delivery. Other prominent Australian companies exploiting the benefits offered by nanotechnology include Agen Biomedical Ltd., Prima Biomed Ltd., and Universal Biosensors Pty Ltd.

The benefits of nanotechnology are not limited to any particular field of biomedicine. Areas that will be revolutionized include detection and analysis, drug delivery, and biocompatible materials in prosthetics and reconstructive surgery. Biosensors for both rapid and sensitive detection and diagnosis will also have a significant impact. Nanotechnology promises to make significant contributions to the animal and human health sector in both diagnosis and therapy. The very nature of nanotechnology dictates the need for cross-disciplinary research and development programs, and research institutions are now active in forming collaborations involving participants from the pure sciences through to engineering, and importantly end users such as clinicians. The key to this is the understanding and exploitation of the surface chemistries of the biological and nonbiological components, and novel properties resulting from their nanofeatures.

Nanotechnology Victoria Ltd. (NanoVic) is a consortium of Victorian Institutes focused on promoting nanotechnology solutions, for public benefit, by meeting the industry needs. The NanoVic BioNanotechnology team has established three vehicles for commercialization of technologies developed by its partners: Interstitial nanosystems (drug delivery), Quintain Nanosystems (diagnostics and imaging), and Envisage Nanosystems (water and environment). This chapter considers NanoVic projects, with applications of nanotechnology within Australia in relation to these three focus areas.

11.2   TRANSDERMAL DELIVERY

It is well recognized that advances in modern drugs have not been matched by equivalent advances in drug delivery systems. Many drugs based on large molecules, such as DNA, proteins, and peptides, cannot be administered orally due to their degradation in the gastrointestinal tract or poor absorption characteristics. Transdermal delivery has emerged as one of the most promising areas to overcome such limitations and represents an alternative form for the delivery of therapeutics through the skin rather than through oral, intravascular, subcutaneous, and transmucosal routes.

Transdermal delivery involves the movement of molecules across the essentially impervious outer layer of the skin, the stratum corneum. There are currently only a limited number of drugs that readily pass through the stratum corneum. This is essentially limited to small molecules (<500 Da) with a high lipophilicity, and is only clinically suitable for drugs that require a low effective dose. Materials for transdermal delivery are generally in the form of patches that are applied directly to the skin. These patches can deliver an agent directly to the lymphatic or circulatory system, or change the permeability or physical properties of the skin to aid in penetration. Currently, transdermal-patch-based delivery has been used with U.S. Food and Drug Administration (FDA) approval for more than 10 therapeutics.

Transdermal delivery has many advantages over conventional methods of drug delivery. Oral administration of drugs is often problematic as drugs may be poorly absorbed or enzymatically degraded in the gastrointestinal system. Hypodermic needles can cause pain and trauma to the recipient, as well as needlestick injuries to medical personnel. Transdermal delivery also increases bioavailability as reagents can be delivered directly to target sites. Transdermal applications also allow for the localized delivery of therapeutics which may be important for skin conditions such as psoriasis or cancers.

Various techniques have been considered for the transdermal delivery of large molecules. Many of these methods increase the permeability of the stratum corneum by disrupting its structure, creating pathways for dermally applied agents to enter the body. Methods for increasing the permeability of the skin include electroporation, iontophoresis, ultrasound, photoacoustic waves, and chemical or lipid enhancers [1,2]. An alternative approach is to create larger delivery pathways using microarray patches. Microarrays consist of arrays of numerous microprotrusions or needles, usually composed of polymer or metal substrates. When applied to the skin, the microarrays create holes in the stratum corneum. The depth of the holes is dependent on the length of the microneedles. Research has demonstrated that arrays of microneedles of 200 to 600 micron lengths can be painlessly inserted into skin [3]. This painless delivery occurs when penetration is limited to the stratum corneum and epidermis, not penetrating greatly into the dermis where blood capillaries and nerves are located.

Several microarray systems have been investigated for transdermal delivery [1,2]. One approach is to apply a patch to increase porosity and then apply a delivery agent directly to the skin for absorption. Another is to load the delivery agent directly onto or within the microprotrusions on the arrays and apply to the skin [4] (see Figure 11.1). Transdermal devices can also be tailored to enable a shock dose or slow release with time.

Transdermal microarrays are manufactured using various microtechnology and nanotechnology fabrication techniques [5]. Typical methods include mold casting with biocompatible polymers and etching of metal substrates. The microarray surface can be further functionalized or structured to allow the covalent binding of delivery agents. Additionally, porous poly-

FIGURE 11.1 Microarray patches for transdermal delivery. (Image courtesy of NanoVic, Australia.)

mers or patches manufactured with a tailored porosity can be used for slow release, with the agent diffusing from the patch over time. Alternatively, the microarrays can be composed of hollow needles, allowing opportunities for controlled delivery or even the sampling of biological fluids. The patch can be further engineered to potentially enable triggered delivery on demand on application of an external stimulus.

The biologicals or pharmaceuticals to be delivered can be nanostructured using various methods including supercritical fluid technologies and mechanical milling (see Figure 11.2). The delivery of nanostructured agents has many advantages by significantly increasing the surface area and enhancing solubility, which when used in combination with transdermal technologies, can potentially minimize the required dose.

NanoVic together with Monash University Victorian College of Pharmacy, Eiffel Technologies Ltd., and MiniFAB (Aust) Pty Ltd., have developed and applied nanotechnology solutions for the transdermal delivery of vaccines, peptide hormones, and other drugs through the stratum corneum of the skin using microarray patches (see Figure 11.3). The project will evaluate and demonstrate the delivery of large peptides, proteins, drugs, and local wound-healing agents for both human and animal health. Current work by this group is investigating the delivery of nanostructured drugs and therapeutics including insulin, ovalbumin, erythropoietin, and novel vaccines.

FIGURE 11.2 Scanning electron microscopy (SEM) image of nanostructured insulin produced using supercritical fluid methods. (Image courtesy of Eiffel Technologies, Australia.)

000013   15 kV   X50.0   0.60 mm

FIGURE 11.3 Penetration of hairless mice skin by a microarray patch. (Image courtesy of Monash University Victorian College of Pharmacy, Australia.)

11.3 DIAGNOSTICS AND SENSING

11.3.1 Nanocomposite Biosensors

Australia faces many challenges in protecting its natural resources, with water being a major focus. The current issues are complex and include increasing salinity levels, nutrient enrichment, and contamination with oils, metals, and other chemicals. Australia's natural resource-based industries, water scarcity challenges, experience working across a diverse range of environmental conditions and industries, outstanding science and research base, and proximity to fast-growing Asian markets make it the perfect place for investment in environmental nanotechnology [6]. It is believed that bionanotechnology can provide important opportunities for the Australian and international water industry to combat some of these issues.

Traditionally, water analysis is based upon wet-chemistry techniques, often determined spectrophotometrically which can be time-consuming and laborious. The techniques often have limited sensitivity and reliability for determining low concentrations. In addition, the collection of samples and transportation to the laboratory can lead to matrix changes or contamination, leading to inaccurate results.

Nanostructured detection surfaces have advantages in sensors to produce rapid response times, real-time measurements, portability for on-site analysis, multiparameter instrumentation, simple design, ease of use, low detection levels, and large working ranges [7,8]. The conducting-polymer–based probes under development by NanoVic and Monash University are based upon proprietary technology. These include bionanofeatured sensing surfaces that are capable of detecting species such as nitrates in environmental samples and sulfites in food and beverage samples.

Nitrate is essential for plant growth and is often added to soil to improve productivity. Water permeating through soil after rainfall or irrigation carries dissolved nitrate with it to groundwater. Bacteria that live in the digestive tracts of newborn babies convert nitrate to nitrite, which in turn reacts with hemoglobin to form methemoglobin, leading to oxygen deficiency in infants under 6 months old. The resulting condition is referred to as methemoglobinemia, commonly called *blue baby syndrome* [9].

Sulfite is a preservative widely used in dried fruits, used to prevent or reduce discoloration and to inhibit the growth of microorganisms in wine [10]. The FDA estimates that 1% of the population is sulfite sensitive, and of those, 5% also suffer from asthma. Symptoms include difficulty in

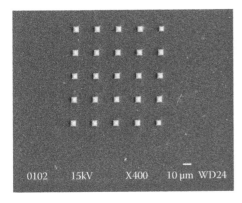 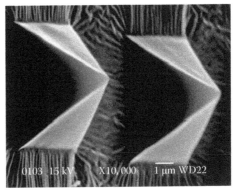

FIGURE 11.4 Microstructured pillars to support diagnostic beads in Diagnostic NanoArrays. (Images courtesy of Swinburne University of Technology, Australia.)

breathing, chest tightness, nausea, hives, or even anaphylactic shock. The Australia New Zealand Food Standards Code states that a mandatory declaration must accompany foods that contain sulfite in concentrations greater than 10mg kg^{-1} (10 ppm).

11.3.2 NANOARRAYS

Nanoarray systems offer a fast, convenient, and accurate method for detecting specific chemical or biological entities. This can include infectious or toxic molecules such as protein markers. In the pharmaceutical industry, applications or nanoarray technologies include drug discovery and DNA analysis and diagnostics [11,12].

Nanoarrays are fabricated via chemical immobilization of a specific library of chemical compounds in the form of diagnostic nanobeads onto microstructured pillars (see Figure 11.4). Due to this unique method of manufacture, a high signal:noise ratio is attainable, with no requirement for sample preparation and high sample throughput, resulting in a low-cost diagnostic system.

One of the many applications of nanoarrays is for the detection of *Salmonellae*. *Salmonella* bacteria are a major cause of food poisoning in Australia and around the world, with approximately 90% of food poisoning cases in Australia and the United States each year being attributed to it. *Salmonella* exists in many different strains or serotypes, each having characteristic antigens. Infection causes diarrhea, abdominal pain, and vomiting and can be more severe with the debilitated, very young, and the elderly [13].

A second application of this technology is for the diagnosis of bovine mastitis, which is caused by microorganisms including bacteria such as *Streptococcus uberis, Streptococcus agalactiae, Staphylococcus aureus*, and *Escherichia coli*. Milk output from infected animals can be reduced by 10% to 26% and mildly infected animals may be unable to produce milk for up to 5 days, leading to substantial income losses for producers. Producers also incur related costs associated with treating the affected animals. Mastitis lowers farm profitability, reduces product quality and quantity, damages exports, and impairs the image and use of milk and milk products in human consumption.

Current methods of detection for both of these applications are based on assays, which require culturing of the sample for a period of 2 to 4 days and multiple tests to identify each different serotype or microorganism. To overcome these problems, producers use antibiotics to keep bacterial levels down, thus creating the problem of antibiotic contamination. The Diagnostic Nanoarray system under development by NanoVic, Swinburne University, and Monash University offers the advantage of multiparameter detection, therefore increasing throughput and reducing costs. Testing is anticipated to produce a more conclusive result within a shorter time period with no requirement for sample preparation.

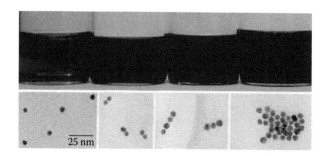

FIGURE 11.5 Chromatic response and evolving gold nanoparticle dimensionality in lysozyme and antilyso-zyme system. (Image courtesy of Professor David Mainwaring, RMIT University, Australia.)

11.3.3 NANOPARTICULATE DIAGNOSTICS

The market for rapid, specific, and deployable diagnostics is rapidly expanding as community demands for health care grow. Selective detection on either nanoparticle or planar platforms offers a significant cost-effective opportunity to develop products for the diagnosis of both pathogenic and genetic diseases.

Gold nanoparticles have been extensively used in immunogold staining methods for microscopy imaging. Nanoparticles can be chemically synthesized using citrate reduction methods, typically producing characteristically red-colored particles in solution with control over size distribution [14]. The aggregation of gold nanoparticles can be detected by monitoring color changes, mass changes, or changes (see Figure 11.5) in magnetism by the use of gold-coated magnetic particles. These systems can be applied for the detection of biological reactions [15,16].

Nanoparticles can be directly coated with biologicals including DNA, PNA, proteins, and Fab fragments using conventional conjugation techniques. Conjugation techniques for biologicals can be noncovalent, based on electrostatic interactions, as well as covalent which are generally based on thiol binding [17]. Upon conjugation of the immobilized biomolecules to other biomolecules, the gold-containing nanoparticles aggregate and a colorimetric, magnetic, or mass change is observed. Those at the Functional Nanosystems Laboratory at NanoVic and The Royal Melbourne Institute of Technology (RMIT) are utilizing these technologies to yield a rapid diagnostic test for Meningococcal disease.

Meningococcal disease is a severe bacterial infection which commonly occurs in children and can cause death within 12 hours of infection [18]. Current diagnosis involves molecular or culture techniques, with test results taking between 2 and 48 hours. Meningococcal disease has a high mortality in children, particularly if diagnosis and antibiotic treatment are delayed or missed [19]. Early recognition is often unreliable because clinical features are often absent or not readily apparent. By applying this technology, nanoparticles can be used to provide a rapid immunodiagnostic agent for the early diagnosis of this disease (see Figure 11.6).

11.4 TARGETED MOLECULAR IMAGING

While medical science continues to make significant progress in the treatment of most major diseases and ailments, early diagnosis of cancer and cardiovascular and neurodegenerative diseases remains the single most important contributor to therapies. The ability to detect cancerous growths in the body before the cells multiply and spread is critical to effective intervention and prevention. Current noninvasive *in vivo* imaging techniques often rely on the oral or intravenous administration of signal amplification agents, which help to produce more detailed and distinctive images of diseased tissue. However early-stage disease diagnosis is often ineffective, due to issues such as the following:

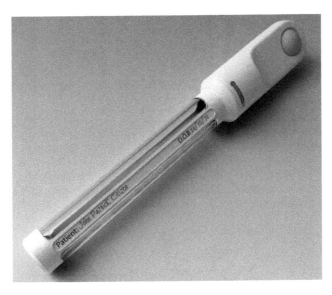

FIGURE 11.6 Point-of-care device format for the diagnosis of Meningococcal disease. (Image courtesy of NanoVic.) **(See color insert following page 112.)**

1. Low sensitivity of many agents causes difficulties in detecting small amounts of diseased tissue.
2. The agents can rapidly degrade, leading to the requirement for more regular and higher dose administration.
3. Poor specificity of agents toward target cells can result in delivery to healthy tissues; this may lead to toxicity problems.

The use of nanoparticles as novel amplification agents is believed to offer many advantages. The ability to functionalize polymer-coated nanoshells, dendrimers, and gold nanospheres could enable specific, site-targeted delivery of agents and drugs. Due to their inherently small size, uptake of these materials into cells of interest is facilitated, leading to lower dose requirements and also significant signal amplification compared to currently used materials. These advantages have already been successfully demonstrated in animal models [19,20], and there is extensive materials development being undertaken internationally to enable the use of these technologies for human benefit. It is also believed that this diagnostic application can be combined with therapeutics by using nanoparticles as a vehicle for delivery of both contrast agents and therapies simultaneously [21,22].

One focus area for NanoVic lies in the field of cardiovascular disease. Vascular disease is one of the leading causes of mortality and morbidity in the Western world. The underlying pathology is atherosclerosis—a progressive chronic inflammatory disease, affecting blood vessels (see Figure 11.7). Numerous risk factors like hypercholesterolemia, smoking, and genetic predisposition are thought to be responsible. These risk factors tend to cause inflammatory plaques within vessel walls. Not uncommonly, the plaques rupture, and depending on the artery involved, a heart attack or stroke can result. Unfortunately, approximately half of these patients do not have any warning symptoms prior to the event.

It is envisaged that nanoparticles that can be conjugated with specific antibodies (see Figure 11.8) will be used as targeted contrast agents and in conjunction with techniques such as magnetic resonance imaging (MRI) and ultrasound scans (USS) it will be possible to identify the formation of inflammatory plaques at an early stage of development. This forms the basis of a project being undertaken by NanoVic, The University of Melbourne, and The Baker Heart Research Institute. The

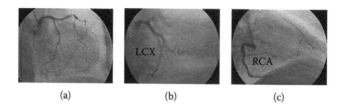

(a)   (b)   (c)

FIGURE 11.7 Conventional angiogram that utilizes an Iodine-based dye injected using a catheter into the coronary artery under X-ray guidance indicating late-stage disease progression. The angiogram shows occluded left anterior descending coronary artery (LAD) consisting of (a) LAD on the right and left circumflex artery (LCx) on the left; (b) LAD blocked, LCx patent; (c) right coronary artery (blood is redirected toward the LAD territory via collaterals due to the occluded LAD). (Images courtesy of Dr. Rajesh Nair, Baker Heart Research Institute, Australia.)

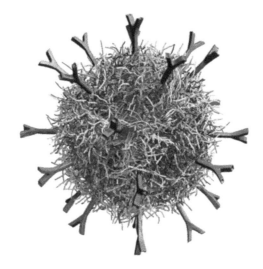

FIGURE 11.8 Graphic representation of targeted, porous nanoparticle. (Courtesy of Centre for Nanoscale Science and Technology, The University of Melbourne, Australia.) It is expected that molecular imaging contrast reagents based on nanoparticle structures will enable early detection of many human diseases, long before onset of symptoms.

use of biodegradable, nontoxic particles for antibody conjugation for targeted molecular imaging will enable direct translation to the clinical setting.

ACKNOWLEDGMENTS

The authors wish to thank the member institutes of NanoVic for making available information prior to publication; Colin Pouton and Paul White, Milton Hearne, Samuel Adeloju (Monash University); David Mainwaring and John Fecondo (RMIT University); Swinburne University of Technology. Microarray patches were produced by MiniFAB and nanostructured drugs by Eiffel Technologies. Thanks also go to Frank Caruso and Angus Johnston (The University of Melbourne) and Rajesh Nair (Baker Heart Research Institute) for provision of images and information.

REFERENCES

1. Nanda A, Nanda S, and Ghilzai N. (2006) Current developments using emerging transdermal technologies in physical enhancement methods. *Current Drug Delivery.* 3: 233–242.

2. Schuetz YB, Naik A, Guy RH, and Kalia YN. (2005) Emerging strategies for the transdermal delivery of peptide and protein drugs. *Expert Opinion in Drug Delivery*. 2: 533–548.

3. Kaushik S, Hord AH, Denson D, McAllister DV, Smitra S, Allen MG, and Prausnitz MR. (2001) Lack of pain associated with microfabricated microneedles. *Anesthetics and Analgesics*. 92: 502–504.

4. Matriano JA, Cormier M, Johnson J, Young WA, Buttery M, Nyam K, and Daddona PE. (2002) Macroflux microprojection array patch technology: A new and efficient approach for intracutaneous immunization. *Pharmacy Research*. 19(1): 63–70.

5. Henry S, McAllister DV, Allen MG, and Prausnitz MR. (1998) Microfabricated microneedles: A novel approach to transdermal drug delivery. *Journal of Pharmacy Science*. 8: 922–925.

6. *Invest Australia, www.investaustralia.gov.au.*

7. Adeloju SB and Wallace GG (1996) Conducting polymers and the bioanalytical sciences: New tools for biomolecular communications—A review. *Analyst*. 121: 699–703.

8. Karube I and Nomura Y (2000) Enzyme sensors for environmental analysis. *Journal of Molecular Catalysis B: Enzymatic*. 10: 177–181.

9. World Health Organisation: Geneva, 2004. *Guidelines for Drinking-Water Quality: Recommendations* Ed.3 Vol. 1: 418–443.

10. Zhao M, Hibbert B, and Gooding JJ. (2006) Determination of sulfite in beer samples using an amperometric fill and flow channel biosensor employing sulfite oxidase. *Analytica Chimica Acta*. 556: 195–200.

11. Dietrich HRC et al. 2004. Nanoarrays: A method for performing enzymatic assays. *Analytical Chemistry*. 76: 4112–4117.

12. Demers L and della Cioppa G. (2003) Dip pen nanolithography as a next generation massively parallel nanoarray platform: A tutorial. *Genetic Engineering News*. 23(15) www.genengnews.com.

13. NSW Food Authority Fact Sheet. 13 July 2006. *Salmonella*. www.foodauthority.nsw.gov.au.

14. Hughes D. (2005) Preparation of colloidal gold probes. *Methods in Molecular Biology*. 295: 155–172.

15. Schofield CL, Haines AH, Field RA, and Russell DA. (2006) Silver and gold glyconanoparticles for colorimetric bioassays. *Langmuir*. 22: 6707–6711.

16. Elghanian R, Storhoff JJ, Mucic RC, Letsinger RL, and Mirkin CA. 1997. Selective colorimetric detection of polynucleotides based on the distance-dependent optical properties of gold nanoparticles. *Science*. 277: 1078–1081.

17. Hermanson GT. (1996). *Bioconjugate Techniques*. Academic Press, San Diego, CA.

18. O'Brien JA, Caro JJ, and Getsios D. (2006) Managing meningococcal disease in the United States: Hospital case characteristics and costs by age. *Value Health*. 9: 236–243.

19. Michalet X. et al. (2005) Quantum dots for live cells, *in vivo* imaging and diagnostics. www.sciencemag.org. 307: 538–544.

20. Hainfeld JF et al. (2006) Gold nanoparticles: A new X-ray contrast agent. *The British Journal of Radiology*. 79: 248–253.

21. Mornet S et al. (2006) Magnetic nanoparticle design for medical applications. *Progress in Solid State Chemistry*. 34: 237–247.

22. Loo C et al. (2005) Immunotargeted nanoshells for integrated cancer imaging and therapy. *NanoLetters*. 5: 709–711.

12 Development of a BioChip for Cardiac Diagnostics

Manoj Joshi, Nitin Kale, R. Lal, S. Mukherji,
and V. Ramgopal Rao
Nanoelectronics Centre and Department of Electrical Engineering,
Indian Institute of Technology—Bombay, Mumbai, India

CONTENTS

12.1 INTRODUCTION: ACUTE MYOCARDIAL INFARCTION (AMI)

Acute myocardial infarction (AMI), commonly known as heart attack, is becoming a major cause of death all over the world. The term *myocardial infarction* is derived from *myocardium* (heart muscle) and *infarction* (tissue death). The phrase "heart attack" is used to refer to heart problems other than AMI, such as unstable angina pectoris. Cardiac diseases are an important cause of mortality today. Earlier it was thought that this disease affects only the population of developed countries in Europe and the United States. However, recent statistical data reveal the prevalence of cardiac diseases among the people primarily in South Asia [1].

The underlying mechanism of a heart attack is the death of heart muscle cells due to lack of oxygen. The heart muscle cells require oxygen for metabolic processes. Oxygen is supplied to the heart by the coronary arteries. If sufficient oxygen is not supplied to the heart muscle cells, the cells die due to a process called infarction. Blood carries oxygen to all organs of the body including the heart muscle. A decrease in blood supply has the following consequences [2]:

1. Heart muscle death (necrosis) occurs if blood flow stops even for a duration of 10 to 15 minutes. The part of the heart muscle that has suffered necrosis does not recover. A collagen scar, which does not have the ability to contract, forms in its place. Thus, the heart ends up permanently weaker as a pump for the remainder of the individual's life.
2. Heartbeats are initiated by electrical impulses, which are in turn initiated by the heart muscle. An injured (but live) heart muscle will initiate impulses at a lower rate, and this would result in a slower heartbeat. The heartbeat can become so slow that the spreading impulse will be preserved long enough for the uninjured muscle to complete contraction. Now the slowed electrical signal, still traveling within the injured area, can reenter and trigger the healthy muscle (termed reentry) to beat again. Thus, the healthy heart muscle does not relax, and does not receive return blood from the veins. If this reentry process results in a sustained heartbeat above 200 beats per minute, then the rapid heart rate effectively stops the heart from pumping blood. Heart output and blood pressure fall to near zero and the individual quickly dies. This is the most common mechanism of sudden death that can result from AMI. The cardiac defibrillator device was specifically designed to stop these too rapid heart rates.

Acute myocardial rupture of a small degree commonly occurs 3 to 5 days after AMI. It may also occur 1 day to 3 weeks later in as many as 10% of all AMI patients. Rupture may occur in the free walls of the ventricles because of increased pressure against the weakened walls of the heart chamber, due to a heart muscle that cannot pump blood out as effectively. Rupture of the walls results in pericardial tamponade (compression of the heart by blood pooling in the pericardium, that is, the heart sac) or sudden death.

12.1.1 History

Before the development of the electrocardiogram (ECG), it was impossible to objectively diagnose AMI. The term *angina pectoris*, to describe chest pain, was coined by William Heberden in 1772. However, little was known about the disease mechanism. As a disease entity, AMI was described by James Herrick in a 1912 article [3]. He is credited as the originator of the "thrombogenic theory"—the theory that AMI is due to thrombosis in the coronary artery. Subsequently, atherosclerosis and plaque rupture were discovered as underlying mechanisms. A major breakthrough in the identification of risk factors was the British doctors' study in 1956, which showed an increased risk of AMI in heavy smokers.

12.1.2 Symptoms

AMI is usually characterized by varying degrees of chest pain, discomfort, sweating, weakness, nausea, vomiting, arrhythmias, and loss of consciousness. Chest pain is the most common symptom of AMI and is often described as tightness, pressure, or a squeezing sensation. Pain may radiate to the jaw, back, epigastrium, and most often to the left arm or neck. Chest pain is more likely caused by AMI when it lasts for more than 20 minutes. The patient may complain of shortness of breath if the decrease in acute myocardial contractility, due to the infarct, is sufficient to cause left ventricular failure with pulmonary congestion or even pulmonary edema.

The symptomatology in women may be somewhat different from that in men [4]. The most common acute symptoms of AMI in women may include dyspnea, weakness, and fatigue. Thus, in women, chest pain may be less predictive of coronary ischemia than in men. Approximately one third of all AMI are silent, without chest pain or other symptoms [5]. This happens more often in elderly patients and patients with diabetes mellitus [6]. Such patients do complain of typical symptoms like fatigue, syncope, and weakness. Approximately half of all AMI patients have experienced warning symptoms such as angina pectoris prior to the infarction.

12.1.3 Diagnostic Criteria

The World Health Organization (WHO) criteria have classically been used to diagnose AMI [7]. A patient is diagnosed with AMI if two (probable) or three (definite) of the following criteria are satisfied:

1. Clinical history of ischemic-type chest pain lasting for more than 20 minutes.
2. Changes in serial ECG tracings.
3. Rise and fall of serum cardiac biochemical markers such as creatine kinase, troponin I, and lactate dehydrogenase isozymes specific for the heart.

There is a considerable variation in the pattern of presentation of AMI with respect to these three criteria. As discussed in the earlier section, one third of patients with AMI do not have the classical chest pain symptom. ECG is the most commonly used tool for diagnosis of AMI diseases. In a cardiac patient's ECG, ST segment elevation and Q-wave are the two factors that are strongly indicative of AMI [8]. However, these indications are seen only in half of the AMI cases. So physicians cannot rely only on the ECG. The third criterion is more reliable as compared to the first two. Today, clinicians rely more on the quantitative measurement of chemical markers released during AMI [9,10]. The WHO criteria were refined in the year 2000 to give more prominence to cardiac biomarkers [11]. According to their new guidelines, a cardiac troponin rise accompanied by either typical symptoms, pathological Q waves, ST elevation or depression, or coronary intervention are all diagnostic of AMI.

12.1.4 Early Detection

After onset of AMI, cardiac tissues release some specific cardiac chemical markers [9] which leads to a rise in their concentration in the blood. After quantifying their concentrations in the blood serum, clinicians can predict the size of infarct and can plan the treatment to be given to the patient. At present there are a number of immunoassays available for the detection of AMI. Conventional pathological investigations such as ELISA (enzyme-linked immunosorbent assay) [10] used for the detection of AMI are time consuming and expensive. To reduce the risk for the cardiac patients, early detection of AMI is absolutely essential, and this is also a challenge for physicians.

Research communities from various fields like very large-scale integrated circuits (VLSI), physics, chemistry, biology, material science, and so on, have come together to develop sophisticated, reliable, and low-cost biosensing devices. Microfabricated biosensors used as point of care assay systems are becoming increasingly popular. Usage of microelectromechanical system (MEMS)–based biosensors is advantageous as they offer a faster response, high reliability, compactness, and also are inexpensive. In such biosensors, the volume of the analyte required for the detection of biological species is also small. Advances in the field of VLSI technology enable us to fabricate a biochip that includes a sensing element embedded within a microfludic system integrated with an electrical readout.

12.2   CHEMICAL MARKERS RELEASED AFTER ONSET OF AMI

Chemical markers are basically macromolecules present in sarcolemmal membrane of myocite. Normally they are not present in the blood. Due to an injury to myocite membrane, they begin to diffuse into the cardiac interstitial and ultimately into microvascular and lymphatic tissues in the region of the infarct [10]. Figure 12.1 shows the various cardiac markers released in the blood after the onset of AMI. Each of these markers attains its peak at a different time after the onset of AMI. Considering this aspect, troponin I, myoglobin, creatine kinase, and its isoforms are commonly quantified in the blood for diagnosis of AMI.

12.2.1   Myoglobin

Myoglobin is a specific cardiac dysfunction marker. It is the first marker to be released in the blood after the onset of AMI. It is a low-molecular-weight (17.8 kD) heme protein present in the myocardial cells. The function of myoglobin is to buffer oxygen concentration in the respiring tissue. Myoglobin has a greater affinity for oxygen than hemoglobin (Hb) [5].

When hemoglobin–oxygen complex reaches the cardiac tissue, oxygen gets transferred to myoglobin. Myoglobin is composed of eight helices, A to H, that are packed in a pattern as shown in Figure 12.2. Heme is held between the E and F helices with a covalent bond. When the muscle suffers an injury or hypoxia, then the tissue cells are affected by necrosis, and myoglobin is released into the bloodstream. It can be detected within 4 to 30 hours after the onset of AMI. After that time, its concentration becomes very low. The presence of myoglobin in the blood may not always be cardiac specific. It is also present in skeletal muscles. However, its specificity to cardiac muscle injury can be increased by measuring it along with carbonic anhydrase (CA-3), which is another cardiac marker. Carbonic anhydrase is mainly present in the skeletal muscles along with myoglobin. In cardiac muscles it is present in trace quantity. Hence, when there is injury to cardiac muscles, only myoglobin is released in the blood [11].

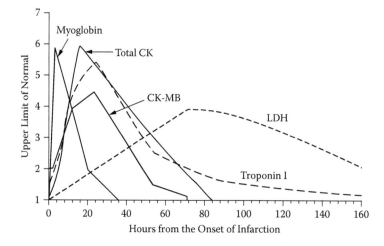

FIGURE 12.1   Plot of the appearance of cardiac markers in the blood versus time after onset of the acute myocardial infarction symptoms. (From Collinson, P.O., Boa, F.G. et al., *Ann Clin Biochem*, 38, 423–449, 2001. With permission.)

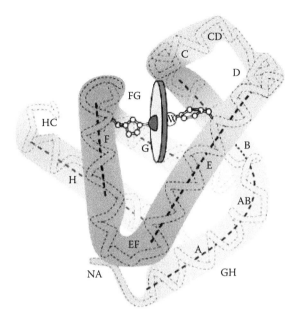

FIGURE 12.2 Structure of cardiac myoglobin shows eight helices A to H of myoglobin. Heme is held between E and F helices. Iron (Fe) ion is at the center of heme where oxygen ion combines. (From Collinson, P.O., Boa, F.G. et al., *Ann Clin Biochem*, 38, 423–449, 2001. With permission.)

The normal range of carbonic anhydrase in blood is 13 to 29 µg/L. A serum concentration of myoglobin greater than 110 µg/L, together with ratio of myoglobin to carbonic anhydrase greater than 3.21 is considered abnormal or indicative of AMI.

12.2.2 CREATINE KINASE (CK) AND ITS ISOFORMS

Creatine kinase (CK) is basically an enzyme. Its serum activity exceeds the normal range within 4 to 8 hours after the onset of the AMI, and declines to normal after 2 to 3 days. CK is also present in the skeletal muscles and maintains ATP level in tissues.

Although elevation of the serum CK is sensitive to enzymatic detector of AMI, it gives false-positive results in patients with muscle disease, alcohol intoxication, and diabetes mellitus. So to make CK more specific to cardiac muscles, three isoforms of CK (MM, BB, and MB) have been identified as biochemical markers of cardiac dysfunction. Only the MB isoenzyme is specific to cardiac muscles [13].

There are two methods for detecting CK in serum: the activity assay method and the mass assay method. Activity assay includes radioimmunoassay. In the mass assay, CK is detected by highly sensitive and specific monoclonal antibodies. In the activity assay, the result is reported in units per milliliter; in the mass assay, the result is reported in nanograms per milliliter. If the ratio of CK-MB mass/CK-activity is about 2.5, then the patient may be suffering from myocardial infarction [13]. However, it is important that the clinician not rely on measurement at a single point of time, but should measure the entire rise and fall trend of CK and CK-MB in a specified period. Even though creatine kinase (CK)-MB isoenzyme is the most widely used marker today for the diagnosis of AMI, it has a limited diagnostic window (i.e., time interval in hours wherein abnormal results are expected after a heart attack) and limited clinical specificity (i.e., the ability of the test to differentiate between cardiac diseases and other conditions).

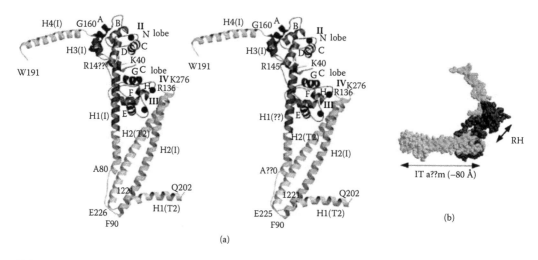

FIGURE 12.3 Stereo view of crystal structure of troponin core domain: (a) TnC and TnT are colored in red and yellow, respectively. TnI is colored in cyan, except for the two stretches of amphiphilic helices (TnC-binding sites), which are dark blue. (b) A space-filling model of the Troponin molecule. RH, regulatory head. (From Takeda, S., Yamashita, A. et al., *Nature*, 424, 35–41, 2003. With permission.) **(See color insert following page 112.)**

12.2.3 TROPONIN

Figure 12.1 shows that troponin is an early marker of AMI. It is released at about the same time as CK-MB (6 to 9 hours); it is also a late marker as it remains elevated for over a week after the infarct. Therefore, it has been advocated that cardiac troponin can replace both CK-MB and LDH for the diagnosis of AMI [14]. Cardiac troponin assays offer clinicians a valuable tool for diagnosing myocardial infarction even at the level of a microinfarction. They also provide independent prognostic information. Aviles et al. suggest that with the advances in the design of assay, many of the technical problems that were encountered earlier appear to have been resolved. Troponin complex is a heteromeric protein (Figure 12.3) that plays an important role in the regulation of skeletal and cardiac muscle contraction. It consists of three subunits: troponin I (TnI), troponin T (TnT), and troponin C (TnC). Each subunit is responsible for a part of troponin complex function (e.g., TnI inhibits ATPase activity of acto-myosin). TnT and TnI are present in both cardiac muscles and skeletal muscles, but in different forms. Only one tissue-specific isoform of TnI is described for cardiac muscle tissue (cTnI) (i.e., cTnI is expressed only in the myocardium).

For more than a decade, the cardiac form of TnI was known as a reliable marker of cardiac tissue injury. Human cardiac troponin I is presented in cardiac tissue by a single isoform with molecular weight 29 kDa. It is considered to be more sensitive and significantly more specific in diagnosis of myocardial infarction than CK-MB, myoglobin, and LDH isoenzymes. cTnI can be detected in the patient's blood 3 to 6 hours after the onset of chest pain, and it reaches a peak level within 16 to 30 hours. cTnI is also useful for the late diagnosis of AMI, because elevated concentrations can be detected in the blood even 5 to 8 days after the onset of AMI.

12.3 WHICH IS THE EARLY MARKER FOR IDEAL DIAGNOSIS?

Early sensitivity and specificity of biochemical markers vary with time from the onset of symptoms and infarct size. Unfortunately, no direct comparison of the most promising markers for early diagnosis (CK-MB, myoglobin) is available at the moment for patients with small infarctions (i.e., those which are more promising to diagnose on the clinical and electrocardiographic basis alone).

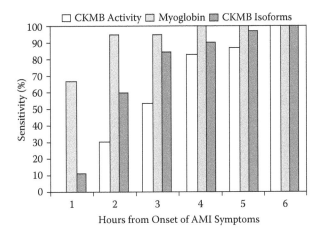

FIGURE 12.4 Sensitivity of CKMB activity, myoglobin, CKMB isoforms within 6 hours from onset of acute myocardial infarction (AMI) of the patients with ST segment elevation treated with thrombolytic therapy. (From Galvani, M., Ferrini, D. et al., *Int J Cardiol*, 65, S17–S22, 1998. With permission.)

With respect to the specificity issue, CKMB isoforms are superior to myoglobin, although it has been found that the ratio MB2/MB1 is frequently elevated in skeletal muscle injury [16]. When considering early sensitivity, Mair et al. suggested that all chemical markers (myoglobin, CKMB mass, CKMB isoforms ratios, cardiac troponin I, cardiac troponin T) useful for early diagnosis are equivalent in AMI patients [17]. Myoglobin is often used to rule out AMI at an early stage, after the onset of chest pain, for more than 10 to 15 minutes. Myoglobin is a low-molecular-weight heme protein that is abundant in cardiac and skeletal muscle, but not in smooth muscle. Thus, it is a sensitive but not specific marker of myocardial infarction. Its presence due to skeletal muscle injury needs to be excluded. Myoglobin is rapidly released from necrotic myocardium with subsequent rapid renal clearance. Decreased renal clearance can result in its elevated levels [18]. The comparison of diagnostic sensitivity of myoglobin and CKMB in the first few hours of the onset of chest pain is demonstrated in the literature [19]. As shown in Figure 12.4, both myoglobin and CKMB isoforms are more sensitive compared to the CKMB activity in the first few hours, but myoglobin was significantly more sensitive than CKMB isoforms in the first 2 hours.

As explained earlier, cardiac troponin I is expressed only in myocardium; hence, it is the most specific biochemical marker released after the onset of AMI. It is released almost at the same time as CKMB (4 to 9 hours), and it can be used as a late marker, as it remains elevated for more than a week after the infarct. Table 12.1 shows the release kinetics of cardiac troponin I and CKMB activity in patients with AMI [20].

As shown in Figure 12.5, the significant appearance of cTnI and CKMB in the blood requires 4 to 6 hours, and they attain their peak value in 12 to 13 hours. However, cTnI takes a long period (~170 hours) compared to the CKMB activity (~51 hours) to return to its normal value.

With this discussion, it can be concluded that cTnI can be used as a late marker of AMI. Quantitative analysis of cTnI also helps to investigate the size of infarct. Also, due to an early appearance of peak and higher rates of rise of myoglobin, it is used as a marker to rule out AMI. A myoglobin–CKMB combination can be used for the confirmation of AMI for the patients with chest pain and elevated ST segment of electrocardiogram.

12.4 POINT-OF-CARE SYSTEM FOR EARLY DIAGNOSIS OF AMI

Point-of-care (POC) or near-patient testing allows diagnostic assays to be performed in locations such as the intensive care unit, where treatment decisions are made, and care is delivered based on the results of these assays. Presently there exist POC immunoassays for several cardiac markers

TABLE 12.1

Release Kinetics of Cardiac Troponin I and CKMB Activity in Patients with Acute Myocardial Infarction (AMI)

Cutoff Value	cTnI (0.1 ug/L)	CKMB Activity (15U/L) (CK>100U/L)
Time to first increased plasma values (h)	4 (2–7.5)	6 (3–9.75)
Time to peak (h)	12.75 (8.15–28)	12.5 (8.75–21)
Return to normal (h)h	170 (99.5–242.2)	51.1 (26.25–74)
Magnitude of increase	119 (22–1333)	9.5 (2.4–24)
Rate of increase	1.23 (0.2–19.8)	12.5 (1.5–50.7)
Percentage (peak)	26.7 (1.2–381)	2.12 (0.2–17.9)

Notes: First appearance in the blood, time to peak values, and time of return to normal are calculated for the onset of symptoms. The magnitude of increase was calculated by dividing peak values by cutoff values. Rate (slope) of increase = (value at peak—baseline value)/(time to peak—time to baseline value). Percentage of increase = (value at peak—baseline value)/baseline value.

Source: Bertinchant, J.-P., Larue, C. et al., *Clin Biochem*, 1996; 29:587–594. (With permission.)

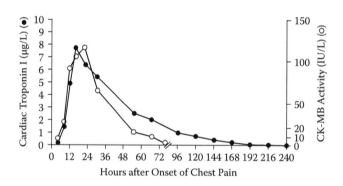

FIGURE 12.5 Cardiac troponin I and CKMB activity release profiles in patients with acute myocardial infarction. (From Bertinchant, J.-P., Larue, C. et al., *Clin Biochem*, 29, 587–594, 1996. With permission.)

including CKMB, myoglobin, troponin I, and troponin T that yield qualitative and quantitative results similar to traditional central lab assays [21]. In the evaluation of emergency room patients with chest pain, POC cardiac markers may improve clinical outcome by reducing the diagnosis time. Existing POC assays that combine myoglobin and CKMB have high sensitivity and specificity for diagnosing AMI and may provide the earliest indication of myocardial injury. POC troponin-I assays provide a better identification of myocardial injury and also provide valuable prognostic information.

Table 12.2 shows the commercially available cardiac marker POC assays with their sensing cutoff values and response time. Most of these assays can give the diagnostic results within 10 to 30 minutes. However, they have some disadvantages [21]:

1. Use of outdated and suboptimal AMI definitions that were based on WHO criteria and only CKMB activity.
2. Lack of clarity over appropriate comparison value or cutoff value, for positive test.
3. Marginal assay performance around discriminatory or diagnostic value.
4. Nonstandard frequency and number of blood collections.
5. Large variation in pretest likelihood of AMI among sample population.

TABLE 12.2

Currently Available Cardiac Marker Point-of-Care Assays

Name	Cardiac Marker	Specimen	Sens Cutoff	Testing Procedure	Time (min)	Reader
Hybritech ICON	CK-MB, (Quan)	300 μl (requires centrifugation)	<2 ng/ml	Requires manual addition of substrate and antibody & 2 wash/drain steps	15–20	Reader console
TROP T 2/ CARDIAC T rapid assay (Roche Diag./ Boerhinger Mannheim Corp)	TnT, (Qual)	150 μl EDTA or heparin whole blood	0.08 ng/ml	Single-step	15–30	Direct visual result
Cardiac STATus (spectral diagnostics)	CK-MB myoglobin, (Qual)	175 μl EDTA whole blood	5 ng/ml 100 ng/ml	Single step plus developer	15	Direct visual result
Cardiac STATus TnI rapid test (spectral diagnostics)	TnI (Qual)	200 μl heparin whole blood	1.5 ng/ml	Single step	15	Direct visual result
TRIAGE® CARDIAC PANEL (Biosite Diagnostics)	CK-MB myoglobin TnI (Quan)	150 μl heparnized whole blood	0.75 ng/ml 2.70 ng/ml 0.19 ng/ml	Single step (blood placed in test cartridge and inserted into automated analyzer)	10–15	Fluorescent detection reader
Alpha DX (first medical)	CK CK-MB myoglobin TnI (Quan)	EDTA whole blood	— — — —	Single step (blood placed in analytical discs and inserted into analyzer)	18–25	Automated analyzer
Stratus CS STAT (Dade Behring Inc)	CK-MB myoglobin TnI (Quan)	3 ml heparnized whole blood	3.5 ng/ml 98 ng/ml* 0.06 ng/ml	Single step analyzer	13–22	Fluorometric analyzer

* Male cutoff value given, female cutoff value 56 ng/ml; Qual = qualitative; Quan = quantitative.

6. Retrospective performance and evaluation of POC assay with no measure of effect on decision-making timing or quality.
7. POC assay performance and interpretation by lab technicians or health-care providers.
8. Exclusion of patients based on early or delayed presentation.
9. Multiple assay and corresponding reference ranges for the same cardiac marker (cTnI, myoglobin).
10. Failure to consider meaningful clinical or resource utilization endpoints.

12.5 MICROFABRICATED BIOSENSORS AS POINT-OF-CARE SYSTEM FOR EARLY DIAGNOSIS OF AMI

In an intensive care unit many physiological parameters are required to be monitored simultaneously and in real time. For example, in the case of AMI, specific protein biomarkers need to be monitored. Knowledge of the levels of these proteins would allow administration of life-saving treatments to patients suffering from AMI. Currently, the most reliable diagnosis of AMI is based on the detection of a temporary significant increase in level of a few proteins in the blood [22,23]. The methods for protein detection involve labeling procedures, which are time consuming [24]. Microfabricated biosensors, however, can allow label-free investigation of processes and real-time

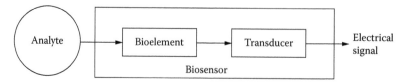

FIGURE 12.6 Schematic representation of a biosensor.

analysis, and hence can play a significant role as a POC system. Use of microfabricated sensors can make it possible to reduce the infarction exclusion time, shorten the time to commence treatment, and therefore reduce the risk for patients.

The history of biosensors started in the year 1962 with the development of enzyme electrodes by Leland C. Clark. Since then, research communities from various fields such as VLSI, physics, chemistry, biology, and material science have come together to develop sophisticated, reliable, and mature biosensing devices. The schematic representation of a biosensor is as shown in Figure 12.6.

Biosensors, like all other sensing devices, can be divided into three main components: a detector that recognizes the signal of interest, a transducer that converts the signal into a more useful output (an electrical signal), and a readout. Biosensors employ biological or biochemical detectors, which can be a single protein, an enzyme, or even whole cells and microorganisms.

Biosensors can be classified based on detector type (immunosensor or enzymatic sensors); transduction principle (amperometric, piezoelectric, or micromechanical); and application (clinical sensors or environmental sensors). Each of this class of sensors for cardiac diagnosis is discussed in the subsequent sections.

12.5.1 Affinity Cantilever-Based Biosensor for Cardiac Diagnostics

Biosensors based on microcantilevers have become a promising tool for detecting bimolecular interactions with a great accuracy. Microcantilevers translate molecular recognition of biomolecules into a nanomechanical motion. This nanomechanical motion can be detected by an optical or a piezoresistive readout detector system. Immobilization of biomolecules on either top or bottom surface of the cantilever generates differential surface stress on the opposite faces of the cantilever. This gives rise to a deflection of the cantilever as shown in the schematic of Figure 12.7.

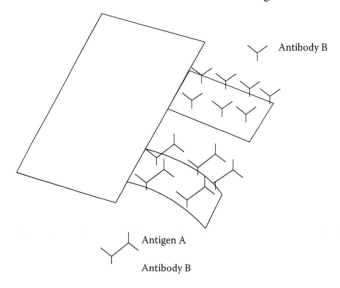

FIGURE 12.7 Cantilever bending due to biomolecular interaction between an immobilized receptor and its target. Only the specific recognition causes a change on the surface stress driving to bending of cantilever.

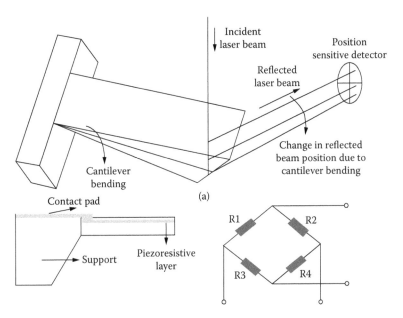

FIGURE 12.8 (a) Optical readout method for cantilever bending evaluation. (b) Piezoresistive readout and the Wheatstone bridge configuration.

Figure 12.8 shows the optical and electrical readout schemes for the detection of deflection due to receptor-target binding. One such application of affinity cantilever for the detection of cardiac markers is reported in [26]. Y. Arntz et al. demonstrated continuous label-free detection of two cardiac biomarker proteins (creatine kinase and myoglobin) using an array of microcantilevers functionalized with covalently anchored anticreatine kinase and antimyoglobin. This method allows biomarker proteins to be detected by measuring the surface stress-induced cantilever deflection. Figure 12.9 shows the cantilever array embedded in liquid cell with a micropump attachment. The deflection of the cantilever is detected optically. A reference cantilever is used along with the main cantilever to eliminate signals due to thermal drift, undesired chemical reactions, and turbulence. Figure 12.10 shows the functionalization of cantilevers and its response. The human cardiac marker specific antimyoglobin and anti-CKMB antibodies are immobilized on the top surface of the cantilever (Figure 12.10a). To achieve selective attachment of target molecules and prevent nonspecific attachment, the bottom surface of the cantilever is blocked using bovine serum albumin (BSA). For getting a better differential deflection signal, both top and bottom surfaces of the reference cantilever are blocked using BSA (Figure 12.10b). As shown in Figure 12.10c, sequential injections of cardiac creatine kinase and myoglobin proteins result

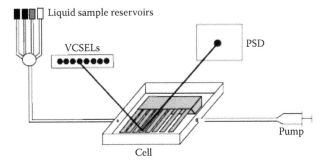

FIGURE 12.9 Experimental setup using a liquid cell with optical readout of cantilever deflections and sample liquid exchange system (VCSEL= vertical cavity surface emitting lasers, PSD = position sensitive detector).

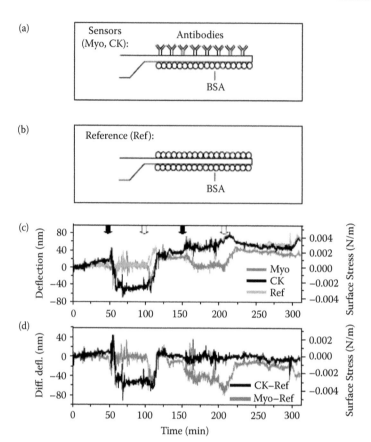

FIGURE 12.10 (a) Two cantilevers in the array are coated with myoglobin (Myo) and creatine kinase (CK) antibodies and passivated on their lower surface with BSA. (b) The reference cantilever is quenched on the two surfaces with BSA. (c) Sequential injections of creatine kinase and myoglobin proteins reveal deflection of the corresponding cantilever due to compressive stress. The experiment is performed in a background of unspecific protein (BSA 100 ug ml[-1]). (d) Differential deflection signals: CK-Ref, Myo-Ref. (From Arntz, Y., Seelig, J.D. et al., *Nanotechnology*, 14, 86–90, 2003. With permission.)

in a deflection of the corresponding cantilever. The direction of cantilever deflection indicates that the surface stress is of a compressive nature. The experiment is performed in a background of unspecific protein (BSA 100 μg ml[-1]). The differential deflection signal and the amount of surface stress generated on the cantilever due to attachments of cardiac myoglobin and CKMB are shown in Figure 12.10d. It should be noted that the stress generated on the cantilever surface due to attachments of targeted cardiac markers is of the order of 5 mN/m. The magnitude of stress is much less than the stress induced due to the other biomolecules such as DNA [27]. Due to low differential surface stress, the deflection of the cantilever is also small. Thus, highly sensitive cantilever structures are required for detecting cardiac markers. However, these structures should withstand the wet-phase immobilization steps and also the turbulence caused by fluid flow in the microfludic system. The response time of such a biosensor is about 10 to 15 minutes (Figure 12.10c). Hence, in the near future, affinity cantilever-based biosensors may be used for real-time monitoring of multiple cardiac markers and may allow an early diagnosis of AMI.

12.5.2 ELECTROCHEMICAL BIOSENSORS FOR CARDIAC DIAGNOSTICS

One of the earliest microfabricated biosensor was in the category of the electrochemical biosensor [28]. Detection of glucose concentration in the blood and hybridized DNA are some of the examples of electrochemical biosensors reported in literature [28–32]. The underlying working principle of such biosensors is that biochemical reactions either produce or consume ions or electrons, which in turn cause a change in the electrical properties of the solution. This change can be sensed and can be used as a measurement parameter. According to measurement parameters, the electrochemical biosensors are classified as [28] conductimetric, potentiometric, and amperometric.

12.5.2.1 Conductimetric

The measured parameter in conductimetric sensors is the electrical conductance/resistance of the solution. When electrochemical reactions produce ions or electrons, the overall conductivity of the solution will change. The change in conductivity is measured, and it can be calibrated to a standard scale. The applied electrical field is generated using a sinusoidal voltage (AC) which helps in minimizing undesirable effects such as Faradic currents, double-layer charging, and concentration polarization [28].

12.5.2.2 Potentiometric

The measured parameter is oxidation or reduction potential of an electrochemical reaction. The working principle relies on the fact that when a ramp voltage is applied to an electrode in the solution, a current flow occurs because of electrochemical reactions. The voltages at which these reactions occur indicate a particular reaction and identify a particular species [28].

12.5.2.3 Amperometric

These are high-sensitivity biosensors that can detect electroactive species present in biological test samples. Because the biological test samples may not be intrinsically electroactive, enzymes are needed to catalyze the production of such species. In these types of sensors, the measured parameter is current [28]. These electrochemical biosensors are compared in Table 12.3, and it can be found that the amperometric electrochemical sensor is the most sensitive among the three.

A disposable amperometric immunosensor for the rapid detection of cardiac myoglobin in whole blood is reported in the literature [33]. As described [33], the carbon working electrode is fabricated using screen printing in a three-step process. The first step involves the printing of a silver conducting path on a PVC substrate. An insulating layer is then deposited, which covers the silver layer between the carbon working area and the connecting area on top. The final layer was the carbon working area (16 mm^2). This was deposited on the top of the silver layer. The electrodes were cured at 90°C for 18 hours.

TABLE 12.3
Different Electrochemical Sensing

Characteristics	Conductimetric	Potentiometric	Amperometric
Measured parameter	Resistance	Voltage	Current
Applied voltage	Sinusoidal (AC)	Ramp voltage	Constant potential (DC)
Governing equation	Incremental resistance	Nerst equation	Cottrell equation
Fabrication	FET + enzyme	FET + enzyme	FET + enzyme
		Oxide electrode	Two electrodes
Sensitivity	Low	Medium	High

Source: Hudson, M.P., Christenson, R.H. et al., *Clinica Chimica Acta*, 1999, 248:223–237. (With permission.)

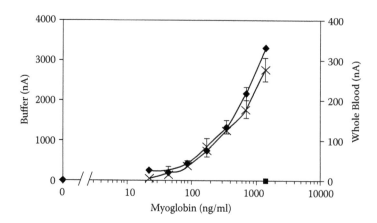

FIGURE 12.11 Indirect one-step sandwich assay with amperometric detection. Comparison of standard curve in buffer versus whole blood. Goat antihuman cardiac myoglobin (antibody A) used as primary antibody and mouse antimyoglobin (antibody B) and goat anti-mouse-IgG AP (antibody C) as detecting antibodies. (From Tina, M., Regan, O. et al., *Analytica Chimica Acta*, 460, 141–150, 2002. With permission.)

A single-step indirect sandwich assay was employed using a polyclonal goat antihuman cardiac myoglobin antibody, with monoclonal mouse antimyoglobin and goat antimouse IgG conjugated to alkaline phosphates (AP) as the detecting antibodies.

Figure 12.11 shows the change in output current with the myoglobin concentration using a buffer solution and whole blood as background during the experimentation. The final sensor requires 30 minutes for incubation, and no cross-reactivity was noted with other cardiac proteins. The overall performance of the sensor, rapid analysis time, wide working range, high precision, and specificity demonstrate its potential usefulness for an early assessment of AMI.

12.6 LAB-ON-CHIP FOR CARDIAC DIAGNOSTIC: OUR APPROACH

For the past few years, the Nanoelectronics Center at the Indian Institute of Technology Bombay has been engaged in the development of a Lab-On-Chip for cardiac diagnostics. In the proposed POC system, affinity cantilevers are used as the sensor element. We have designed, simulated, and fabricated such cantilevers using various materials such as silicon dioxide, silicon nitride, and SU-8. For electrical detection, p-type polysilicon was used as a piezoresistive layer. Mechanical and electrical properties of these cantilevers were investigated using the atomic force microscope (AFM). The top surface of these cantilevers was selectively immobilized with antibodies. Antibody-specific antigens were allowed to react with the cantilever in a liquid cell of the AFM setup, so as to detect the cantilever bending optically. For electrical detection, microcontroller-based signal conditioning and a digital readout circuit were developed.

12.6.1 MATERIAL DEVELOPMENT

Hotwire chemical vapor deposition (CVD) was chosen as the technique for depositing thin films of doped microcrystalline silicon (μSi:H) and silicon nitride. Hotwire CVD deposits films at high deposition rate. Using this technique, one can deposit thin films at a low substrate temperature, thus enabling the usage of polymers as a structural material.

Silicon nitride films were deposited at a substrate temperature of 300°C, at various gas pressures and also with or without nitrogen dilution. The films were characterized for their deposition rate, refractive index, and etch rate in buffered hydrofluoric acid (BHF). We optimized the deposition parameters for low etch rate as $SiH_4:NH_3$ flow rate ratio = 1:20; gas pressure = 50 mTorr; and no usage of nitrogen dilution [34].

Doped p-type microcrystalline silicon films were developed for achieving a high gauge factor. The deposition rates of these films were in the range of 3.6 to 4.0 A/sec. These films have a grain size of 15 to 30 nm [35]. We have achieved a gauge factor of above 20 for films deposited at a substrate temperature of 150°C.

12.6.2 DESIGN AND SIMULATION

We first show the effect of material properties on the deflection of a cantilever due to myoglobin-induced surface stresses (typically 6 mN/m). We also show the importance of locating the immobilization layer and an embedded piezoresistive layer on the same side of the neutral axis [36].

The study has been carried out using the Coventorware MEMS tool. The cantilever structure is 200 µm long and 30 µm wide. The thickness of individual layers is given in Figure 12.12. The piezoresistive layer is assumed to be single crystal silicon (Young's modulus ~170 GPa and piezoresistive coefficients as given in [36]). It is assumed that the cantilever is slightly oxidized to form an immobilization layer. From this figure, it is clear that both deflection and ΔR/R, in the case when the immobilization layer and the piezoresistive layer are placed on opposite sides of the neutral axis, are much lower compared to the case when these layers are on the same side of the neutral axis. Second, there is hardly any increase in ΔR/R as we scale down the Young's modulus of the structural layer when the immobilization layer and the piezoresistive layer are placed on the opposite side of the neutral axis. Thus in biosensing systems, it is preferable to place the immobilization layer and the piezoresistive layer always on the same side of the neutral axis.

We also observe that both deflection and ΔR/R increase as the Young's modulus decreases in the case when the immobilization layer and the piezoresistive layer are on the same side of the neutral axis. However the ΔR/R and the deflection do not scale as effectively as the Young's modulus. For example, as we reduce Young's modulus of the structural layer from 169 GPa (Silicon) to 5 GPa (polymers) (i.e., about 1/35th), the deflection and ΔR/R have increased only by a factor of about 2. This is due to the nature of surface stresses and high Young's modulus of the thin piezoresistive layer (compared to the structural layer) that is located away from the neutral axis. It has resisted deflection and the development of a scaled higher strain in the piezoresistive material.

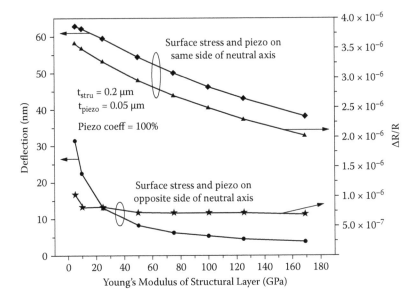

FIGURE 12.12 Deflection and change in resistance as a function of Young's modulus of the structural layer. Immobilization layer and piezoresistor are on the same side and opposite side of the neutral axis.

 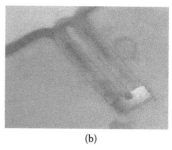

(a)   (b)

FIGURE 12.13   Affinity cantilevers to fit in to atomic force microscopy (AFM) nose. (a) Scanning electron microscope (SEM) image of silicon dioxide cantilever; (b) optical image of SU-8 cantilever with gold patch at the tip for optical detection.

 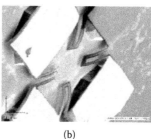

(a)   (b)

FIGURE 12.14   Scanning electron microscope (SEM) images of piezoresistive (a) silicon dioxide cantilever and (b) silicon nitride cantilever.

12.6.3   FABRICATION OF CANTILEVERS

To detect the cantilever bending optically, we decided to use the AFM setup and its liquid cell accessories. The main bottleneck of this experiment was to fabricate cantilevers that would fit in the nose of the AFM setup.

Figure 12.13 shows the pictures of the single-layer silicon dioxide and SU-8 cantilevers that can fit in the AFM nose. The surface stress developed due to cardiac markers is in the order of a few mN/m. Because the surface stress is small, cantilever structures with high sensitivity are essential to obtain a measurable deflection. One way to increase the sensitivity is to have thin single-layer cantilevers. Selective immobilization of biomolecules on either top or bottom surface of the cantilever is an important issue. We developed immobilization techniques that can selectively immobilize biomolecules on only one surface of the cantilever.

To detect the cantilever bending electrically, cantilevers with embedded piezoresistors were fabricated in our laboratory. The reference cantilever and the measurement cantilever are employed in a half-bridge configuration.

Figure 12.14a shows the silicon dioxide cantilevers with gold as the strain-sensitive material. The cantilevers are curled upward due to compressive stress. Figure 12.14b shows the silicon nitride with p-type polysilicon embedded as a piezoresistive layer. Downward curling of these cantilevers is due to the tensile stress in the films.

12.6.4   IMMOBILIZATION OF BIOMOLECULES

Antibody immobilization on cantilever surface is preceded by its surface modification (i.e., making the surface amenable to antibody immobilization). For the optical detection method, it is essential to selectively immobilize antibodies only on either the top or the bottom surface of the cantilever. A detailed study of selective immobilization is reported in [37]. For the immobilization of antibodies

(a)   (b)

FIGURE 12.15 Micrograph of SU-8 cantilever used for the optical detection with incubation of HIgG followed by fluorescein isothiocyanate (FITC) tagged goat anti-HIgG observed under (a) optical microscope and (b) fluorescent microscope.

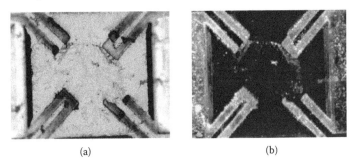

(a)   (b)

FIGURE 12.16 Micrograph of piezoresistive silicon nitride cantilevers used for the electrical detection with incubation of HIgG followed by fluorescein isothiocyanate (FITC) tagged goat anti-HIgG observed under (a) optical microscope and (b) fluorescent microscope.

on various materials, different protocols were developed and characterized [38–41]. Human immunoglobulin (HIgG) antibodies were immobilized on the top surface of the cantilevers using these immobilization protocols. This was followed by incubation with fluorescein isothiocyanate (FITC) tagged antihuman-HIgG. The cantilever surface was then observed under an optical and a fluorescent microscope. Examples of antibody immobilization on SU-8 affinity cantilever and silicon nitride cantilever are shown in Figure 12.15 and Figure 12.16, respectively.

Obtaining a large deflection due to myoglobin induced stresses is of paramount consideration. This is possible if we employ polymeric materials. One such material is SU-8, basically a negative photoresist with excellent structural and chemical properties. However, immobilization on polymers requires that their surface be modified via high-energy plasma or by the use of strong acids or bases. A novel SU-8 surface modification technique is described in the literature [40]. In this method, amino (NH_2^+) groups are grafted onto the SU-8 surface.

Cardiac markers such as myoglobin and troponin are protein molecules. They can be immobilized by the methods described in this section.

12.6.5 CANTILEVER CHARACTERIZATION

Sensitivity of the affinity cantilever depends on how it performs in the mechanical and the electrical domains. Hence, before using a cantilever for any biosensing application, it is essential to investigate its performance in these domains.

12.6.5.1 Mechanical Characterization

An AFM setup was used to find the mechanical properties of the cantilever. The force–distance spectroscopy feature of AFM was used to measure the *spring constant* of the cantilever structure.

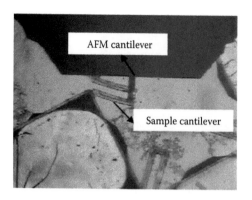

FIGURE 12.17 Atomic force microscopy (AFM) cantilever deflecting sample cantilever taken from the charge-coupled device (CCD) camera. **(See color insert.)**

The Young's modulus of the cantilever material can be calculated from the measured spring constant and the dimensions of the cantilever. In this measurement, the tip of the AFM cantilever is aligned near the free end of the sample cantilever (Figure 12.17), and then the AFM cantilever is swept in the Z-direction to bend the sample cantilever. The standard procedure for force–distance spectroscopy using an AFM setup was used to obtain the force–distance curves. The slope of these curves was used to calculate the spring constant and the Young's modulus of the structural material [42]. The measured values agreed with the simulated values.

12.6.5.2 Electrical Characterization

The expected change in the resistance is of the order of a few parts per million (ppm) for myoglobin-induced surface stress. To detect such a small change in resistance, sensitive signal conditioning and display circuits were developed [43]. The microcontroller MSP430f161x was used to achieve a high resolution. Faraday's cage was used to eliminate the noise within the measurement setup. Change in resistance as low as 10 ppm was measurable with the setup developed in our laboratory.

12.7 CONCLUSIONS

A microfabricated biosensor is an important component of POC devices. The overall performance of such a sensor depends on the various functional domains of the microsystem. Careful selection of materials and optimization of material properties is an important prerequisite for the development of a biomicrosystem. Microfabrication process parameters are the key factors to get the required functionality of the biosensor. Because these biosensors are supposed to work in the liquid media, a thorough understanding of the behavioral aspects of the sensor in liquid media is essential. Microsystem packaging and related experimentation in a controlled environment are some of the other important requirements.

Development of Lab-on-Chip for cardiac diagnostic is in progress. Extensive studies are required in the area of biomolecule immobilization, stability, and sensitivity of the device. The day is not far when the development of a BioMEMS-based lab-on-chip will bring a revolution in the field of cardiac diagnostics.

REFERENCES

1. Nishtar S. Prevention of coronary heart disease in south Asia. *The Lancet*, 2002;360:1015–1018.
2. Myocardial Infarction. Wikipedia, the free encyclopedia, http://en.wikipedia.orgl/wiki/Myocardial_infarction.
3. Herrick JB. Clinical features of sudden obstruction of the coronary arteries. *JAMA*, 1912;59:2015–2019.
4. McSweeney JC, Cody M, et al. Women's early warning symptoms of acute Acute Myocardial infarction. *Circulation*, 2003;108(21):2619–2623.
5. Spodick DH. Decreased recognition of the post-Acute Myocardial infarction (Dressler) syndrome in the postinfarct setting: Does it masquerade as "idiopathic pericarditis" following silent infarcts? *Chest*, 2004;126 (5):1410–1411.

6. Davis TM, Fortun P, et al. Silent Acute Myocardial infarction and its prognosis in a community-based cohort of Type 2 diabetic patients: The Fremantle Diabetes Study. *Diabetologia*, 2004;47(3):395–399.

7. Gillum RF, Fortmann SP, et al. International diagnostic criteria for acute Acute Myocardial infarction and acute stroke. *Am Heart J*, 1984;108:150–158.

8. Zimetbaum PJ, Josephson ME. Use of the electrocardiogram in acute Acute Myocardial infarction. *N Engl J Med*, 2003;348(10):933–940.

9. Masahike HN, et al. Usefulness of serum Troponin T levels on day 3 or 4 in predicting survival after Acute Myocardial Infarction. *Am J Cardiol*, 2001;87:284–297.

10. Collinson PO, Boa FG, et al. Measurement of cardiac Troponin. *Ann Clin Biochem*, 2001;38:423–449.

11. Alpert JS, Thygesen K, et al. Acute Myocardial Infarction redefined—A consensus document of The Joint European Society of Cardiology/American College of Cardiology Committee for the redefinition of Acute Myocardial infarction. *J Am Coll Cardiol*, 2000;36:959–969.

12. Stryer L. *Principle of Biochemistry.*

13. Apple, FS et al. Simultaneous rapid measurement of whole blood myoglobin, creatine kinase MB, and cardiac troponin I by the triage cardiac panel for detection of myocardial infarction. *Clinical Chemistry*, 1999;45:199–205.

14. Wu AHB, Apple FS, et al. Recommendations for use of cardiac markers in coronary artery diseases. National Academy of Clinical Biochemistry Standards of Laboratory Practice. *Clin. Chem.*, 1999;45:1104–1112.

15. Takeda S, Yamashita A, et al. Structure of the core domain of human cardiac troponin in the Ca^{2+}-saturated form. *Nature*, 2003;424:35–41.

16. Wu AH, Wang XM, et al. Creatine kinase MB isoforms in patients with skeletal muscle injury ramifications for early detection of acute myocardia infarction. *Clin Chem*, 1992;38:2, 396–400.

17. Mair J, Morandell D, Genser N, et al. Equivalent early sensitivities of myoglobin, creatine kinese MB mass, creatine kinase isoform ratios, and cardiac troponins I and T for acute myocardial infarction. *Clin Chem*, 1995;41:1266–1272.

18. Roberts R. Myoglobinemia as an index to myocardial infarction. *Ann Intem Med*, 1977;87:788–798.

19. Galvani M, Ferrini D, et al. New markers for early diagnosis of acute myocardial infarction. *Int J Cardiol*, 1998;65:S17–S22.

20. Bertinchant J-P, Larue C, et al. Release kinetics of serum cardiac troponin I in ischemic myocardial Injury. *Clin Biochem*, 1996;29:587–594.

21. Hudson MP, Christenson RH, et al. Cardiac markers: Point of care testing. *Clinica Chimica Acta*, 1999;248:223–237.

22. Engel G, Stanley GR. Feasibility and reliability of rapid diagnosis of myocardial infarction. *Am J Med Sci*, 2001;322–339.

23. Panteghini M. Recent approaches in standardization of cardiac markers. *Clin. Chim*, 2001:311–319.

24. Walker JM. Protein protocols on CD-ROM. ISBN, 2002;1-58829-054-9.

25. Carrascosa LG, Moreno M, et al. Nanomechanical biosensors: A new sensing tool. *Trends Anal Chem*, 2006;25:196–206.

26. Arntz Y, Seelig JD, et al. Label-free protein assay based on a nanomechanical cantilever array. *Nanotechnology*, 2003;14:86–90.

27. McKendry R, Zhang J, et al. Multiple label-free biodetection and quantitive DNA binding assays on a nanomechanical cantilever array. *Proc Natl Acad Sci USA*, 2002;9783–9799.

28. Sethi RS, Lowe CR. Electrochemical microbiosensors. *IEE Colloquium on Microsensors*, 1990;911–915.

29. Higson SPJ, Reddy SM, et al. Enzyme and other biosensors: Evolution of a technology. *Eng Sci and Educ J*, 1994;435–445.

30. Fraser DM. Glucose biosensors—The sweet smell of success. *Med Device Technol*, 1994;5:44–47.

31. Fraser DM. Biosensors in critical care. *Med Device Technol*, 1995;6:36–40.

32. Northrup MA, Gonzalez C, et al. A MEMS-based miniature DNA analysis system. *Proceedings of the 8th International Conference on Solid-State Sensors and Actuators*, 1995;764–767.

33. Tina M, Regan O, et al. Direct detection of myoglobin in whole blood using a disposable amperometric immunosensor. *Analytica Chimica Acta* 2002;460:141–150.

34. Rokade H. HWCVD Silicon nitride based microsensor. M.Tech. Project Report, Reliability Engineering Department, Indian Institute of Technology Bombay, July 2005.

35. Sahu, et al. Internal stress in CAT-CVD films. Third International Conference on Hotwire CVD process, Utrecht, 2004.

36. Nitin S, Kale V, Rao R. Design and fabrication issues in affinity cantilevers for bioMEMS applications. Accepted, to appear in *IEEE/ASME J Micromech Sys*, 2006.

37. Joshi M, Rao R, Mukherji S. AFM characterization and selectivity of immobilization of antibodies in bio-MEMS. Proceedings of International Bioengineering Conference (IBEC), September, 2004.

38. Joshi M, Goyal M, Pinto R, Mukherji S. Characterization of anhydrous silanization and antibody immobilization on silicon dioxide surface. IEEE-EMBS International Summer School on Medical Devices and Biosensors, July 2004.

39. Joshi M, Singh S, Swain B, Patil S, Dusane R, Rao R, Mukherji S. Anhydrous silanization and antibody immobilization on hotwire CVD deposited silicon oxynitride films. IEEE India Annual Conference, INDICON, 2004.

40. Joshi M, Pinto R, Rao VR, Mukherji S. Silanization and antibody immobilization on SU-8. Accepted for publication, *Appl Surface Sci*, 2007;253:3127–3132.

41. Joshi M, Kale N, Mukherji S, Dusane RO, Rao VR, Lal R. A dry method for surface modification of SU-8 for immobilization of biomolecules using hotwire induced pyrolytic process. Indian provisional patent application number 1264/MUM/2004.

42. Serre C. et al. Determination of micromechanical properties of thin films by beam bending measurements with an AFM. *Sensors and Actuators* 1999:134–137.

43. Lukachan G. Measurement setup for characterizating piezoresistance of affinity cantilever. M.Tech. Project Report, Electrical Engineering Department, Indian Institute of Technology Bombay, 2005.

13 Synthesis, PhysicoChemical Properties, and Biologic Activity of Nanosized Silver Particles

Alexandra A. Revina
A.N. Frumkin Institute of Physical Chemistry and Electrochemistry,
Russian Academy of Sciences, Moscow, Russia

CONTENTS

13.1 Introduction 161
References 167

13.1 INTRODUCTION

During the last decade in different fields of science, medicine, and engineering, nanosized metallic particles have been used with great success. The importance of nanosized particles was stressed in 1959 by Richard Freiman, winner of the Nobel Prize in physics, in his lecture at the California Institute of Technology: "There is Plenty of Room at the Bottom in Miniaturization," during which he told of the immense possibilities of materials and devices of atomic and molecular size. He pointed out that in small assemblies of atoms, quantum effect must be taken into account, and that physical laws do not exclude the possibility of developing such systems. Beginning from the 1990s, the amount of publications on nanotechnology and its use in different fields increased sharply [1–5]. Buchachenko [2] noted the increased importance of chemistry for the development of nanotechnology. Even before the beginning of the "Nanoboom," different small-sized systems were investigated by scientists, such as microorganisms, aerosols, catalysts, carbon black, zeolytes, ultradisperse powders, thin films, and many others. Many archeologic discoveries showed the use of colloid systems in the ancient world (for example, the use of "Chinese ink" in ancient Egypt). Thus, the modern concept of nanotechnology includes also many "old" fields of science [4,5].

In 1857 Michael Faraday synthesized gold colloids with a particle size below 1 μ (down to 10 nm). Bach and Balashova [7] investigated electrochemical properties of platinum soles in the presence of oxygen and hydrogen and showed that (as predicted by Frumkin) these soles have the same behavior as electrodes from platinized platinum.

The use of pulse and steady-state radiolyses for the investigation of short-living and metastable reaction centers of biological and catalytical systems containing metal ions gave the possibility to establish a general concept for the preparation of materials with nanosized structures [8–11]. A new method for a radiation–chemical synthesis of stable nanoparticles of metals in liquid media was developed with the use of reverse micellular systems for the reduction of metal ions, and a γ-radiation ^{60}Co

$$H_2O \longrightarrow e^-_{aq}, H^\bullet, OH^\bullet, H_2, H_2O_2$$

$$Ag^+ \xrightarrow{\ e^-_{aq}\ } Ag^0 \xrightarrow{\ Ag^+\ } Ag_2^+ \xrightarrow{Ag_2^+} Ag_4^{2+} \longrightarrow \cdots Ag_n^{m+}$$

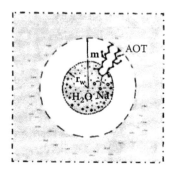

FIGURE 13.1   Ag$^+$ reduction with products of water radiolysis and a reverse micelle AOT/bis(2-ethylhexyl) sodium sulfosuccinate.

subsequent formation of nanoaggregates under the influence of an ionizing radiation was developed in 1987 [11,15].

It is exactly in natural biological systems that nanosized entities are formed. A nanotechnologic approach while investigating natural biological active compounds with metals of variable valency showed the influence of solvent organization on redox reactions resulting in the formation of nanoaggregates. It was found that in reverse micellular solutions Ag(NO$_3$)$_{aq}$/surface active compound/n-hydrocarbon in presence of the flavonoid pigment quercetin, a sequence of events occur which are very frequent in natural systems with metal ions: dissolution, formation of complexes, transport, aggregation, and destruction. It was indeed possible to develop with these components a technological method for the synthesis of metal nanoparticles [12–14].

During radiation chemical synthesis (RCS) of nanosized silver particles, silver ions Ag$^+$ are reduced by hydrated electrons e$^-_{aq}$ or other radical-type entities with reducing properties formed during ionizing irradiation (γ-radiation from ^{60}Co) in an aqueous-organic micellular solution Ag$^+$/ H$_2$O/AOT/isooctane. Figure 13.1 represents schematically a reverse micelle and the reactions occurring during radiolysis leading to the formation of nanosized particles Ag$_n^{m+}$. The radiation dose was 1 Mrad (10 KGr) [11].

Stable silver nanoparticles (SNPs) in reverse micelles have characteristic optical absorption spectra in the ultraviolet-visible (UV-Vis) region. This gives the possibility to detect such entities in liquid media. Figure 13.2 and Figure 13.4 show changes of the absorption spectra of SNP with time at room temperature. The results of data measured by correlation photon spectroscopy (CPS) are shown in Figure 13.3 and Figure 13.4.

In contrast to the radiation–chemical synthesis, the reduction of Ag$^+$ ions in micelles takes place in the presence of air oxygen. As a reducing agent, the flavonid queceitin was used. After introducing a certain amount of an aqueous silver salt solution into a stirred 0.15 M solution of AOT in an apolar solvent containing quercetin (40 to 200 μM) a gradual increase in the color density was observed. The spectra of solutions with SNP show distinct absorption bands in the range of 390 to 440 nm.

In order to obtain composite materials containing SNP, the interaction of SNP with different adsorbing materials was investigated by methods of spectrophotometry, CPS, and thin-layer chromatography. The results of such measurements with different carbonaceous materials are shown in Figure 13.5. It can be seen from these results that the amount of SNP adsorption depends on the nature of the adsorbent.

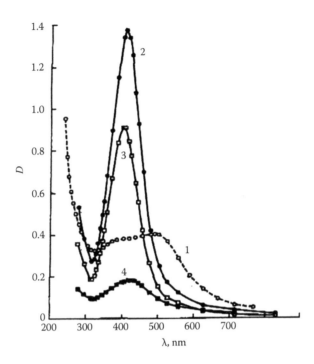

FIGURE 13.2   Optical absorption spectra of stable silver nanoparticles (SNPs) at a concentration of Ag$^+$ ions in the micellular solution 8.1 mM and a hydration degree $\omega = [H_2O]/[AOT]$, $\omega = 10$ and a AOT concentration on isooctane 0.15 M.

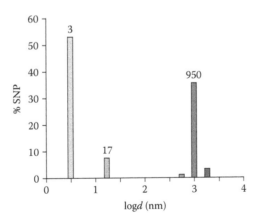

FIGURE 13.3   Histograms of the size distribution of silver nanoparticles (SNPs) in the range of 3 to 300 nm 2 weeks after their synthesis. The digits above the columns indicate the size of the SNP in nanometers. (From W.M. Gelbart, B.S. Avinoam, R. Redier, (Eds.), *Micelles, Membranes, Microemulsions and Monolayers*, Springer-Verlag, Heidelberg, 1994; E.M. Yegorova, A.A. Revina, *Colloids and Surfaces. A: Physicochem. and Eng. Aspects*, 168, 87, 2000. With permission.)

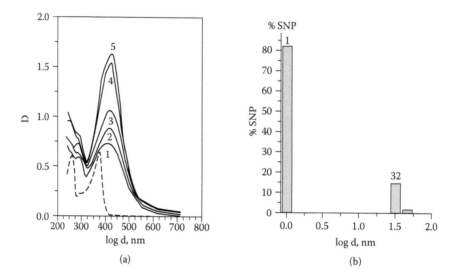

FIGURE 13.4 (a) Evolution of optical absorption spectra during the formation of silver nanoparticles (SNPs), measured: (1) 15 minutes, (2) 40 minutes, (3) 3 hours, (4) 2 days, (5) 2 weeks after introducing $AgNO_3$ into the micellular quercetin solution. (Dashed line, pure quercetin solution.) (b) Size distribution of SNPs 2 weeks after synthesis.

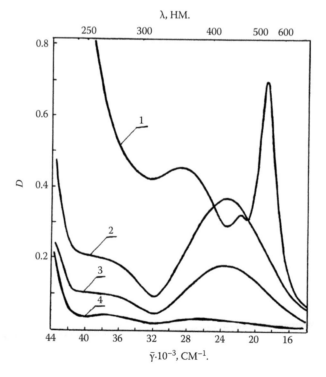

FIGURE 13.5 Optical absorption spectra of a micellular solution with silver nanoparticles (SNPs): (1) immediately after the SNP synthesis (γ-radiation, 2 Mrad), (2) 2 hours and (3) 42 hours after the contact with carbon cloth, (4) 2 hours after contact with carbon black BAU-1.

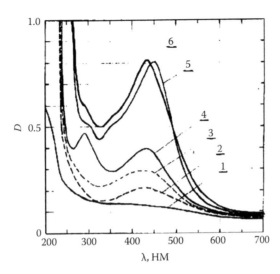

FIGURE 13.6 Changes in the optical absorption spectra for cellophane containing silver nanoparticles (SNPs) and contacting with solutions of natural sample: (1) pure sample, (2) sample after irradiation during the SNP synthesis (1.5 Mrad), (3–6) sample 2 after a 12-hour contact in a SNP solution containing 300 μM of (4) morin, (5) rutin, and (6) quercetin.

The method of radiation–chemical synthesis has an important advantage. When using compact materials with fine micropores, silver ions freely enter into these micropores and are there reduced *in situ* by radiation penetrating into the bulk of the material [15]. Using this method, filters were produced containing microporous ceramic and polymeric materials. This method gives also the possibility to introduce SNP into polymeric films (cellophane a.o.). The optical and electrical properties of such films were investigated in the literature [17]. With an increasing radiation dose, the optical absorption of such films in the region of maximal absorption for SNP particles ($\lambda_{max} = 410$ to 440 nm) and the electrical conductivity of such films increases also. These results correlate with results obtained by AFM.

From Figure 13.6, it can be seen that the presence of flavonoids (depending on their structure) can be detected by a change in the intensity of the absorption band characteristic for SNP. It must be noted that films not containing SNP do not adsorb molecules of the pigment.

Great attention is devoted to investigations of the biocid activity of SNP and silver ions. Introducing aqueous suspensions of SNP into infected wounds results in a considerable concentration reduction (or complete elimination) of microorganisms. SNP seems to be more effective than the same concentration of silver ions. As an example in Table 13.1, the results of comparative measurements of the biologic effect of SNP and silver ions on water infected with Coliphags MS-2 (a model for virusological water contamination) is shown.

Different versions of varnishes and colors with virusocidic and antibacterial properties were developed [19,20]. A point of interest is the effectiveness of the transfer of biocidic and antibacterial properties from SNP to composite materials containing such particles. An investigation of antibacterial activity of filters with SNP showed that at the surface of filtering materials they actively eliminate different bacteria [15]. So, after introducing intesting bacteria into water, their concentration in the presence of filters with adsorbed SNP diminishes immediately 15 times from the initial value of 10^7 cells/L. After 30 minutes, they diminish 300 times and after 3 hours they completely disappear, whereas in control experiments with filters without SNP their concentration remains unchanged. At present, such filters based on porous polyethylene (developed at the Karpov Institute, Moscow) and modified with SNP are used in communal systems for water conditioning.

The prospects for widespread use of SNP in biotechnology and the food industry are connected not only with their bacteriocidic properties, but also with the fact that these particles remain

TABLE 13.1

Comparative Dynamics of Coliphag MS-2 Inactivation under the Influence of Different Concentrations of Silver

| | | 1 × 10⁻⁴ g-ion/l | | | | 6 × 10⁻⁵ g-ion/l | | | |
| | | SNP | | AgNO₃ | | SNP | | AgNO₃ | |
Time of Sampling	Control KOE/ml	KOE/ml	% Inactive	KOE/ml	% Inactive	KOE/ml	% Inactive	KOE/ml	% Inactive
Initial	185,000	25,000		100,000		50,000		59,000	
1 h	182,000	23,000	90.8	70,000	70	7,000	86	26,200	55.59
2 hhca	179,000	700	99.72	8,000	92	2,000	96	14,800	77.91
3 aca	172,000	620	99.75	6,200	93.8	900	98.2	5,900	90
4 hca	172,500	500	99.8	5,000	95	900	98.2	5,000	91.52

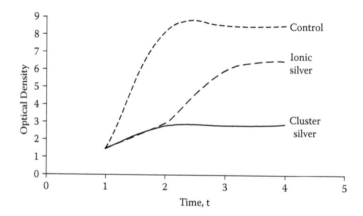

FIGURE 13.7 Dynamics of the increase of optical density for a solution of a culture of Candida Utilis: control experiment and after introducing silver ions and silver nanoparticle (SNP) clusters (both in concentrations 8 to 86 mg/L).

strongly fixed on the filter surface, in contrast to silver ions that easily desorb from the filter. In particular, SNP can be useful for preventing the development of microorganisms in filters used in breweries for separating the biomass containing microbes from the reaction product.

In order to assess the possibility of using SNP in the food industry, it is necessary to investigate the influence of different concentrations of SNP and silver ions on yeast cells. It can be seen in Figure 13.7 that in the phase of their linear growth, the development of yeast cells is inhibited both by silver ions and by SNP. With an increasing amount of these cells, the biocidic effect of silver ions is much lower, whereas SNP completely inhibit their growth.

Observations of samples of Cutilis, prepared after a 2-hour washing of yeast suspensions with deionized water and a subsequent incubation period of 1 hour at 200°C in the presence of 45 µg AgNO₃ showed the presence of nanoparticles (d = 2 nm) on the outer surface of the cells (Figure 13.8a). The insertion of SNP into the suspension of yeast cells resulted in a destruction of the cell membrane.

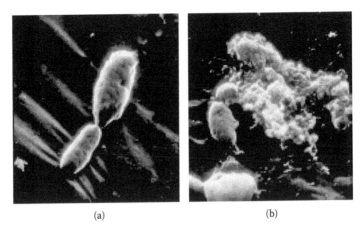

(a)   (b)

FIGURE 13.8   Electron micrograph (EM) of the cells *Candida utilis*: (a) in the presence of $AgNO_3$ and (b) in the presence of silver nanoparticles (SNPs).

Thus, these results indicate that the mechanism of influence of ionic silver on different microorganisms and yeast cells is different from the corresponding mechanism in the case of silver nanoclusters.

REFERENCES

1. Zh. I.Alferov, *Semiconductor Physics and Engineering* (Russia), 3, 2 (1998).
2. A.L. Bochachenko, *Vestnik Russ. Acad. Sci.* 71, 544 (2001).
3. L.I., Trusov, V.F. Petrunin, UFN 133. No. 4, 653 (1981).
4. B.D. Summ, N.I. Ivanova, *Uspekhi Khimii*, 69, 995 (200).
5. R.A. Andrievski, *Sci. Sintering.* 32, 155 (2000).
6. N.A. Bach, N.A. Balaschowa, *Nature.* 137, 617 (1936); *Acta Physicochimica* USSR. 7, 866 (1937).
7. A.N. Frumkin, *Trans. Farad. Soc.*, 69, 204 (1936).
8. R. Tausch-Treml, A. Henglein, J. Lilie Ber. *Bunsenges.*, 82, 770 (1978).
9. B.G. Ershov, N.L. Sukhov, D.A. Troitsky, *Izvestiya of the Russ. Acad. Sci.*, 8, 1930 (1989).
10. B.G. Ershov, E. Janata, A. Henglein. *J. Phys. Chem.* 339 (1993).
11. A.G. Dokuchaev, T.G. Myasoyedova, A.A. Revina, *High-Energy Chemistry* (Russia), 5, 353 (1997).
12. W.M. Gelbart, B.S. Avinoam, R. Redier (Eds.), *Micelles, Membranes, Microemulsions and Monolayers.* Springer-Verlag, Heidelberg. 1994.
13. E.M. Yegorova, A.A. Revina, *Colloids and Surfaces. A: Physicochem. and Eng. Aspects.* 168, 87 (2000).
14. A.A. Revina, Ye.M. Yegorova, A.D. Karatayeva, *Zh. Phys. Khim.*, 73, 1897 (1997).
15. A.A. Revina, Russian Patent 2212268.
16. A.A. Revina, A.G. Dokuchaev, Ye.B. Khaylova, M.G. Teodoradze, *High-Energy Chemistry* (Russia), 43, 11 (2000).
17. Ye.M. Yegorova, *Kolloidniy Zhurnal* (Russia), 64, 134 (2002).
18. A. Maitra, *J. Phys. Chem.* 88, 5122 (1984), 4, 28 (2001).
19. A.A. Revina, Ye.M. Yegorova, B.B. Kudryavtsev, Chemical Industry (Russia).
20. Ye.M. Yegorova, A.A. Revina et al., Vestnik Moscow State Univ., Series 2, Khemistry, 12, 33 2 (2001).
21. A.A. Revina, Ye.K. Baranova et al., Paper on the First Seminar on Application of Nanotechnology to Biologic Systems, Moscow, 2003, Abstracts, p. 53.
22. Ye.K. Baranova, A.L. Mulyukin et al., in the Internet Journal http://zhurnal.ape.relarn.ru/articles/2005/139.pdf.

14 Nanocrystalline Silicon for Biomedical Intelligent Sensor Systems

Alexandra Shmyryeva and Elena Shembel
National Technical University of Ukraine—
Kiev Polytechnic Institute, Kiev, Ukraine

CONTENTS

14.1   INTRODUCTION

Significant progress in modern medicine is impossible without the development of new medical technologies using the latest results of fundamental investigations in the related fields of physics, chemistry, electronics, mathematics, and biology. The development of a wide spectrum of diagnostic systems enabling control of the key parameters of organisms and the diagnosis of concrete diseases is based on using biomedical sensors with microelectronic structure [1–3].

A very important role of human vital function is undoubtedly an effective protection from the different toxic substances that are practically found in all fields of industry and farm production. In some cases, as a result of chemical transformations in the organism, substances more toxic than the initial products can be formed (lethal synthesis). Various food additives can lead not only to an effective protection from intoxication but also to a lethal outcome. Their use is clearly regulated by normative documents that state their permissible concentrations (usually with a measurement error no more than 2%). Mycoplasmas used in biotechnology for virus vaccine production can cause respiratory, urogenital, autoimmune diseases, arthritis, and immune deficit disorders [2], as well as activate many viruses including oncogenic viruses and human immunodeficiency virus (HIV). Undoubtedly, ideal methods of diagnostics of mycoplasma infections should give the possibility to provide fast and reliable measurements. They should be simple in operation, available, and allow a large-scale application.

A brief analysis of the properties of some toxic substances indicates the necessity to develop new electronic devices for the measurement of their concentration and to develop sensitive cells—biosensors [3].

The base of all investigations in this field is the elucidation of informative parameters of biochemical processes, the development of sensitive devices cells (sensors)—transducers transforming these parameters into electrical signals, the development of electronic systems for a quantitative determination, and registration of these parameters [2–4]. Recently, more and more attention is devoted to the development of experimental methods of analysis where the sensors should be characterized by simplicity, have a sufficient level of sensitivity and selectivity, and also the possibility of miniaturization. They must have the ability to work independently without an operator and have the possibility to be integrated with systems of automated data accumulation and processing. Achievements in the field of microelectronics have stimulated the development of biosensors of new design—biochips, combining sensor systems transducers, analog/digital [A/D] converters, and a microprocessor to analyze the obtained data. Usually, the development of such intelligent sensor systems is based on using a microelectronic technology, where silicon is the main material [1,5].

Due to the development of silicon thin films of various modification, especially, nanocrystalline which has new functional properties [5,6], the possibilities of biomedical sensor systems with unique properties are expanded. Thermal and optical effects, intrinsic to many biochemical reactions, can be the physical–chemical basis for the development of biosensor systems for information measurement. In this chapter, some versions of biosensor systems and the basis properties of nanocrystalline silicons will be considered.

14.2 BIOSENSORS

Biosensors are a version of chemical sensors in which biochemical reactions are used for the recognition of either individual biomolecules or biological supermolecular structures in order to provide information about the presence of definite "recognized" molecules and their amount. The term "biosensor" implies a device in which the sensitive layer comprising such biological material as ferments, living tissues, bacterium, yeast, antigens/antibodies, liposomes, organelles, receptors, DNA, and living organisms reacts directly with the components to be determined and generates signals functionally connected with their concentration.

The chemical and physical phenomena that are the basis of the biochemical reaction of energy conversion into luminous radiation are now well known. Chemi- and bioluminescence are now widely used in the biochemical laboratory and clinical investigations. Increasing the chain of biochemical stages preceding the stages with bioluminescence, one can obtain new analytical possibilities [9].

For a successful development of this bioelectronic field, it is necessary to develop new highly sensitive photodetecting devices for the spectrum range of 300 to 500 nm. The development of such devices is based on the use of new thin-film nanocrystalline silicon alloys connected with dielectrical layers with an increased photosensitivity and detecting ability. A microelectronic technology gives the possibility to develop portable biochemiluminometers and to extend the field of their application.

The main problems in the field of biosensors for determination of the concentration of different substances are connected with the development of stable biochemical cells allowing multiple use. Biochemical reactions in such cells lead to the formation of reaction products that then, with the help of transducers, are transformed into electrical or optical signals. For example, the heat evolving during interaction with microorganisms is measured with the help of microminiature thin-film thermoresistors. Optical radiation evolving during interaction of toxins with luminescent bacteria is measured with the help of photodetectors. The micromovements of living organisms (e.g., dafnies) when changing the toxicity of the surrounding medium are measured with the help of optoelectronic sensors.

14.3 FUNCTIONAL PROPERTIES OF NANOCRYSTALLINE SILICON

A stimulus for the development of microelectronic sensors is the synthesis of new semiconductive materials having special functional properties. Among such materials, porous (PS) and nanoporous (NPS) silicon are of great interest. Until recently, PS attracted the attention of investigators, mainly because of its luminescent properties [6]. However, a wide study of its other properties has opened the prospects for many other applications, including solar cells, biotechnology, sensors, humidity, gas detectors, and photodetectors. Solar cells and light-emitting diodes based on PS have been already developed [1,2,11,14,15].

Among the methods for preparing PS layers, anodic electrolytic etching of monocrystalline silicon in acidic solutions and chemical etching are the two most commonly used [1,12,16]. The first method allows the properties of porous silicon layers to be changed within the wide ranges, but it also has some essential disadvantages, including the difficulty to implement it with the standard silicon technology, an incomplete use of the silicon surface, and the impossibility of a large-scale treatment of products. Properties of nanoporous silicon that can be produced by chemical etching of monocrystalline silicon have not been studied from the point of view of its application in sensor systems. Nanocrystalline silicon is produced by vacuum spraying followed by laser recrystallization, ion implantation [17], or by changing the technological conditions of deposition.

The microstructure determines the majority of physical properties of silicon [18,19]. However, during studies of nanocrystalline materials, as a rule, the main emphasis is placed on the scale factor, whereas their properties will depend on the structure and properties of the interfaces and the ratio of crystalline, amorphous, or dielectric phases. In particular, many properties of PS, NPS, and nanocrystalline silicon (NS) are determined by quantum-dimension effects [18,20]. But, nevertheless, there is no clear understanding of the microstructure and its connection with luminescent, transport, and other properties. From the numerous works about the use of porous silicon in sensors, it follows that the main problem is connected with an instability of its physical and optical characteristics [21].

For the development of different types of sensors, it is necessary to produce stable films of nanoporous and nanocrystalline silicon with controlled properties and with a possibility of their optimization for specific applications in sensor systems.

14.3.1 INFLUENCE OF PREPARATION TECHNOLOGY ON THE PROPERTIES OF NANOPOROUS SILICON

Both electrolytic and chemical etchings of silicon are alike in many respects. They are multistage processes including a great variety of series and parallel reactions. The two main reactions proceed consecutively. The first is directly connected with a charge transfer through the silicon/solution interface and proceeds with using h^+ holes and can be presented as [1,6]

$$Si + 2h^+ \rightarrow Si^{2+} \tag{14.1}$$

The Si^{2+} ions formed are unstable, and they are either oxidized by hydrogen ions

$$Si^{2+} + 2H^+ \rightarrow Si^{4+} + H_2 \tag{14.2}$$

or participate in a reaction of disproportionation [6]:

$$Si^{2+} + Si^{2+} \rightarrow Si + Si^{4+} \tag{14.3}$$

As a result of the reaction (Equation 14.3), secondary corpuscular silicon is formed in quantities equal to half of the participating silicon ions, and its second half reacts with F^- ions forming strong complex ions $[SiF_6]^{2-}$ in the solution. Both reactions (Equation 14.2 and Equation 14.3) proceed in parallel. The share of each of them is determined by the experimental conditions. Among the great variety of semiconductor materials, only silicon is characterized by the possibility of ion disproportionation with mutual exchange of electrons and formation of neutral atoms and ions of higher valence. This serious chemical factor distinguishes silicon from other semiconductors and enables the formation of a porous layer. Germanium, closest to silicon by its properties, can also dissolve with the formation of two- and four-valence ions. However, compounds of two-valence germanium are less stable than two-valence silicon, and the possibility of their disproportionation is completely suppressed by an oxidation of Ge^{2+} ions by hydrogen ions.

On the silicon surface, randomly distributed spots with increased stability to dissolution (surface nanocrystallites) are formed. The formation of such surface heterogeneities of quantum-dimension scale is the first stage of formation of porous silicon. Subsequently, silicon will dissolve only in the spaces between such nanocrystalline aggregates (i.e., a formation and deepening of pores will begin). Simultaneous with this pore formation, a growth of secondary silicon continues on the whole surface of *por*-Si, including on pore walls. Its high specific resistance together with quantum-dimension effects in thin pores provides an increased resistance to dissolving and promotes the growth of large pores in the bulk of the substrate. Deposition of secondary silicon on pore walls leads to the formation of small-size pore branches. In accordance with this concept, in *por*-Si two systems of pores are formed: larger "main" pores of micrometer width, going into the bulk of the substrate bulk for the tens of micrometers and well visible in optical microscopes, and rather shorter "nanopores" branching from the larger pores. Such a character of the porous structure enables the transport of initial reagents and reaction products in the whole volume of *por*-Si and corresponds to the results of microscopic investigations of the *por*-Si surface [6]. The formation on amorphous silicon substrates of porous layers with luminescent properties in the visible part is of an indirect confirmation of the possible formation of secondary silicon aggregates within its crystalline structure.

An investigation of the structure, thickness of nanoporous silicon layers, and their photoluminescent properties has shown that all samples independent of the value of the porous silicon area were characterized by a high degree of homogeneity in contrast to layers prepared by chemical or electrochemical etching. The thickness of the prepared PS layers is less than 20 nm, which indicates that they may be considered as nanoporous silicon.

In Figure 14.1, the images of different silicon modifications, including nanoporous silicon prepared by chemical treatment of monocrystalline silicon of p-type, monocrystalline silicon, nanocrystalline silicon prepared by the methods of electron beam evaporation, and porous silicon prepared by an anodizing method, are shown. It can be seen that that there is a substantial difference in the configuration, size, and surface state of these nanocrystals.

For a more detailed study of NPS, the method of scanning tunnel microscopy was used. An evaluation of the thickness of NPS has shown that d < 20 nm. This value correlates with the height of columns observed by the method of atomic force microscopy (AFM). In this case, visible differences in the surface structure between samples produced under similar conditions are not observed.

Samples with nanoporous and nanocrystalline silicon were investigated by the method of "cylindric mirror-type" electron Auger spectroscopy with a spectrometer LAS-2000 (Riber, Bezons, France).

During formation of a layer of nanoporous silicon on the surface of monocrystalline silicon, both the amount and the distribution profile of oxygen on the oxidized surface change. In addition, the content of hydrogen in the near-surface region increases, essentially providing a significant decrease of the amount of recombination centers.

During etching of porous silicon, partially oxidized layers enriched by hydrogen are formed on the surface. When heating such a surface in contact with aluminum, hydrogen is released, for example, by the following reaction:

$$SiO_2(...H, OH...) + 2Al \rightarrow AlO_3 + Si + 2H\uparrow \qquad (14.4)$$

Monoatomic hydrogen released in reaction (Equation 14.4), penetrates the bulk of silicon where it can effectively passivate defects on grain boundaries and thus decrease the recombination of minority charge carriers.

It is well known that hydrogen at room or even lower temperature can penetrate deep into silicon evidently as the result of an interaction with crystalline lattice defects. Investigations of nanoporous

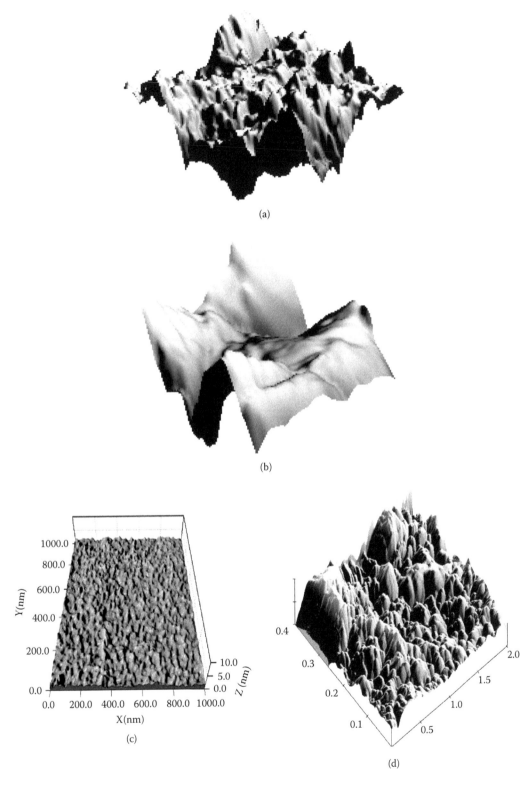

FIGURE 14.1 Atomic force microscopy (AFM) images of the surface structure of nanoporous silicon (a), monocrystalline silicon (b), nanocrystalline silicon (c), and porous silicon (d).

silicon layers showed that the distribution of hydrogen-containing defects AH (where *i* is the quantity of hydrogen atoms in a defect) for *z* depth from the etched surface deep into crystal is described by the exponential dependence $[AH_i] \approx \exp(-iz/L)$, which can be used for an experimental determination of the parameters *L* and *i*. *L* is the diffusion length of hydrogen, referred from the crystal boundary that is moving during etching. The value of *L* is the same for all defects in the investigated sample and is equal to [1]

$$L = \left(\frac{V}{2D_H} + \sqrt{\frac{1}{L_0^2} + \frac{V^2}{4D_H^2}} \right)^{-1}$$

(14.5)

where *V* is the rate of etching, L_0 is the diffusion length of hydrogen in the initial crystal, and D_H is the hydrogen diffusion coefficient. It can be seen from Equation 14.5 that measurements of *L* give, in principle, a possibility to determine the diffusion coefficient of hydrogen, the values of which quoted in many works differ by some orders of magnitude. However, in deriving Equation 14.5, some assumptions were made which in certain cases can be true for n-type Si (where the interaction between hydrogen and doping admixture is comparatively weak), but in p-type Si are applicable only at rather great distances from the surface, where the hydrogen concentration is small. A stable hydrogenation of the silicon surface layer provides a passivation of the surface recombination centers and enables the development of photodetectors highly sensitive to ultraviolet and optical irradiation.

14.3.2 Photoluminescent and Photoelectrical Characteristics

Photoluminescence in the visible spectrum range with a maximum at the wavelength 640 to 670 nm and ultraviolet spectrum range at 350 nm is one of the interesting properties of porous and nanoporous silicon [1,2,4,21]. For an investigation of the photoluminescence spectra of nanoporous silicon, a portable spectrometer SL40-2 and a photodiode TCD 1304 (Toshiba, Tokyo, Japan) with a resolution up to 1.5 to 2 nm were used. As can be seen from the photoluminescence spectra shown in Figure 14.2, the radiation intensity depends on the crystallite size. A dimension quantization of charge carriers in PS increases the optical width of the prohibited zone, for example, by the following value [22]:

$$\Delta E = \frac{3\pi^2 \hbar^2}{\mu q^2} - \frac{3e^2}{\varepsilon q}$$

(14.6)

where μ is the relative value of the effective mass, and ε is the dielectric permeability.

The second term in Equation 14.6 describes the energy decrease due to coulombic interaction of photogenerated holes and electrons. It should be noted that owing to the comparatively small effective mass of holes, as well as to a tortuosity of the valence zone, the contribution of holes into the value of ΔE value is evidently dominating, and according to the literature [21], is about 60%.

Using the measured intensities of photoluminescence of the prepared nanoporous silicon films, the optimal composition for chemical treatment and etching for the further investigations have been selected.

During the formation of porous and nanoporous silicon on substrates of monocrystalline silicon, not only a change of the structural properties (resulting in a change of the prohibited zone's width and the appearing of quantum-dimension effects) takes place, but also the formation on the surface of new silicon compounds with an increased content of hydrogen and amorphous silicon takes place. Such complex structures have new electrophysical, photoelectrical, thermal–physical,

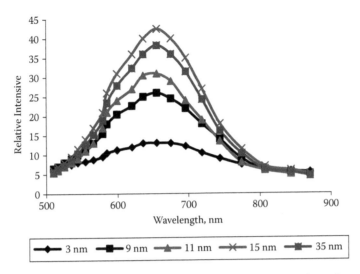

FIGURE 14.2 Luminescent spectra of nanoporous silicon films with different sizes of nanocrystals.

electro- and photoluminescent properties. This gives the possibility of development of the new types of semiconductor devices, particularly, photodetectors for the short-wave part of the spectrum.

Further investigations have the goal to investigate catalytic and adsorption characteristics of nanoporous silicon in order to develop selective sensors for gas and humidity concentration, sensors for pressure and force, biosensors for determination of organic impurities, concentration of complex organic substances, drugs, toxic substances based on optical, and measurement of photoluminescent and photoelectrical characteristics. Investigations of properties of nanoporous silicon prepared by chemical treatment of monocrystalline silicon performed up to now confirm the possibility of achieving stable and controllable characteristics for this multifunctional material and the possibility of developing highly sensitive photodetectors for visible and ultraviolet radiation exceeding the parameters of other known analogues.

The possibility to control the photosensitivity of nanocrystalline silicon films using laser recrystallization allows the development of highly sensitive sensors for ultraviolet radiation which are needed in many spheres of life, including biomedicine.

REFERENCES

1. Shmyryeva A.N., Melnichenko N.I., 2006, Sensors systems with application of nanoporous silicon, *Electronics and Communication*, Part 1, 17–22.
2. Starodub N.F., Starodub V.M., 2005, Biosensors based on porous silicon photoluminescence. Application for environmental quality monitoring, *Sensor Electronics and Microsystem Technologies*, 1, 63–74.
3. Varfolomeyev S.D., Biosensors, 1997, *Soros Educational Journal*, 1, 45–49.
4. Starodub N.F., Starodub V.M., 2004, Biosensors based on porous silicon photoluminescent. General characteristics and application for medical diagnostics, *Sensor Electronics and Microsystem Technologies*, 2, 63–83.
5. Zimin S.P., 2004, Porous silicon—material with new properties, *Soros Educational Journal*, 8, 1, 101–107.
6. Svechnikov S.V., Sachenko A.V., Sukach G.A. et al., 1994, Light-radiative layers of porous silicon: production, properties and application, *Optoelectronics and Semiconductive Equipment*, 27, 3–28.
7. Vladimirov Yu.A., 2001, Activated chemi-luminescence and bioluminescence as the instrument in medicine-biological investigations, *Soros Educational Journal*, 7, 1, 16–23.
8. Yevdokimov. Yu.M., Bundin V.S., Ostrovsky M.A., 1997, Biosensors and sensor biology, *Sensor Systems*, 11, 4, 374–387.
9. Mirzabekov A.D., 2003, Biochips in biology and medicine of XXI century, *Bulletin of Russian Academy of Science*, 73, 5, 412–416.

10. Ganshin V.M., Danilov V.S., 1998, Bacterial biosensors with bioluminescence output of information, *Sensor Systems*, 12, 1, 56–68.
11. Melnichenko N.N., Shmyryeva O.M., 2003, Application of sub-micron layers of porous silicon in solar cells, *Electronics and Communication*, 20, 204–207.
12. Kulinich O.A., 2004, Investigations of the photoluminescence properties of chemically modified surface region of silicon in the structures silicon oxide-silicon, *Sensor*, 2, 13–16.
13. Karachentceva L.A., Litvunenko O.A., Stronskaya E.I., 2003, Investigation of local chemical states in the structures of macro porous silicon, *Theoretical and Experimental Chemistry*, 39, 2, 77–81.
14. Melnichenko M.M., Svezhentseva K.V., Shmyryeva O.M. et al., 2005, Peculiarities of porous silicon formation on the texture surface of photoelectrical transducers, *Scientific News of NTUU "KPI"*, 2, 20–24.
15. Zimin S.P., Bragin A.N., 2004, Tensoresistive effect in porous silicon layers with different morphology, *Physics and Technics of Semiconductors*, 38, 5, 594–597.
16. Astrova E.V., Belov S.V., Lebedev S.V., 1994, Some properties of the structures based on porous silicon, produced by the method of painting etching, *Physics and Technics of Semiconductors*, 28, 2, 332–337.
17. Koval V.M., Shmyryeva A.N., 2005, Nanocrystal silicon with the controlled semi-conducting properties, *Scientific News of NTUU "KPI"*, 4, 14–28.
18. Guschina N.V., Dneprovsky V.S., Divydenko E. Yu. et al., 1994, Optical non-linearity and effect of dimension quantization in porous silicon, *JTP*, 106, 6, 1830–1838.
19. Kulinich O.A., Glauberman N.A., Sadova N.N., 2004, About connection between photoluminescent properties of porous silicon and its actual structure, *Surface, X-ray, Synchronic and Neutron Investigations*, 7, 96–98.
20. Salcedo W.J., Fernandez F.R., Rubim J.C., 2005, Polarization effects on the Raman and photoluminescence spectra of porous silicon layers, *Symposium Amorphous and Nanocrystalline Silicon Science and Technology*, San Francisco, 717–723.
21. Fischer M., Hillerich B., Kozlowski F., 2001, Long-time stability of photoluminescence in porous silicon, *Thin Solid Films*, 372, 209–211.
22. Agarwal V. et al., 2000, Analysis of the shape of PL spectra and its temperature dependence in self-supporting porous silicon, *Physica Status Solidi*, 182, 385–388.
23. Fraiden D., 2005, *Modern Sensors. Handbook*, Moscow, Technocphere, 592.
24. Vasilyev R.B., Ryabova L.I., Rumyantceva M.N. et al., 2005, Gas sensitivity of interface in semi-conduction materials, *Sensor*, 1, 21–50.

15 Wetting the Surface: From Self-Cleaning Leaves to Energy Storage Devices

V.A. Lifton and S. Simon
mPhase Technologies, Inc., Little Falls, New Jersey, USA

CONTENTS

15.1 INTRODUCTION

The surface is an important part of virtually every physical object and often plays an overriding role in many processes, beyond just connectivity and structural support, but more deeply into areas involving chemical and biological interactions. In some instances, the surface provides an easy entry into the biological or chemical systems; in others it protects the elements surrounded by the surfaces. An ever-increasing amount of surface-related research and development (R&D) work is being performed in nearly every field of science. In this overview, we focus our attention on a single aspect of surface properties: its topographical heterogeneity that gives rise to interesting wetting properties of naturally occurring biological objects. We then compare how this phenomenon is implemented in a variety of man-made objects. It will also be shown how surface roughness modifies wetting of solid surfaces, which is not confined to the droplets rolling off the surfaces but can also be implemented in such a seemingly unrelated subject as a novel energy storage device.

Surface topography is known to significantly alter the wetting properties of solid surfaces [1–3]. It changes the way liquid (water being one of the most commonly tested liquids) behaves when in contact with a surface of another object. Surfaces are generally grouped into two main categories: hydrophilic (surfaces that are wet by the liquid) and hydrophobic (surfaces that are not wet by the liquid). Quantitatively, a contact angle is often used to characterize the wetting behavior. It represents a balance of forces acting on a liquid when in contact with a solid surface. A surface is called hydrophilic when the contact angle is less than 90°, and hydrophobic when the contact angle is greater than 90°. Recently, surfaces where contact angles typically exceed 130° to 150° and even

170° started to gain much attention [4]. These surfaces are now called superhydrophobic surfaces and are the main focus of this overview.

Hydrophobic and hydrophilic surfaces are abundant in nature and in man-made materials, both organic and inorganic. The exploration and synthesis of superhydrophobic surfaces have remained elusive for quite some time, but with the advances made in nanotechnology, they have started to become more widely available and better understood [5–8]. In parallel, work in the biological sciences has made similar advances in surface characterization, so that some peculiar wetting effects (or nonwetting effects, to be exact) in plants such as lotus, rice, and many other species of plants, as well as peculiar water-collecting abilities of some beetles living in the driest of places, deserts, have been explained [9–11].

The observations of interest involve the "self-cleaning" lotus leafs and water-harvesting beetles. The term "self-cleaning" comes from the fact that lotus leaves appear to remain clean under the conditions of their natural habitat, which in many instances are swampy areas that do not appear to be particularly clean or free of soil and clay-based contaminants. A remarkable ability of the plant to remain clean has been described by numerous investigators and is a result of the surface of the leaf being highly hydrophobic such that when a droplet of water falls on the leaf, instead of remaining immobile on the leaf, it rolls off cleaning the leaf of dirt or other particles adhering to the surface [9,10].

The Stenocara beetle living in the Namib Desert has developed an elegant way to harvest water from humid air. Through the process of condensation, small water droplets accumulate on the back of a beetle and when grown sufficiently large via coalescence, roll off the back of the exoskeleton of the beetle and channel themselves directly into the mouth of the beetle [11].

A common phenomenon invoked in both cases to explain the observations is the superhydrophobic surfaces. Close examination of these examples revealed that the superhydrophobicity is caused by the heterogeneous micro-/nanostructures on the surfaces of both objects. If we examine recent progress in understanding and synthesis of superhydrophobic coatings, we notice that they all share a similar trait: a combination of micro- and nanostructures judiciously placed on their surfaces and an overlay of a hydrophobic coating on these structures. We will now describe the basic mechanisms responsible for this remarkable behavior and compare them with the man-made structures mimicking these properties. The following highlights only a small portion of the published literature, giving only the most pertinent citations for the interested reader.

For brevity, the superhydrophobic surfaces most commonly have the following features [4,12]: a hydrophobic coating (often, waxy substances in Nature and self-assembled monolayers or Teflon-like coatings in man-made materials); micro- (approximately 1 to 10 μm) and nano- (approximately 10 to 500 nm) scale roughness. As will be seen in the following discussion, these two features are prevalent in all of the examples cited.

15.2   SUPERHYDROPHOBIC SURFACES IN NATURE

Although many objects in Nature exhibit hydrophobic properties, here we will consider only those that come about because of the surface texture and hydrophobic coatings.

15.2.1   Lotus Leaf

The lotus leaf has been the source of fascination and admiration by many generations of people (Figure 15.1). In was not until the late 1990s that the remarkable self-cleaning action was investigated in detail. It has also been known for some time that its surface is covered with a waxy substance that gives a low surface energy surface to reduce water spreading and is at least partially responsible for its hydrophobic properties. However, in order to explain remarkably high contact angles measured on the leafs' surfaces and extremely low droplet sliding angle (on the order of a few degrees, very shallow tilt), some other mechanisms need to be present. Barthlott and Neinhuis observed that the

FIGURE 15.1 Overview photo of a leaf of lotus. (Blossey, R. 2003. *Nature Materials* 2:301–306. With permission.) **(See color insert following page 112.)**

FIGURE 15.2 Detailed images of the microstructures found on the surface of the lotus leafs. (a) and (c) are the untreated, dry lotus leafs; (b) and (d) are heat-treated at 150°C for 1 hour—lotus leaf to melt any waxy compounds on the surface. (From Cheng, Y.T., Rodak, D.E., Wong, C.A., Hayden, C.A. 2006. Effects of micro- and nano-structures on the self-cleaning behaviour of lotus leafs, *Nanotechnology* 17:1359–1362, Institute of Physics. With permission.)

leaf consists of the micrometer-scale papillae with the diameter of 5 to 9 µm, which in turn consists of branched nanostructures with the diameter on the order of 100 to 200 nm in size (Figure 15.2) [9,10]. Therefore, it appears that this multilevel roughness (micro- and nanoscale) gives rise to the enhanced hydrophobicity of the leaf's surface. Empirical observations indicate that surface roughness need not to be regular or periodic, as long as it conforms to the micro- and nanoscale lengths required. Several research groups have directly replicated the leaf structure by soft and nanoimprint lithography and observed superhydrophobic behavior on such surfaces [13–20].

15.2.2 STENOCARA BEETLE

This species are known for their remarkable ability to harvest water from humid air. All the more impressive are their exoskeleton surface structures that give them this ability. Parker and Lawrence

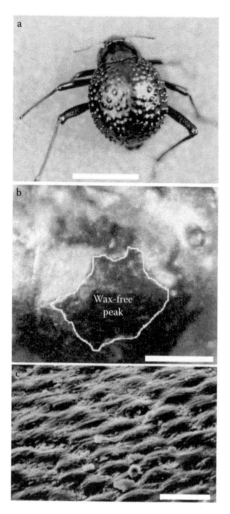

FIGURE 15.3 Details of the water-capturing, hydrophobic surface of the desert beetle *Stenocara* sp. (a) Adult beetle. (b) A close up of a "bump" on the fused overwings. Special staining marks waxy depressed areas as colored, whereas bumps remain unstained (black). (c) SEM of the depressed areas. Scale bars: (a) –1 mm; (b) –0.2 mm; (c) –10 μm. (From Parker, A.R., Lawrence, C.R. 2001. *Nature* 414:33–34. With permission.) **(See color insert.)**

[11] have shown that its back consists of an array of bumps 0.5 mm in diameter. A close examination of these bumps revealed that the tops of the bumps are smooth, without any surface films, and are hydrophilic. The sides of the bumps and the "flat" surface surrounding the bumps are covered with microscopic roughness, on the order of 10 μm, with a "wax" coating rendering the surfaces superhydrophobic. The suggested mechanism on how this phenomenon works is the following. In the morning fog, small droplets condense on the hydrophilic "bumps." As the droplets grow, gravity forces them to roll down the bumps and down the hydrophobic surface toward the beetle's mouth. In this case, there is the need for two types of surfaces: hydrophilic for water condensation and hydrophobic for water transport. Superhydrophobicity is also required to assure that water droplets will move without stopping and pinning. Figure 15.3 shows photos of the beetle as well as the details of its hydrophobic structures. Direct mimicking of such structures of the beetle has recently been described [21].

In addition, some insects such as cicada and termites are thought to develop regular nanostructured patterns on their wings to ensure water repellency and nonwetting [22]. The structures consist of ordered hexagonal arrays 150 to 350 nm in height and spaced some 200 to 1000 nm from each other. The structures appear to be multifunctional, with smaller size arrays serving also as antireflective coatings and larger structures providing mechanical rigidity. Note that the antireflective properties would be significantly degraded if water accumulated on the surface, thus changing the reflective index. It is quite plausible that the superhydrophobic surface helps maintain the required optical properties. Potentially, water striders also developed superhydrophobic legs to help support their weight while walking on water [4].

In many cases, superhydrophobic surfaces have evolved in order to provide "an instant" dewetting (water runoff) as soon as the plant or a bird (e.g., duck) or an insect emerges from underwater. Quere [4] suggested that butterflies may have developed it in order to prevent dew condensation from sticking and "welding" their wings normally closed together during the night.

15.2.3 CARNIVOROUS PLANTS

Drosera is a carnivorous plant that has superhydrophobic surfaces. The surface of the leaves of the Drosera looks like an entangled web of submicron hydrophobic fibers. In this example, bends in the fibers provide "roughness" as well modulation to submicron and micron sizes. There are sev-

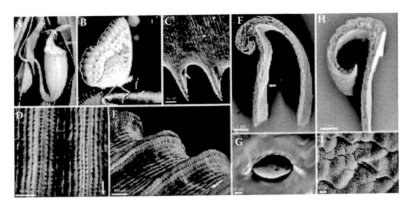

FIGURE 15.4 *Nepenthes* pitcher and its morphology. (A) Pitcher. (B) Butterfly sitting on the surface of the plant. (C) Details on the inner structure with visible tooth-like protrusions and nectar pores. (D and E) Detailed morphology of the internal surface of the plant, notice microscale surface roughness superimposed on macroscale surface waviness. (F and H) Cross-section of the trapping part of the plant with the arrow indicating transition from the digestive zone. (G) Inner digestive gland. (I) Morphology of the waxy inner wall. (Bohm, H.F., Federle, W. 2004 (Sept. 28). Proceedings of the National Academy of Sciences 101(39):14138–14143. With permission.)

eral theories why it developed such surfaces. One theory suggests that it may play a role in helping trap an insect as a potential meal, by making the leaf's surface extremely slippery to the feet of the insect, hindering its escape from the digestive liquid of the plants. Another hypothesis is that it needs such surfaces to help spread its seeds by the water currents. The seeds have been described as having a "rough, waxy" surface, perhaps a reference to a superhydrophobic surface. Clearly, the structures responsible for the superhydrophobic behavior in this example are different from the superhydrophobic structures found on lotus leaf. A nanostructured superhydrophobic surface prepared from nanofibers with high static contact angle has been reported by Shang et al. [23].

Another carnivorous pitcher plant, Nepenthes, appears to have superhydrophobic surfaces as well [24]. Again, we see a combination of microscale roughness and waxy substance covering it (Figure 15.4). At present, unanimity has not been reached on the exact action of this phenomenon. One hypothesis as in the case of Drosera plant, is that the superhydrophobic surface improves trapping efficiency by making surfaces more slippery. Another study on Nepenthes pitcher suggested aquaplaning as a cause for better sliding of the insect into the digestive tract of the plant. Both waxy coating and microscale roughness have been noticed as well. Authors also described a region where liquid droplets accumulate and suggested that a trapped insect then slides (aquaplanes) down this surface on a thin layer of liquid [25]. Alternatively, we may propose that no aquaplaning [26] occurs, but rather an insect rolls down the superhydrophobic surface because no droplet pinning can occur on such a surface. A remarkable ability of the plants' leafs to self-clean, just like lotus, is cited in all accounts, and once again, superhydrophobic surfaces play an overriding role in this aspect.

Clearly, we presented only a handful of superhydrophobic structures found in Nature. The list is sure to continue to grow [27]. Consider the fact that even the pioneering work by Neinhuis and Barthlott [9] contains the description of some 200 plant species exhibiting hydrophobic behavior. More work is needed to identify which ones possess superhydrophobic structures and their origins and intended functions.

15.3 MAN-MADE STRUCTURES AND MATERIALS

Mankind has always been fascinated with finding ways to replicate Nature's evolution, and hence the term "biomimetics" or "biomimicry." In the following section, we will describe mostly recent attempts in synthesis and application of superhydrophobic surfaces. It must be kept in mind that

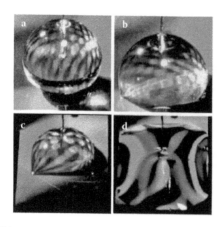

FIGURE 15.5 Man-made tunable nanostuctured surfaces that can be switched from completely non-wetting to completely wetting and form the basis for novel electrochemical energy storage devices among other potential applications. Frames from the videorecording demonstrate electrically induced transitions between different wetting states of a liquid droplet on the nanostuctured surface (a) with no voltage applied, a droplet of ionic liquid (molten salt) forms a highly mobile ball that can move freely across the surface; (b) upon application of voltage pulse, the droplet undergoes a sharp transition to the immobile droplet state on a fully wetted surface; (c) an immobile droplet of cyclopentanol spreads and fills most of the surface area (d) upon application of a voltage pulse. (Langmuir, T.N., Krupenkine, J.A., Taylor, T.M. Schneider, S., Yang. 2004. American Chemical Society 20(10):3824–3827. With permission.) **(See color insert.)**

while some of the works were derived directly from the attempts to mimic Nature, some others were performed "independently," only later to realize its close resemblance to the naturally occurring structures.

15.3.1 SUPERHYDROPHOBIC SURFACES IN MICROFLUIDICS

Microfluidic devices receive significant attention due to their promise to change chemical detection in small volumes for rapid chemical, biochemical, and drug discovery analysis. However, the ability to alter and manipulate fluid flow is degraded in microchannels where viscous drag conditions prevail. This is one area where superhydrophobic surfaces may find their application. In a report by Krupenkin et al., the authors showed how a drop of liquid could be switched between two distinct states on a superhydrophobic nanostructured surface [28]. There, a mobile droplet (Cassie-Baxter state) freely moving around the surface can be turned into an immobile droplet firmly attached to the nanostructure by fully penetrating it (Wenzel state). The switching has been achieved through an electrowetting phenomenon [29] (Figure 15.5). The method gives an elegant and straightforward way to manipulate and move liquids on the solid surface using digital electronics [30,31]. Some other research groups and commercial enterprises are commercializing similar approaches (or at least electrowetting aspects of it) for applications involving various devices, most notably for lab-on-a-chip devices used for biomedical applications [32,33].

Other microfluidic applications can be envisioned, such as drag reduction of underwater vehicles with the superhydrophobic "skin," cooling structures that selectively transport heat-transfer fluid to "hot" spots on computer chips, and many others, including self-adaptive surfaces [34] and reversible electrically tunable superhydrophobic surfaces [35].

15.3.2 ENERGY STORAGE DEVICES

In this section, we reference the coauthors' efforts in applying superhydrophobic surfaces with tunable wetting to create novel nano- and microscale electrochemical energy storage devices [36,37]. We know them in everyday life as "batteries." The emphasis is placed on this application as it points to not only using phenomenon of superhydrophobic surfaces found in Nature but also shows that biomimicry is not limited to direct replication of the observed phenomenon. It is possible to implement this finding in areas seemingly far removed from the original discovery. Our ongoing development capitalizes on the fundamental work of Krupenkin et al. [28] on the dynamic tuning and control of fluids on nanostructured surfaces to create a special class of batteries known as reserve-type cells.

The key function of such a battery is to keep liquid electrolyte away from solid electrode materials until energy generation is required, thus assuring that no dissipation occurs while not in service (and hence, the term "reserve"). In this design, we chose a superhydrophobic surface to perform that function. It is accomplished through the use of the "nanograss" surfaces. The surfaces consist of an array of cylindrical posts, 1 to 5 μm apart, 5 to 10 μm tall, and 100 to 300 nm in diameter (Figure 15.6). When treated with an appropriate fluorocarbon polymer or other hydrophobic material, such a surface demonstrates superhydrophobic behavior, distinguishing itself from a regular surface by the substantially higher contact angle (i.e., 150° versus 90°) of a liquid suspended on it and the absence of droplets' pinning to the surface. Pinning is the direct result of the contact angle hysteresis, where values of the receding and advancing contact angles differ significantly. A droplet will remain mobile in

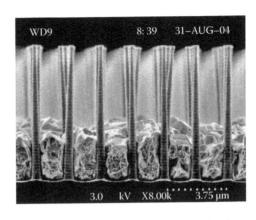

FIGURE 15.6 Electroplated zinc anode deposited on the bottom of the tunable nanostructured surfaces.

the absence of pinning. Therefore, a larger reservoir of electrolyte does not stick but instead remains highly mobile. Moreover, a reservoir on such a surface is supported by the very tips of the nanoposts and does not penetrate into the space between them. It is worth noting that further experimentation has indicated that nanoposts are not the only surface structures capable of supporting the batteries' architecture, and the team is conducting ongoing research that studies nanostructured surfaces with other geometries that appear suitable for a wider range of electrolyte chemistries, for example, primary and secondary (rechargeable) Li-based chemistries for increased energy density.

When suitable electrolyte is placed on a nanograss surface, it remains suspended above the electrodes and never comes into contact with them. Because no contact between the electrolyte and electrode occurs, no energy is being generated, and therefore, no dissipation can occur. Thus, the battery is assured to have its full charge (capacity) when triggered by an external event or by a user. In order to trigger or activate this battery, one needs to bring the electrolyte and electrode together. The use of electrowetting makes it possible to rapidly change the contact angle of the solid–liquid interface by applying an external voltage pulse to the liquid. Figure 15.7 shows a schematic of such a battery as well as its performance in storing and generating energy. Electrowetting has been successfully used to create a variety of optical devices (e.g., lenses and diffraction gratings) and is now being combined with nanostructures to create novel batteries and battery architectures with unique characteristics.

15.3.3 SELF-CLEANING WINDOWS AND UTILITARIAN GLASS STRUCTURES

Similarly, one can think of creating superhydrophobic surfaces on common items made of glass, such as automobile windows, house windows, and even glass shower doors, frequently exposed to water or rain. Clearly, water droplets impinging upon such surfaces will not stick but will rather roll off, helping to improve visibility, while carrying away dirt or deposits to increase the apparent cleanliness of the glass surface. Superhydrophobic property is the key attribute needed for this application, as well as the means to prevent droplets from pinning to the surface of glass when the individual water beads coalesce as they roll across the surface of the glass. Several attempts are currently under way to develop such glass materials while retaining optical transparency by creating micro- and nano-sized roughness on glass surfaces, which is then functionalized with self-assembled monolayers to render the surface superhydrophobic and to maintain reasonable transparency [38–41]. An obvious difficulty in creating such materials and coatings is in their inherent low abrasion resistance, whether to a household wipe, a car's wiper, a cleaning solution, or even repeated exposure to rain and dust of a windshield while driving a car. Although glass is chemically and

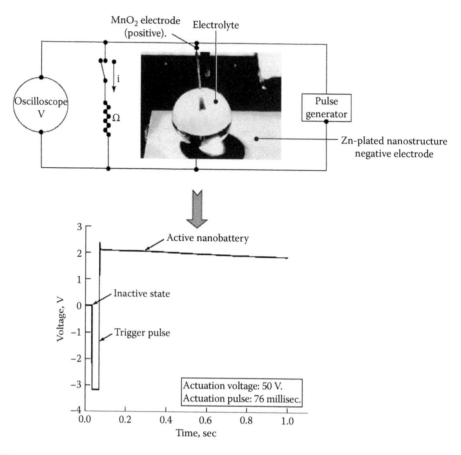

FIGURE 15.7 Schematic of the set-up and experimental details of the first proof-of-principle demonstration of the novel energy storage device based on tunable superhydrophobic nanostructured surfaces.

mechanically stable, nanostructured surface layers with the ultra-thin coatings deposited over them are quite the opposite and will quickly degrade. Work is currently under way to address these issues as well [42–45].

Superhydrophobic coatings are employed in a similar function to prevent water, rain, and snow accumulation on satellite dish receiver antennas and radomes. There, a fine mist of silica particles and hydrophobic material, mostly fluorinated silane, is spread directly on the surface of the structure. Silica particles form "roughness" and silane assembles into hydrophobic coating covalently attached to the silica surface, thus making the surface superhydrophobic.

In summary, we described several examples of how surface topography or roughness on the nanoscale modifies the wetting behavior of such surfaces, making them highly nonwettable. Remarkable, self-cleaning properties of lotus leafs, water-harvesting ability of a desert beetle, and trapping efficiency of carnivorous plants have been described as examples of nanostructured superhydrophobic surfaces evolved in Nature. Several man-made materials are cited as having surface properties similar to those in Nature. Although the origin of these examples may have been from different scientific disciplines that were not initially based on mapping observations from naturally occurring phenomena, it is nevertheless striking on the similarity with a number of nano- and microstructures found in Nature. The ongoing challenge for the scientific and engineering communities will be in extracting technical nuggets in understanding these phenomena and in creating useful and practical designs for the advancement of real-world systems.

ACKNOWLEDGMENTS

Special thanks are extended to the Alcatel-Lucent/Bell Labs team for the R&D efforts in developing the micropower cell architecture described in this chapter.

REFERENCES

1. Wenzel RN. Surface roughness and contact angle. *J. Phys. Colloid Chem.* 53:1466–1467, 1949.
2. Cassie ABD, Baxter S. Wettability of porous surfaces. *Trans. Faraday Soc.* 40:546–551, 1944.
3. de Gennes PG, Brochard-Wyart F, Quere D. *Capillarity and Wetting Phenomena: Drops, Bubbles, Pearls, Waves*, 2003, Springer.
4. Quere D. Non-sticking drops. *Rep. Prog. Phys.* 68:2495–2532, 2005.
5. Cassie ABD, Baxter S. Large contact angles of plant and animal surfaces. *Nature* 155:21–22, 1945.
6. Fogg GE. Diurnal variation in a physical property of leaf cuticle. *Nature* 154:515, 1944.
7. Gao L, McCarthy T. "Artificial Lotus leaf" prepared using a 1945 patent and a commercial textile. *Langmuir* 22:5998–6000, 2006. For some of the critique of the lack of historical perspective on "super-hydrophobic" phenomenon.
8. Callies M, Quere D. On water repellency. *Soft Matter.* 1:55–61, 2005.
9. Neinhuis C, Barthlott W. Characterization and distribution of water-repellent, self-cleaning plant surfaces. *Ann. Bot.* 79:667–677, 1997.
10. Barthlott W, Neinhuis C. Purity of the sacred lotus, or escape from contamination in biological surfaces. *Planta* 202:1–8, 1997.
11. Parker AR, Lawrence CR. Water capture by a desert beetle. *Nature.* 414:33–34, 2001.
12. Otten A, Herminghaus S. How plants keep dry: A physicist's point of view. *Langmuir* 20:2405–2408, 2004.
13. Gao L, McCarthy T. The "Lotus Effect" explained: Two reasons why two length scales of topography are important. *Langmuir* 22:2966–2967, 2006.
14. Oner D, McCarthy T. Ultrahydrophobic surfaces: Effects of topography length scales on wettability. *Langmuir* 16:7777–7782, 2000.
15. Cheng YT, Rodak DE, Wong CA, Hayden CA. Effects of micro- and nano-structures on the self-cleaning behavior of lotus leaves. *Nanotechnology* 17:1359–1362, 2006.
16. Feng L, Li S, Li Y, Li H, Zhang L, Zhai J, Song Y, Liu B, Jiang L, Zhu D. Super-hydrophobic surfaces: From natural to artificial. *Adv. Mater.* 14:1857–1860, 2002.
17. Sun T, Feng L, Gao X, Jiang L. Bioinspired surfaces with special wettability. *Acc. Chem. Res.* 38:644–652, 2005.
18. Liu Y, Chen X, Xin JH. Super-hydrophobic surfaces from a simple coating method: A bionic nanoengineering approach. *Nanotechnology* 17:3259–3263, 2006.
19. Furstner R, Barthlott W, Neinhuis C, Walzel P. Wetting and self-cleaning properties of artificial super-hydrophobic surfaces. *Langmuir* 21:956–961, 2005.
20. Zhai L, Cebeci FC, Cohen RE, Rubner MF. Stable superhydrophobic coatings from polyelectrolyte multilayers. *Nano Lett.* 4:1349–1353, 2004.
21. Zhai L, Berg MC, Cebeci FC, Kim Y, Milwid JM, Rubner MF, Cohen RE. Patterned superhydrophobic surfaces: Toward a synthetic mimic of the Namib Desert beetle. *Nano. Lett.* 6:1213–1216, 2006.
22. Watson GS, Watson JA. Natural nano-structures on insects-possible functions of ordered arrays characterized by atomic force microscopy. *Appl. Surf. Sci.* 235:139–144, 2004.
23. Shang HM, Wang Y, Takahashi K, Cao GZ, Li D, Xia YN. Nanostructured superhydrophobic surfaces. *J. Mater. Sci.* 40:3587–3591, 2005.
24. Bohm HF, Federle W. Insect aquaplaning: *Nepenthes* pitcher plants capture prey with the peristome, a fully wettable water-lubricated anisotropic surface. *PNAS* 101:14138–14143, 2004.
25. Gaume L, Gorb S, Rowe N. Function of epidermal surfaces in the trapping efficiency of *Nepenthes alata* pitchers. *New Phytologist* 156:479–489, 2002.
26. Some of us may have, unfortunately, experienced aquaplaning or hydroplaning when one's car starts skidding off the flooded roadway. In this case, the car tires lose traction with the road when they become separated from the pavement by a thin sheet of water.
27. The most up-to-date list of "Lotus effect" related work is located at http://www.nees.uni-bonn.de/lotus/en/lotus_effect_multimedia.html.

28. Krupenkine TN, Taylor JA, Schneider TM, Yang S. From rolling ball to complete wetting: The dynamic tuning of liquids on nanostructured surfaces. *Langmuir* 20:3824–3827, 2004.
29. Electrowetting is a phenomenon of controlling (changing) surface energy of the liquid–solid interface by applying electrostatic charges to the interface between phases in contact.
30. Kim J, Kim CJ. Nanostructured surfaces for dramatic reduction of flow resistance in droplet-based microfluidics. *Proc. IEEE Micro Electro Mech. Syst.* 479–482, 2002.
31. Ichimura K, Oh S-K, Nakagawa M. Light-driven motion of liquids on a photoresponsive surface. *Science* 288;1624–1626, 2000.
32. Wheeler AR, Moon H, Kim CJ, Loo JA, Garrell RL. Electrowetting-based microfluidics for analysis of peptides and proteins by matrix-assisted laser desorption/ionization mass spectrometry. *Anal. Chem.* 76:4833–4838, 2004.
33. www.nanolytics.com.
34. Minko S, Muller M, Motornov M, Nitschke M, Grundke K, Stamm M. Two-level structured self-adaptive surfaces with reversiblyt properties. *J. Am. Chem. Soc.* 125:3896–3900, 2003.
35. Krupenkin TN, Taylor JA, Hodes M, Kolodner PR, Wang EN, Salamon TR. Reversible wetting–dewetting transitions on electrically tunable superhydrophobic nanostructured surfaces. *Langmuir* 23:9128–9133, 2007.
36. Lifton VA, Simon S, Frahm RE. Reserve battery architecture based on superhydrophobic nanostructured surfaces. *Bell Labs Tech. J.* 10:81–85, 2005.
37. Lifton VA, Simon S. Reserve battery architecture based on superhydrophobic nanostructured surfaces. *Proc. NSTI.* 2:726–729, 2005.
38. Nakajima A, Hashimoto K, Watanabe T, Takai K, Yamauchi G, Fujishima A. Transparent superhydrophobic thin films with self-cleaning properties. *Langmuir* 16:7044–7047, 2000.
39. Shang HM, Wang Y, Limmer SJ, Chou TP, Takahashi K, Cao GZ. Optically transparent superhydrophobic silica-based films. *Thin Solid Films* 472:37–43, 2005.
40. Ogawa K, Soga M, Takada Y, Nakayama I. Development of a transparent and ultrahydrophobic glass plate. *Jpn. J. Appl. Phys.* 32:L614–L615, 1993.
41. Feng L, Zhang Z, Mai Z, Ma Y, Liu B, Jiang L, Zhu D. A super-hydrophobic and super-oleophilic coating mesh film for the separation of oil and water. *Angew. Chem. Int. Ed.* 43:2012–2014, 2004.
42. Nakajima A, Abe K, Hashimoto K, Watanabe T. Preparation of hard super-hydrophobic films with visible light transmission. *Thin Solid Films* 376:140–143, 2000.
43. Kemell M, Farm E, Leskela M, Ritala M. Transparent superhydrophobic surfaces by self-assembly of hydrophobic monolayers on nanostructured surfaces. *Phys. Stat. Sol. A.* 203:1453–1458, 2006.
44. Nakajima A, Fujishima A, Hashimoto K, Watanabe T. Preparation of transparent superhydrophobic boehmite and silica films by sublimation of aluminum acetylacetonate. *Adv. Mater.* 11:1365–1368, 1999.
45. Duparre A, Flemming M, Steinert J, Reihs K. Optical coatings with enhanced roughness for ultrahydrophobic, low-scatter applications. *Appl. Opt.* 41:3294–3298, 2002.

16 Nanotechnology in Drug Delivery for Malaria and Tuberculosis Treatment

Hulda S. Swai, Paul K. Chelule,
Boitumelo Semete, and Lonji Kalombo
Polymers and Bioceramics, Council for Scientific and
Industrial Research (CSIR), Pretoria, South Africa

CONTENTS

16.1   INTRODUCTION

Nanotechnology has gained ground in the twenty-first century and is rapidly growing due to the ability to manipulate and harness properties of assemblies that are at the nanosize scale of various biomolecules. In addition, nanotechnology allows scientists to alter the chemical, physical, and biological properties of these assemblies, allowing for their synthesis at a controlled size range of 1 to 500 nm. When combined with other biotechnology tools such as bioinformatics, imaging, and systems biology, nanotechnology holds great promises to address challenges faced in the field of drug development and delivery. Nanotechnology may be used to modify other factors such as localized drug delivery, site-specific activity, periods of maintenance within the body, patient compliance, as well as adverse drug effects. If these factors are not controlled, they may eventually lead to lowered drug efficacy. The challenges for site-specific delivery are the cellular barriers, which prevent the drug from reaching the desired target [1], thus presenting a need for a carrier system that will overcome this effect.

A wide variety of natural polymers have been investigated for drug delivery. The limitation of these natural polymers, however, is the uncertainty of their purity as well as their efficiency [2]. Synthetic polymers have also been explored, along with various systems such as hydrogels, microspheres, and nanoparticles. Based on the schematic illustration indicated in Figure 16.1, it can be concluded that nanoparticles have the best delivery efficiency followed by microparticles.

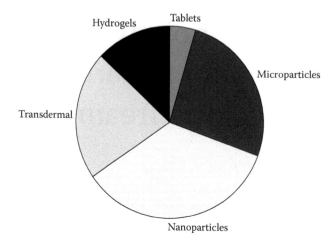

FIGURE 16.1 Drug release efficiency by various polymeric systems. (Adapted from Ravi Kumar MNV, Kumar N, *Drug Dev. and Ind. Pharm.* 27:1–30, 2001. With permission.)

Since the last two decades, synthetic polymers have been used more frequently than the natural polymers, as they have fewer limitations than the latter. However, ways of overcoming these limitations are becoming available of late, due to research inputs in this field. These will be discussed further in subsequent sections. Due to the challenges that have been encountered in the field of drug development and the treatment of diseases such as malaria and tuberculosis, which are of a major burden in Africa, various nano-based carriers have been developed. The rest of the chapter will, therefore, focus on the use of polymeric nanodrug delivery systems developed to date, to address challenges related to treatment of these diseases [4,5].

Nanoparticles, which are submicron-sized (less than 1 micron) polymeric colloidal particles, are used to encapsulate therapeutic agents within their polymeric matrix, or adsorbed or conjugated onto the surface. The major advantage of nanoparticulate drug delivery systems is that the release of the active agent can be controlled. Thus, the release may be cyclical, or it may be triggered by the environment or by another external event [6]. The nanopolymeric drug delivery systems have proved to have the following advantages [7]:

- Ease of administration.
- Localized delivery/drug targeting.
- Protection of the cargo from degradation before reaching the target site.
- Intracellular delivery.
- Provision of stealth functions.
- Prolonged delivery periods.
- Decreased drug dosages with concurrent reduction in possible undesirable side effects common to most forms of systemic delivery.
- Improved patient compliance and comfort.

16.2 MALARIA

Malaria is the leading cause of death in sub-Saharan Africa, even surpassing human immunodeficiency virus and acquired immunodeficiency syndrome (HIV/AIDS) [8]. The annual worldwide mortality rates due to malaria range from 0.5 to 3 million people, and up to 3000 children in Africa die every single day due to this infectious disease. Malaria contributes about 2.3% of the disease burden in Africa [9]. Increased malaria prevalence and higher patient mortality can be attributed to a number of factors. The major ones include increased drug resistance by malaria parasites,

TABLE 16.1
Various Classes of Antimalarials and Their Parasite-Target Activity

Class	Examples	Activity
4-Aminoquinolines	Chloroquine, amodiaquine	Erythrocytic schizonts
8-Aminoquinolines	Primaquine, tafenoquine	Gametocytocidaltissue schizonts
Aryl-amino alcohols	Quinine, mefloquine, halofantrine, pyronaridine	Erythrocytic schizonts
Folate inhibitors	Pyrimethamine, proguanil, sulfadoxine, chlorproguanil, dapsone	Tissue/erythrocytic schizonts
Peroxides	Artemether, artesunate dihydroartemisinin, arteether	Tissue and erythrocytic schizonts
Antimicrobials	Tetracycline, doxycycline, clindamycin, azithromycin	Erythrocytic schizonts
Napthoquinones	Atovaquone	Erythrocytic schizonts

Sources: See References [12] and [13].

especially *Plasmodium falciparum* [10], which decrease the efficacy of some antimalarial drugs and subsequently increase noncompliance in patients. For those drugs that may still be effective, noncompliance to the drug regimen is the leading cause of treatment failure. Most antimalarials are purchased over the counter and not sold as a prescription drug, especially in developing countries where they are available in convenience stores with the retailer having little or no knowledge of proper dosage. In this case, the drugs are unlikely to be taken correctly or patients may not adhere to treatment. In some cases, patients may take antimalarial drugs when they feel sick but may not have malaria. Therefore, noncompliance becomes an issue when patients take these drugs without completing the full course and thus increase the chances of becoming drug resistant. Furthermore, there are reports of resistance emerging from newer artemisinin usage *in vitro* [11]. Artemisinins are usually administered as intramuscular formulations that are erratically absorbed and hence become less bioavailable in the tissues, giving suboptimal concentrations and poor therapeutic outcome.

Antimalarial drugs are currently classified into six groups according to their chemical structure or function. A comprehensive review on these drugs is available [12,13] (Table 16.1). Antimalarials are known to target different stages of the malaria parasite (i.e., tissue and erythrocytic scizonts) and gametocytes as shown in Table 16.1. The older drugs (4- and 8-aminoquinolines and folate inhibitors), which were cheaper and readily available in the poor countries, are either ineffective or too toxic (e.g., chloroquine, primaquine, pyrimethamine, and sulfadoxine). In the case of treatment with chloroquine, the oral dose is usually 10 mg/kg start dose, then three doses of 5 mg/kg each, over 36 to 48 hours [14]. From this it can be observed that frequent dosages have to be given (due to the poor bioavailability of the drug) in order to maintain the minimum inhibitory concentration (MIC). For example, the mean 50% inhibitory concentration (IC_{50}) for chloroquine, quinine, tafenoquine, primaquine, dihydroartemisinin, mefloquine, desbutylhalofantrine, and proguanil were 29.7, 64.8, 64.6, 242.7, 0.2, 5.7, 3.0, and 58.6 ng/ml, respectively, when tested *in vitro* on one *P. falciparum* clone from Thailand [15]. The concentration of drugs taken is usually high and can cause toxic side effects. For example, at therapeutic doses, chloroquine can cause dizziness, headache, disturbed visual accommodation, dysphagia, nausea, malaise, and pruritus of palms, soles, and scalp [16]. It can also cause visual hallucinations, confusion, and psychosis on some occasions.

When used as a prophylactic at 300 mg of the base/week, it can cause retinal toxicity after 3 to 6 years (i.e., after an accumulated concentration of 50 to 100 g of chloroquine). Intramuscular injections of chloroquine can cause hypotension and cardiac arrest, particularly in children [10]. Although chloroquine is still used in many parts of the world as a first-line drug, it has been banned in several countries in sub-Saharan Africa where resistance has reached levels above 50% [14]. This high rate of chloroquine resistance is partly due to antimalarials not being used in combination so as to improve efficacy and reduce chances of resistant parasites emerging. Amodiaquine, in the same class as chloroquine, has a potential for resistance and liver toxicity. Other drugs brought in to replace chloroquine, such as mefloquine, primaquine, atovaquone, proguanil, doxycycline, and sulfadoxin-pyrimethamine, are not faring better and may only treat certain species of *Plasmodium*. In fact, it has been reported that *P. falciparum* is resistant to almost all of these drugs, and the spread of this highly resistant strain could be a recipe for disaster [16]. Furthermore, the response to quinine (aryl-amino alcohol), a drug useful in treating severe falciparum malaria (which is resistant to chloroquine), has been reducing steadily and may require additional artemisinin dosage to eliminate the parasites from the body.

Newer drugs, termed artemisinin-based combination therapies (ACTs) are the best available in the market and can lower resistance while enhancing efficacy [17]. This drug regimen has been adopted as the first line of treatment in 43 countries in the world. However, they are very expensive, costing up to 20 times more than the commonly available antimalarials. This makes treatment difficult to implement in developing countries. Most artemisinin-derived drugs are poorly soluble in water; thus, they have to be given in high doses (which may give rise to drug toxicity) to maintain the MIC levels and increase bioavailability in the body. In addition, they are usually administered intradermally which limits their use in rural villages where no specialized and skilled medical personnel are available. The normal dosage for artemether is 3.2 mg/kg intramuscularly as a loading dose, followed by 1.6 mg/kg daily until oral therapy or a maximum of 7 days. This method and length of treatment period may lead to increased noncompliance in patients. Artesunate, the oral artemisinin derivative, is available and is usually administered as 5 mg/kg on the first day followed by 2.5 mg/kg on the second and third days with 15 mg/kg of mefloquine taken as a single dose on the second day [14]. This treatment regimen over a maximum of 7 days is quite long and often reduces compliance in patients. There is thus a need for a sustainable delivery system in which antimalarials can be given, for example, in a nanoencapsulated formulation, which can address dose frequency, treatment duration, improved bioavailability, and efficacy. In this system of delivery, drugs may be given in smaller efficient dosages with frequencies in excess of a week, maximizing compliance and successful therapeutic outcome.

16.3 TUBERCULOSIS

Tuberculosis (TB), once regarded as a disease of the mid-twentieth century, is still a leading cause of death of young adults worldwide [18]. Despite the discovery of the TB bacillus in 1882, and that of anti-TB drugs since 1944, efforts to control TB globally have so far failed. Approximately 2 billion people (i.e., one-third of the world's population) are infected with the TB bacterium. Up to 70% of adults in Sub-Saharan Africa are infected, and once infected, a person stays infected for life [18]. In fact, 70 to 85% of TB patients are HIV infected [19]. More than a third of HIV-infected people will develop TB, mainly due to a suppressed immune system. The immune deficiency state makes it possible for opportunistic infections to thrive. TB is not just one of the opportunistic infections associated with HIV/AIDS—it is the most lethal coinfection [19]. Every year, 8 million people worldwide develop active TB and 3 million die from it, while more than 400,000 new cases of multidrug resistant TB (MDR-TB) are diagnosed [20]. The first-line drugs currently administered to treat TB include rifampicin (RIF), isoniazid (INH), pyrazinamide (PYR), and ethambutol (ETB). The current doses administered and the MIC_{90} of these drugs are indicated in Table 16.2.

This fixed dose combination has proved to be effective when taken according to prescription. However, patient noncompliance due to the patient not taking the medication as prescribed has led to the escalation of TB cases and emergence of MDR-TB, particularly in Africa, leading to treatment failure. Furthermore, clinical management of these diseases is limited because of toxic side effects of the drugs, degradation of drugs before reaching their target site, low permeability, and poor patient compliance [21]. Thus, the drawbacks of conventional chemotherapy necessitate

TABLE 16.2

Properties of First-Line Antituberculosis Drugs

First-Line Drugs	Mode of Action	Metabolism	Daily Dose	MIC$_{90}$
RIF (MW = 822.9)	Inhibits assembly of bacterial DNA and protein into mature virus Inhibits initiation of RNA synthesis	Deacetylation	10–12 mg/kg	0.2 ug/ml
INH (MW = 137.1)	Inhibits synthesis of cell-wall components	Acetylation and hydroxylation	5 mg/kg	0.3 ug/ml
PYR (MW = 123.1)	Disrupts membrane potential Inhibits membrane transport functions	Hydrolysis	25 mg/kg	8 ug/ml
ETB (MW = 277.23)	Inhibits cell-wall synthesis	Metabolized by hepatic enzymes	15–20 mg/kg	6 ug/ml

Sources: See References [21], [22], and [23].

the development of a delivery or carrier system that can release drugs slowly over extended time periods and reduce the current costs of treatment.

16.4 NANOTECHNOLOGY IN DRUG DELIVERY

Nanotechnology for drug delivery offers a suitable means of delivering small-molecular-weight drugs as well as macromolecules such as proteins, peptides or genes to specific tissues and intracellular compartments, particularly in cases of DNA delivery [24]. Particles in the submicron range possess very high surface-to-volume ratios, thus allowing for an intimate interaction between the surface of the particles and the mucus of various tissues. Additionally, carriers in the particulate form reportedly diffuse further into the mucus layer, enabling them to reach the cells. The particle size, surface properties, and relative hydrophobicity are the main factors affecting the particles' effectiveness in prolonging their transit time in the gastrointestinal (GI) tract and additionally protecting the active agents from degradation [25].

Polymers, used to make nanoparticles, can be divided into two classes: natural and synthetic. Natural polymers (i.e., proteins or polysaccharides) include chitosan, alginate, starch, and albumin. The advantages of using natural polymers are that they are nontoxic and can easily be biodegraded; however, they are not widely used to make nanoparticles because they vary in purity and often require rigorous optimization before successful encapsulation of drugs can be achieved. Smart, synthetic polymers that have been widely used include polylactic acid (PLA), poly(glycolic acid) (PGA), poly-lactic-glycolic acid (PLGA), poly-e-caprolactone (PCL), and poly(alkylcyanoacrilate) [24]. Currently, it is possible to generate degradable nanoparticles with responsive polymers that have accessible surface functionality for targeting, steric shielding for biocompatibility (bioavailability), therefore making them stealth, and with encapsulated polymer prodrug for controlled release of specific drugs [26], thus allowing for the development of smart therapeutics. Although the application of nanoparticles for targeted drug delivery is well received, much work still needs to be performed due to an array of receptors/signals/molecules expressed on infected cells as well as a number of physiological signals emitted by the infected cells or tissue. Understanding these processes will improve the development of targeted nanoparticles for drug delivery. It is envisaged that these polymeric nanosystems will enable entry and retention of the antibiotics in the cells. Nanoparticles also facilitate the subcellular distribution and activity of the antibiotic in the infected cells, thus addressing the major shortfalls of failed malaria and TB therapy. In addition to these advantages, the carrier systems will reduce unwanted systemic side effects associated with conventional free antibiotics [27].

The anti-TB and antimalarial drugs can be incorporated either in their native forms or as conjugated polymer prodrugs. Controlled release is dependent on nanoparticle biodegradability and drug diffusion from the nanoparticles. In the case of hydrophobic drugs such as rifampicin and halofantrine (anti-TB and antimalarial drugs, respectively), the attachment of hydrophilic polymer chains greatly enhances water solubility once released, leading to better penetration into infected cells and ultimate clearance of the infection.

Limited reports on polymeric encapsulation of antimalarials are available and are based on animal models. For example, halofantrine was formulated in nanocapsules with an oily core, prepared from PLA that was surface modified with polyethylene glycol chains for longer residence in the circulatory system [25]. An encapsulation efficiency greater than 90% was attained. When this formulation was administered to *Plasmodium*-infected mouse model, the drug concentration in plasma was increased sixfold in comparison with free drug throughout the experimental period of 70 hours. Furthermore, no toxic effects associated with halofantrine were observed when it was nanoencapsulated while the same dose of free drug was toxic. Other antiparasitic drugs like pentamidine and atovaquone have been delivered using nanoparticles with success in leishmania- and toxoplasma-infected mice, respectively [28,29]. The ED_{50} (effective dose for 50% of the population exposed to the drug or a 50% response in a biological system that is exposed to the drug) of nanoencapsulated pentamidine was six times lower than that of free pentamidine, demonstrating the ability of nanocapsules to increase the bioavailability and decrease drug toxicity of pentamidine in treatment of leishmaniasis. Similarly, atovaquine when administered in nanosuspensions to *Toxoplasma*-infected murine model demonstrated excellent therapeutic effect following clearance of parasites from the system [29]. In this study, two groups of mice were treated separately with free 100 mg/kg atovaquone suspension orally and different concentrations of atovaquone nanosuspensions (ANSs) (0.1, 1.0, and 10.0 mg/kg of body weight), administered as a single intravenous dose every other day for 16 days. Survival rates of mice orally treated with atovaquone suspension (100 mg/kg every other day) were significantly lower than those of mice treated intravenously with ANSs at a dose of 10 mg/kg. Nanoencapsulation of atovaquone made it possible to administer this drug orally and also increased the plasma concentrations of the drug 2.4 times that of the free drug.

It would seem, from the report on antiparasitics [29], that it is possible to encapsulate antimalarials into polymeric nanoparticles for targeted delivery and release into body organs. In fact, triclosan, a broad-spectrum antimicrobial agent shown to be effective in treating malaria, has very low aqueous solubility and can potentially give rise to shortfalls in formulation and bioavailability. This drug has been loaded into chitosan-coated nanocapsules with encapsulation efficiency of 93 to 98% [30]. Chitosan-coated nanocapsules were obtained by adding triclosan-loaded submicron emulsions, previously prepared by solvent displacement method, to chitosan solution. The slow release studies were done *in vitro* by reverse dialysis bag technique. The drug-loaded nanocapsules allowed slower and more controlled drug release (over 6 hours) than the free drugs, which rapidly diffused across the dialysis membrane within 60 minutes [30]. This nanoencapsulated drug delivery system is in the process of development before it can be used to deliver trichlosan into human tissues.

16.5 RESEARCH IN PROGRESS

16.5.1 INTRODUCTION

Our team synthesized nanoparticles formulated using U.S. Food and Drug Administration (FDA)-approved biodegradable and biocompatible polymers with a therapeutic agent encapsulated (in this case antituberculosis drugs [ATDs]) into the polymer for targeted and sustainable drug delivery, which addresses patient noncompliance, toxicity, dose frequency, length of treatment, and low bioavailability of drugs for the treatment of TB and also reduces the current cost of treatment. Once this technology is optimized, we intend to apply the same technique for encapsulation of antimalarials.

Several attempts have been made to encapsulate TB drugs in micro-/nanoparticles with a degree of success in mice models [31]. Recently, Pandey et al. [31], developed an oral anti-TB drug-loaded nanoparticle that can release anti-TB drugs continuously for 10 days in a murine model. The slow release of drugs was possible due to the bioadhesive properties of PLGA polymers, which enable them to adhere to the small intestinal mucosa. Also, unlike the microparticles, nanoparticles can be taken up via various routes through the intestinal mucosa, as discussed in Section 16.5.3. Following oral admission of anti-TB drug-loaded nanocapsules to TB-infected mice, no tubercle bacilli could be detected in the tissue after five oral doses of treatment [31]. It can be concluded that nanocapsules loaded with anti-TB drugs form a sound basis for reduction in dosing frequency for better treatment of TB. The authors are now conducting a clinical trial based on animal-trial findings.

16.5.2 SYNTHESIS OF NANOPARTICLES

PLGA particles of an average size of 250 nm (Figure 16.2) have been prepared and characterized in our laboratory. In the preparation process, the double emulsion solvent evaporation freeze-drying technique was utilized. In addition, several techniques such as nanoprecipitation and spray drying have been explored. The scanning electron microscopy (SEM) images of the particles prepared using the above techniques are presented in Figure 16.3A and Figure 16.3B, respectively. As illustrated in Figure 16.3C, our team has also prepared alginate/chitosan particles via ionotropic gelation resulting in particles of an average size of 200 nm. An encapsulation efficiency of 55 to 65% for INH and rifampicin has, to date, been achieved.

16.5.3 MECHANISM OF ACTION

When orally taken, nanoparticles reach the intestinal mucosa where they are taken up paracellularly via the M cell, intracellularly via epithelial-cell–intestine mucosa and into the lymphatic system via Peyer's patches [32,33]. Due to the rapid clearance by macrophages once internalized, the nanoparticle degradation rate can be manipulated by varying the ratio and increasing the hydrophilicity of the polymer surface. Preliminary *in vitro* release assays performed at this stage by Pandey et al. [31], in mice, show a slow and prolonged release of drugs when encapsulated into PLG nanoparticles. It was observed that free drugs are released almost immediately after absorption and cleared within 24 hours, and the release of encapsulated drugs begins after 6 hours of internalization and lasts over 192 hours [31].

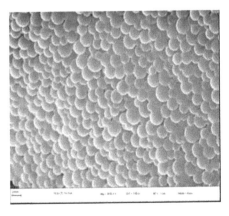 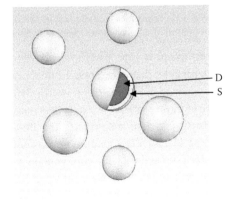

FIGURE 16.2 Left: Scanning electron micrograph of poly-lactic-glycolic acid (PLGA) nanoparticles containing encapsulated isoniazid (INH) and rifampicin drugs with particle sizes of 200 to 400 nm (scale bar is 200 nm). Right: Tuberculosis-encapsulated drugs showing the drug (D) shaded and the nanopolymeric shell (S) unshaded.

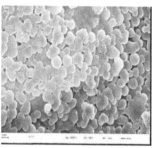 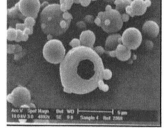 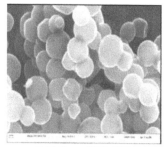

A: Nanoprecipitation   B: Spray drying of emulsion   C: Chitosan-TPP+PVA

FIGURE 16.3 Scanning electron micrograph of nanoparticles prepared by (A) nanoprecipitation resulted in rifampicin, encapsulation efficiency of 65%. (B) Spray drying resulted in isoniazid encapsulation efficiency of 55%. (C) Alginate/chitosan particles prepared by ionotropic gelation.

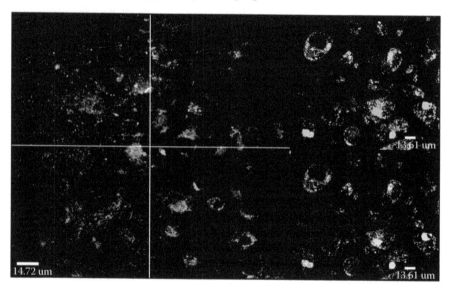

FIGURE 16.4 Nanocapsule transport/uptake by CaCo-2 cells viewed under a confocal microscope. (A) Z-stack of cells treated with nanoparticles over a 5 minute incubation period. (B) Appearance of cells depicting lysotracker stain in cells treated with rhodamine-labeled particles. (C) Same image showing colocalization of nanoparticles with lysosomes as indicated by the orange staining at 60 minutes postadministration. Rhodamine dye was used as a fluorophore. **(See color insert following page 112.)**

We performed *in vitro* particle uptake assays in CaCo-2 cells, which are cell models for the colon with particles that were labeled with rhodamine dye. Based on the confocal analysis of particle uptake as depicted in Figure 16.4, it was evident that the size of the particles (mean 250 nm) enables the uptake of these particles by cells, thus making intracellular delivery a feasible objective.

The second system we investigated involves the use of amphiphilic block copolymers of polyethylene glycol (PEG) and polypropylene sulphate (PPS), to form micelles (23 nm) with a hydrophobic core and rubbery-core nanoparticles of PPS stabilized with copolymers of PEG and poly (propylene oxide), which are referred to as Pluronics [34,35], to form solid particles within a nano-size range (30 to 100 nm). Rifampicin was encapsulated into both of these particles, and an encapsulated efficiency of 86% and 90% was achieved, respectively. As depicted in Figure 16.5A, the *in vitro* release of rifampicin encapsulated in the micelles and solid nanoparticles was slower and prolonged for a period of 10 days when compared to free rifampicin. As indicated in Figure 16.5B, at an acidic pH of 1.2, mimicking stomach conditions, only a very low concentration of

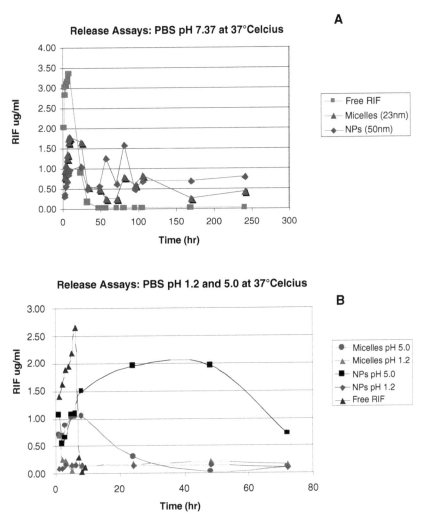

FIGURE 16.5 *In vitro* release profile of rifampicin (RIF) at different pH values in phosphate-buffered saline (PBS).

the drug was released. This indicates that with this system, the first-pass metabolism in the stomach can be minimized, due to the low concentration of drugs released. We intend on extending the study to mice followed by nonhuman primates and finally to human clinical trials. Once this drug delivery system is optimized for TB treatment, we will explore ways of using it to deliver malaria and HIV drugs.

16.6 CONCLUSION

Due to the slow degradation of polymers, which results in a slow release mechanism of the carrier systems, it is envisaged that using nano-based delivery systems will prolong the bioavailability of the drugs, allowing their administration only once in 7 days, instead of the current daily administration. Using this approach, we aim to reduce the dose currently administered, as well as the dose frequency, thereby minimizing the cost of TB treatment, which will ultimately ease the economic burden, estimated to be $16 billion per annum. Once the system has been optimized for TB drugs, it is envisaged that the technology will be applied for malaria, HIV/AIDS, as well as other diseases, which suffer from patient noncompliance and are major burdens in Africa.

ACKNOWLEDGMENTS

We would like to acknowledge the financial assistance of the Department of Science and Technology (South Africa) and CSIR Biosciences and Grobler at North West University, Potchfestroom Campus, for assistance in confocal work. We would also like to thank Professor Khuller, our project collaborator from India, for the provision of data on mice experiments, and Professor J. Hubbell for providing the facility and infrastructure for performing the encapsulation of Rifampicin into PEG/Pluronic-PPS systems.

REFERENCES

1. Pinto-Alphandery H, Andremont A, Couvreur P. Targeted delivery of antibiotics using liposomes and nanoparticles: Research and applications. *Int. J. Antimicrob. Agents,* 13:155–168, 2000.
2. Ratajska M, Boryniec S. Physical and chemical aspects of biodegradation of natural polymers. *React. and Functional Polym.,* 38:35–49, 1998.
3. Ravi Kumar MNV, Kumar N. Polymeric controlled drug-delivery systems: Perspective issues and opportunities. *Drug Dev. and Ind. Pharm.,* 27:1–30, 2001.
4. Panyam J, Labhasetwar V. Biodegradable nanoparticles for drug and gene delivery to cells and tissues. *Adv. Drug Deliv. Rev.,* 5524:3329–3347, 2003.
5. Davis BG, Robinson MA. Drug delivery systems based on sugar-macromolecule conjugates. *Curr. Opinion in Drug Discov. and Dev.,* 5:279–288, 2005.
6. De las Hera Alarcon C, Pennadam S, Alexander C. Stimuli responsive polymers for biomedical applications. *Chem. Soc. Rev.,* 34:276–285, 2005.
7. Davis SS. Biomedical applications of nanotechnology—Implications for drug targeting and gene therapy. *TIBTECH,* 15:217–223, 1997.
8. Health-related issues for the African continent, *African Health Care,* November 2003.
9. World Health Organization. Malaria in Africa, Roll Back Malaria Partnership, WHO, 2001.
10. Baird KJ. Effectiveness of antimalarial drugs. *N. Eng. J. Med.,* 135:1565–1577, 2005.
11. Krishna S, Woodrow CJ, Staines HM, Haynes RK, Mercereau-Puijalon O. Re-evaluation of how artemisinins work in the light of emerging evidence of *in vitro* resistance. *Trends Mol. Med.,* 12:201–205, 2006.
12. Antimalarial drugs: www.malariasite.com/malaria/chloroquine.htm. Last updated April 14, 2006.
13. Robert A, Benoit-Vical F, Dechy-Cabaret O, Meunier B. From classical antimalarial drugs to new compounds based on the mechanism of action of artemisinin. *Pure. Appl. Chem.,* 73:1173–1188, 2001.
14. Ginsburg H. Should chloroquine be laid to rest? *Acta Tropica,* 96:16–23, 2005.
15. Ohrt C, Willingmyre GD, Lee P, Knirsch C, Milhous W. Assessment of Azithromycin in combination with other antimalarial drugs against *Plasmodium falciparum in vitro. Antimicrob. Agents and Chemother.,* 46:2518–2524, 2002.
16. White NJ, Nosten F, Looareesuwan S, Watkins WM, Marsh K, Snow RW, Kokwaro G, Ouma J, Hien TT, Molyneux ME, Taylor TE, Newbold CI, Ruebush II TK, Danis M, Greenwood BM, Anderson RM, Olliaro P. Averting a malaria disaster. *Lancet,* 353:1965–1967, 1999.
17. Mutabingwa TK. Artemisin-based combination therapies (ACTs): Best hope for malaria treatment but inaccessible to the needy. *Acta Topica,* 95:305–315, 2005.
18. World Health Organization. Global Tuberculosis Control report, WHO, March 2002.
19. Villacian JS, Tan GB, Teo LF, Paton NI. The effect of infection with mycobacterium tuberculosis on T-cell activation and proliferation in patients with and without HIV co-infection. *J. Infection* 2005.
20. UNAIDS. The impact of HIV and AIDS on Africa, UNAIDS, 2004.
21. Smith PJ, van dyk J, Fredericks A. Determination of rifampicin, isoniazid and pyrazinamide by high performance liquid chromatography after their simultaneous extraction from plasma. *Int. J Tuberc. Lung Dis,* 3:S325–S328, 1999.
22. Agrawal S, Singh I, Kaur KJ, Bhade S, Kaul CL, Panchagnula R. Bioequivalence trials of rifampicin containing formulation: Extrinsic and intrinsic factors in the absorption of rifampicin. *Pharmacol. Res.,* 50:317–327, 2004.
23. World Health Organization. Fixed dose combination tablets for the treatment of tuberculosis, WHO, April 1999.

24. Bugunia-Kubik K, Sugisaka M. From molecular biology to nanotechnology and nanomedicine. *Biosys.,* 65:123–138, 2002.
25. Biganzoli E, Cavenaghi LA, Rossi R, Brunati MC, Nolli ML. Use of CaCo-2 cell culture model for the characterization of intestinal absorption of antibiotics. *Il Farmaco,* 54:594, 599, 1999.
26. Hans ML, Lowman AM. Biodegradable nanoparticles for drug delivery and targeting. *Curr. Opinion Solid State and Mater. Sci.,* 6:319–327, 2002.
27. Mosqueira VCF, Loiseau PM, Bories C, Legrand P, Devissaguet J-P, Barrat G. Efficacy and pharmacokinetics of intravenous nanocapsule formulations of halofantrine in *Plasmodium berghei*-infected mice. *Antimicrob. Agents and Chemother.,* 48:1222–1228, 2004.
28. Durand R, Paul M, Rivollet T, Houin R, Astier A, Deniau M. Activity of pentamidine-loaded nanoparticles against *Leishmania infantum* in a mouse model. *Int. J. Parasitol.,* 27:1361–1367, 1997.
29. Scholer N, Krause K, Muller RH, Borner K, Hahn H, Liesenfeld O. Atovaquone nanosuspensions show excellent therapeutic effect in a new murine model of reactivated toxoplasmosis. *Antimicrob. Agents Chemother.,* 34:1771–1779, 2001.
30. Maestrelli F, Mura P, Alonso MJ. Formulation and characterisation of trichlosan submicron emulsions and nanocapsules. *J. Microencapsulation,* 21:857–864, 2004.
31. Pandey R, Zahoor A, Sharma S, Khuller GK. Nanoparticle encapsulated antitubercular drugs as a potential oral drug delivery system against murine tuberculosis. *Tuberculosis,* 83:373–378, 2003.
32. El-Shabouri MH. Positively charged nanoparticles for improving the oral bioavailability of cyclosporin-A. *Int. J. Pharm.,* 249:101–108, 2002.
33. Salem II, Flasher DL, Duzguzenes N. Liposome-encapsulated antibiotics. *Methods in Enzymol.,* 391:261–291, 2005.
34. Napoli A, Tirelli N, Kilcher G, Hubbell JA. New synthetic methodologies for amphiphillic copolymers of ethylene glycol and propylene sulfide. *Macromolecules,* 34:8913–8917, 2001.
35. Rehor A, Hubbell JA, Tirelli N. Oxidation-sensitive polymeric nanoparticles. *Langmuir,* 21:411–417, 2001.

17 Nanophotonics for Biomedical Superresolved Imaging

Zeev Zalevsky, Dror Fixler, Vicente Mico,
and Javier García
School of Engineering, Bar-Ilan University, Ramat-Gan, Israel

CONTENTS

17.1 INTRODUCTION

The resolution of every imaging system is limited by the diffraction spot size that equals $\lambda/(2 \, NA)$, where λ is the optical wavelength and NA is the numerical aperture of the lens. This limit is obtained due to the wave nature of light, and it is considered as an unsurpassable bound for most practical cases. Nevertheless, the resolution limit is obtained under certain assumptions, such as source monochromaticity, unpolarized light, short time imaging, and so on. In fact, for many practical situations, additional a priori information can assist in surpassing the classical resolution limit. This assertion is supported by the classical works of Toraldo di Francia and Luckosz [1–3], describing a system by the numbers of degrees of freedom it can transmit, instead of the space-bandwidth product. Later on, the signal-to-noise ratio (SNP) of the image was also included in the computation [4]. Thus, owing to the invariance of the information throughput of a system, the spatial resolution of a system can be enhanced, at the expense of other degrees of freedom [5]. As examples, the resolution in one axis can be enhanced by sacrificing the resolution on the orthogonal direction [2,6], or the two orthogonal polarizations can be used for doubling the resolution in the case of a polarization-independent object [7]. Methods based on extrapolation of the spectrum (e.g., [8]) try to squeeze the SNR degree of freedom [5].

Obviously, the most appealing degree of freedom that can be used is the temporal one, due to the high amount of available bandwidth for the cases where the images are essentially static. Eventually, the maximum amount of information that can be multiplexed is given by the time-bandwidth product, related to the illumination source temporal spectrum. For a wavelength-independent object, several methods have been devised to encode the spatial resolution in different wavelengths, obtaining superresolution [9,10]. With a different approach, but also exploiting the temporal bandwidth of the source, it is possible to encode the spatial resolution in the coherence of the light [11,12]. Both wavelength and coherence encoding methods are especially suitable for a one-point spatial resolution channel, like a single-mode fiber. For the case of systems with a significant spatial resolution, the most widely studied methods are based on illuminating the object with tilted waves, in such a way that the high spatial frequencies of the sample are downconverted into low spatial frequencies and, thus, can pass through a limited resolution system [13,14]. Passing high frequencies through the optical system is not enough, as the phase of the frequency bandpass must also be recovered. One way to surpass this problem is to record the phase by interferometry [15–17], although it requires a reference arm aside of the imaging system. Another widely used method employs moving gratings so that the different tilted beams coming from the grating are encoded in a phase that varies linearly with time [18]. A main problem of these systems is that they need a high-resolution pattern (namely, a grating) to be projected onto the sample with high resolution. In general, this means a high-quality system preceding the system to be enhanced.

Several efforts have been made to overcome the detector pixel size limitation when several displacement frames of the same sample have been captured [13,19–21]. Most of the methods are based on interlacing or averaging the frames after a subpixel resolved registration. Nevertheless, an increase of resolution by a factor of 2 is usually considered the maximum realistic improvement, owing to the loss of information in each frame [13]. In the literature [21], a technique using a mask with subpixel details attached to the detector is proposed, in order to obtain a higher-resolution improvement. The implementation of this method is very costly because of the need to prepare a high-resolution mask and to set it in physical contact with the sensor (or fabricate the mask over it).

Digital holography is a widely useful technique to record the complex amplitude distribution of an object. In a basic configuration, the interference pattern composed by object and reference wave is recorded onto an image sensor (typically a charge-coupled device [CCD]), and the reconstruction is performed by computer using discrete Kirchhoff-Fresnel propagation equations. The size of the CCD pixels is usually bigger than the photographic grain size that composes the holographic plate. The worse resolution of digital imaging media as compared to conventional holographic media limits the maximum angle between interfering beams to a few degrees. Thus, in many cases, the zero order and both diffracted images are overlapping and no reconstruction can be done. The use of an in-line setup permits the measurement of the object complex amplitude distribution by using phase-shifting interferometry [22,23]. In a similar way, Iemmi et al. [24] applied a point diffraction interferometer to digital holography using phase-shifting steps controlled by means of a liquid crystal device (LCD). This procedure allows both amplitude and phase object distributions to be obtained and the three-dimensional object to be visualized by simple propagation of the calculated distributions using computer tools.

One of the most useful applications of digital holography is microscopy. Digital in-line microscopy with numerical reconstruction [25,26] provides a powerful technique to lensless imaging when a low-density object is imaged. As the interference is done between the object diffracted beam and the undiffracted one, the illuminated object should block only a fraction of the cone of radiation recorded at the CCD. Typical objects that satisfied the previous conditions are microspheres, cells, plankton, and other biological specimens. For not-so-transparent objects, the reference beam must be introduced separately (not through the object), and it allows the possibility for phase recovering by phase-shifting techniques [27].

On the other hand, digital holographic microscopy involving image formation with lenses is a common technique that permits the recovery of the phase distribution by adding a reference beam onto the CCD at the image plane. Basically, two ways are commonly used to perform digital holography. One involves off-axis addition of the reference wave and some optical postprocessing stage for suppressing the undesired diffraction orders [28–30], and the second one, which we will follow, is performed in on-axis configuration and using phase-shifting technique to recover phase information.

17.2 PATTERN PROJECTION FOR SUBPIXEL RESOLVED IMAGING IN MICROSCOPY

In this section, we project periodic patterns (grating) on the inspected sample in order to obtain geometrical superresolution in microscopic imaging systems (i.e., we overcome the resolution limit dictated by the geometry of the detection array). Instead of projecting the patterns, the nanophotonic element can also be attached to the detector array and realize the same functionality [31].

17.2.1 THEORY

For the mathematical model, we assume an M pixels sensing device that samples the scene N times. In each sample, the device is shifted with respect to the image a distance of $\Delta x/N$, where Δx is the pitch of the detector's pixels. At the same time, a periodic pattern, with period Δx, is used to illuminate the sample. This pattern is displaced synchronously with the detector. For CCD sensors, each pixel integrates all the light impinging upon it within the cycle. We will deal with the one-dimensional mathematical derivations, although the extension to two dimensions is straightforward. The light distribution just before the detector for a subpixel displacement $n\Delta x/N$ is

$$s(x)g\left(x-n\frac{\Delta x}{N}\right) \tag{17.1}$$

where $n = 0, 1, 2, ..., N–1$. Denoting by $p(x)$ the spatial shape of the pixel (a rectangle function for instance), the integrated signal for pixel m at subpixel displacement n is

$$\hat{y}_{m,n} = \int_{-\infty}^{+\infty} s(x)g\left(x-n\frac{\Delta x}{N}\right)p\left(x-m\Delta x-n\frac{\Delta x}{N}\right)dx \tag{17.2}$$

with $m = 0, 1, 2, ..., M–1$. Notice that Equation 17.2 implies a synchronous subpixel movement of the pattern and the detector. This can be easily accomplished if instead of projecting the pattern on the object, the pattern is set directly in contact with the detector. Then the movement of the pattern will follow the movement of the detector. A simpler way to achieve the scanning, without physical attachments on the detector, is to displace the object, keeping static the rest of the system.

Owing to the periodicity of the pattern, $g(x) = g(x – m\Delta x)$. Moreover, we can join the two indexes m and n in a single index as $k = n + mN$, that runs continuously from 1 to NM, covering all pixels and subpixel displacements. Thus, we can rewrite:

$$\hat{y}_k = \int_{-\infty}^{+\infty} s(x)g\left(x-k\frac{\Delta x}{N}\right)p\left(x-k\frac{\Delta x}{N}\right)dx \tag{17.3}$$

For simplicity, we can redefine:

$$c(x) = g(x)p(x) \tag{17.4}$$

This gives the unit cell of the pattern modified by the pixel. In the simplest case, where the pixel is just a rectangle function, $c(x)$ represents a period of the pattern. With this definition,

$$\hat{y}_k = \int_{-\infty}^{+\infty} s(x)c\left(x - k\frac{\Delta x}{N}\right)dx \qquad (17.5)$$

Equation 17.5 states a convolution operation between the original signal and the unit cell of the combination given by the product of the projected pattern and the sensing array. The convolution is sampled at the rate given by the subpixel displacements. And, according to the sampling theorem, these values are sufficient to fully recover a spatial distribution sampled at rate of $\Delta x/N$. The convolution in Equation 17.5 can be expressed in the Fourier domain as a product. Naming $\{S_k\}$, $\{Y_k\}$, and $\{C_k\}$ the discrete Fourier transforms (DFTs) of the sampled input and output and unit cell distributions,

$$Y_k = S_k C_k \qquad (17.6)$$

The sampled original signal can be recovered from the samples $\{y_k\}$ by means of an inverse filtering that removes the convolution effect of $c(x)$. This operation is easily performed in the Fourier domain; the restored high-resolution image can be obtained as follows:

$$\hat{s}_k = DFT^{-1}\left\{\frac{Y_k}{C_k}\right\} \qquad (17.7)$$

In order to summarize, the full process consists of the following steps: (a) The object is illuminated with a pattern such that on the detector plane its period coincides with the detector pixel pitch. (b) A set of images, each for a subpixel object displacement, is recorded at the detector's resolution (M pixels). (c) The images are interlaced to give an image with subpixel pitch sampling (MN pixels). (d) The influence of the pattern is removed by inverse filtering, rendering the final reconstructed image.

At this point, it is worth considering the gain obtained by using the pattern projected on top of the object. In principle, the same method could be used without the pattern, the only difference being that the output image would be a convolution with the pixel function [$c(x) = p(x)$], typically a rectangle. The main drawback of this approach is that (according to Equation 17.6) the frequencies are attenuated with the Fourier transform of the pixel function. Normally, this means a strong attenuation of high frequencies and the presence of zeros in the spectrum where the signal cannot be recovered. In general, a twofold improvement in resolution is considered the maximum achievable [13]. On the other hand, using a projected pattern with subpixel details will enhance the high frequencies with respect to the conventional illumination case, making possible the recovery of the signal at higher resolutions. As an extreme case example, if the pattern contains a single transparent spot, the resolution obtained will be the size of this spot. Also, a proper design of the subpixel structure can avoid the presence of zeros in the Fourier transform of the unit cell [21], avoiding the need of any regularization when applying Equation 17.7.

In the above analysis, we neglected the effect of diffraction. This approximation can hold as long as the diffraction spot size in the detector plane is smaller than the target resolution after the method is applied. This is when the resolution limit is given by the pixel size and not by diffraction.

17.2.2 MATERIALS, METHODS, AND EXPERIMENTS

The principle of the proposed method is demonstrated by means of two experiments: one in transmission microscopy and the other in fluorescence microscopy. For the purpose of comparison, we use the binning capability of the camera to simulate a large pixels detector. This way a reference high-resolution image is also available.

The experiments were done on biological samples as beads and human cells that were imaged using the same epifluorescence microscope (BX61, Olympus, Japan) as before, with 20× and 4× (0.6 and 0.13 NA) LCPlanFl objective (Olympus, Japan) and 488 nm of Ar Ion laser (Spectra-Physics, California). Images were collected by the same photometric CCD as before.

For the experiments with cells, human Jurkat T-lymphoblast cell line was grown in a humidified atmosphere containing 5% CO_2, in RPMI 1640 medium (Biological Industries, Israel); supplemented with 10% (v/v) heat-inactivated fetal calf serum (Biological Industries, Israel), 2 mM L-glutamine, 10 mM Hepes buffer solution, 1 mM sodium pyruvate, 50 U/ml penicillin, and 100 μmg/ml streptomycin. The cells were washed twice with incomplete RPMI 1640 medium, without phenol red, containing 10 mM HEPES buffer solution. An aliquot of 100 μl of cell suspension (5 × 106 cells/ml) was added to 50 μl of fluorescein diacetate (FDA) staining solution (Sigma, St. Louis, Missouri, F7378) dissolved in phosphate-buffered solution (PBS) to a final concentration of 2 μM in PBS, and incubated at room temperature for 5 minutes. At the end of incubation, cells were loaded onto a microscope slide and measured.

For the experiments with beads, polyscience fluorescent beads Fluoresbrite (fluorescein) of 10 ± 0.1 μm diameter, with FI values of 2000 to 50,000 molecules of equivalent soluble fluorochromes (MESF) were used. These were obtained from Bangs Laboratories, Inc. (Fishers, Indiana).

17.2.2.1 Microscopic Superresolution Measurements of Beads

The periodic pattern projection method was first tested on 10 ± 0.1 μm diameter beads using the 4× lens. The diffraction resolution limit is thus about 2 μm in the sample plane, corresponding to 8 μm in the sensor plane. Thus, the diffraction spot size is slightly larger than the physical pixel size (6.45 μm). A two-dimensional grating with 200 line pairs/mm was projected on the beads at such a magnification that 80 line pairs/mm periodicity approximately was generated on the sample. The grating is formed by square tessellation of replicas of a unit cell consisting on a transparent square on a black background. The projection system magnification was fine-tuned until a full grating period covered eight pixels in the captured image. The profile of the grating is known from previous measurements and is used for the inverse filtering needed for the reconstruction. For the proof of concept, the captured images were binned into macro pixels of 8 × 8 original pixels. The scan was two dimensional with eight steps in each axis, and thus, a total of 64 images were taken prior to the digital processing. Figure 17.1 shows the high-resolution images prior to binning while a long exposure time was applied.

Figure 17.2 demonstrates the low-resolution image when the 8 × 8 binning was done and the exposure time was reduced by a factor of 64. Indeed many spatial features have disappeared.

Figure 17.3 illustrates the reconstruction using the suggested approach via illumination with periodic pattern. The processing consists of interlacing the 64 images according to the shift of the sample for each one of them and then applying the inverse filtering correction.

These operations are very fast to perform and do not compromise the speed of the final imaging. Note that the resolution is comparable to that of Figure 17.1. The square artifacts (8 × 8 pixels) in some parts of the image are due to nonuniform transmittance of the projected grating. Also, the nonuniformities in the mechanical scanning of the sample may introduce a periodic

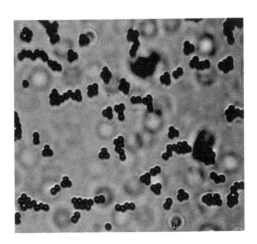

FIGURE 17.1 High-resolution image of polyscience fluorescent beads obtained with long exposure time.

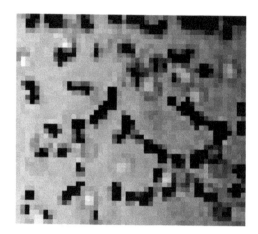

FIGURE 17.2 Image obtained via an 8 × 8 binning. Exposure time is reduced from that in Figure 17.1 by a factor of 64, although the resolution is not enough for resolving the beads.

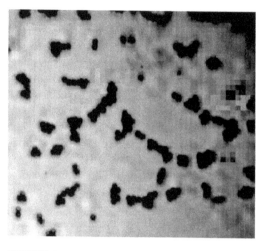

FIGURE 17.3 Reconstruction obtained with the suggested approach including the application of illuminating with a periodic pattern.

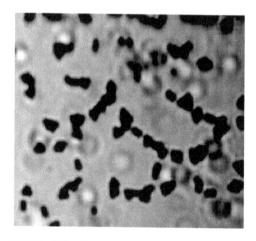

FIGURE 17.4 Reconstruction from a set of 64 images, as in Figure 17.3, but without the grating illumination.

pattern in the resulting image. Therefore, aside from the interlacing and unit cell effect compensation, a simple digital processing was performed in order to obtain higher uniformity, consisting of removing the periodic patterns appearing in the spectrum of the reconstructed image. A more complete processing can be made by taking an image over a uniform sample to calibrate the local grating transmittance.

For comparison, in Figure 17.4 we present the reconstruction obtained from the set of 64 images as in Figure 17.3, but without projecting the two-dimensional grating. Despite the increased resolution with respect to Figure 17.2, the beads are not separable as in Figure 17.3. This operation is equivalent to "micro scanning" [21], and its resolution enhancement is limited by the fill factor of the detector's pixels. When the projected grating is used, this limitation in the resolution improvement factor no longer exists, and it is then directly proportional to subpixel scanning steps.

17.2.2.2 Microscopic Superresolution Measurements of Human Cells Stained with Fluorescein

In this subsection, we describe the experiment performed with Jurkat T-lymphoblast cell lines while the binning and scanning were done in one dimension only. The scanning was one-dimensional, and thus only eight images were captured to fit the binning factor of 8. A projection grating of 600 lines per mm was used, fitting the magnification to match on the camera one period of the grating to eight camera pixels. The diffraction resolution limit for the 20× lens is in this case about 0.5 μm in the sample plane, corresponding to 10 μm in the sensor plane. Figure 17.5 shows the high-resolution image obtained with a long exposure time. This image was obtained without binning. The

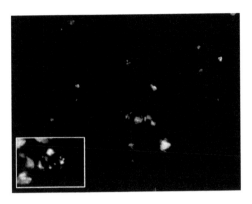

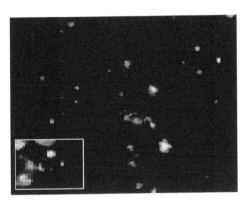

FIGURE 17.5 High-resolution image of Jurkat T-lymphoblast cells obtained with long exposure time. The inset displays a magnified portion of the image with high-resolution details for comparison.

FIGURE 17.6 Image obtained via binning of 8 × 1. Exposure time is reduced from that in Figure 17.5 by a factor of 8, although the resolution is not enough for resolving the details.

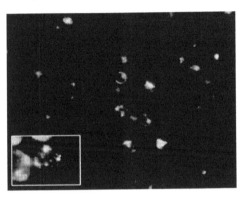

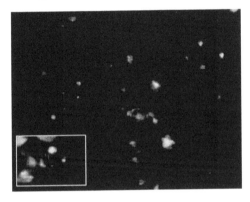

FIGURE 17.7 Reconstruction obtained with the suggested approach—illuminated with a grating. The artifacts (8 × 1 pixels) in some parts of the image are due to nonuniformities of the grating.

FIGURE 17.8 Reconstruction from a set of eight images as in Figure 17.7, but without the grating illumination.

inset displays a magnified portion of the image with high-resolution details and will be used for reconstruction quality comparison.

Figure 17.6 displays the image obtained via 8 × 1 binning. Exposure time is reduced by a factor of 8, although the resolution is not enough for resolving the details.

In Figure 17.7, one can see the reconstructed results obtained with the suggested approach including the grating projection. Note that the resolution is comparable with that of Figure 17.6. The artifacts (8 × 1 pixels), in some parts of the image, are due to nonuniformities of the grating. The reconstruction included simple digital enhancement of the image in order to improve its spatial uniformity, as described in the previous subsection.

Figure 17.8 shows the reconstruction from a set of eight images as in Figure 17.7, but this time without the grating illumination. Despite the increased resolution with respect to Figure 17.7 due to the "microscanning" operation, the high-resolution details are not as clearly resolved as in Figure 17.7.

17.3 PHASE-SHIFTING INTERFEROMETRIC MICROSCOPY

The experiments done in this section demonstrate the reconstruction of the phase information with subwavelength accuracy. To obtain that, the mirror of the presented interferometric setup should perform subwavelength scanning. However, the same concept may be obtained without the mirror scanning if the axial information is encoded by different wavelengths. Such a concept can be realized using the same setup but illuminated with a white light source and therefore having a white light interferometer.

17.3.1 THEORETICAL ANALYSIS OF THE OPTICAL SYSTEM

The optical setup is shown in Figure 17.9. A Mach-Zehnder interferometric configuration is used to implement the technique. A collimated input laser beam illuminates the interferometric system.

A first beam splitter (BS1) is used to split the collimated incoming laser light into two optical beam paths. In the first one (the imaging arm), the sample is imaged by the microscope objective into the CCD after reflection in a mirror (M1). In the other optical path (the reference arm), the light is directed toward the CCD through a piezoelectrical mirror (PZT). A simple lens is placed in the reference arm at the appropriate distance d from the CCD to ensure equal divergence of both interference beams. A second beam splitter (BS2) mixes the two beams into the image plane and produces the interferometric recording. Notice that no tilt is introduced in one of the optical beams versus the other one, resulting in an on-axis hologram at the CCD plane.

This setup, in static mode, gives no possibility to recover the phase information of the sample. To obtain this information, one can introduce a tilt in one of the mirrors (usually in the reference mirror) of the interferometer for off-axis hologram recording. This procedure enables the recovery of both phase and amplitude distribution by simple filtering and relocation process of the first order diffracted by the hologram. Sufficient carrier frequency must be introduced to avoid the overlapping between the different orders diffracted by the hologram.

An extensively used method to recover phase and amplitude information is the phase-shifting technique. A time-varying phase shift is introduced in one of the optical beams of the interferometer. Thus, recording a set of holograms with different phase between the beams, and later digital processing, can give the phase object distribution. We introduce the phase shift in the reference beam by means of a PZT mirror. Several algorithms exist for phase extraction depending on the phase steps and the number of images [32]. We use a continuous-phase shifting produced by external modulation of the PZT mirror power supply by means of a frequency function generator. In particular, we applied a sawtooth profile with higher amplitude than that necessary to produce a 2π phase displacement in the reference beam. This ensures that at least one phase period of the reference beam is covered with a constant and linear phase displacement. Thus, by storing in the memory of the computer a sequence of on-axis holograms, one can recover the phase and amplitude information of the sample under test with a simple numerical reconstruction procedure [32].

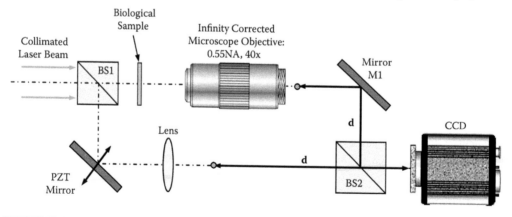

FIGURE 17.9 Experimental setup. The object is imaged onto the charge coupled device (CCD) by the microscope lens. A lens placed in the reference arm compensates the divergence introduced by the microscope lens.

In the following lines, some mathematical expressions of the proposed method are presented. The CCD image sensor records a hologram of a microscopic object. Let us denote as $f(x,y)$ the input object complex amplitude distribution. Then, the amplitude distribution onto the CCD done by the imaging arm, under quadratic approximation, is

$$U_{imag}(x,y) = \left[f\left(-\frac{x}{M}, -\frac{y}{M}\right) \otimes disk\left(\Delta v\, r\right) \right] e^{j\frac{k}{2d}\left[x^2+y^2\right]} \tag{17.8}$$

"with M being the imaging magnification factor, $k = 2\pi/\lambda$ the wave number, λ the wavelength and d the distance between the laser beam image plane through the microscope objective and the CCD. r is the polar coordinate defined as $r = \sqrt{x^2 + y^2}$ and $disk(\Delta v \cdot r)$ is the point spread function (PSF). of the optical microscope lens. The symbol $\otimes$ denotes the mathematical operation of convolution.

From Figure 17.9, the intensity distribution recorded at the CCD comes from the addition of a reference beam to the amplitude given by the previous equation. The reference can be expressed as

$$U_{ref}(t) = C\, e^{j\frac{k}{2d}\left[x^2+y^2\right]} e^{j\phi_n(t)} \tag{17.9}$$

where C is a constant related with the real amplitude of the reference beam, and $\phi_n(t)$ is the temporal phase shift introduced by the PZT mirror. The spherical phase factor comes from the lens placed at the reference beam to equal beam convergence. Then, the recorded intensity at the CCD is

$$I_{CCD}^m(x,y;t) = \left| \left[f\left(-\frac{x}{M}, -\frac{y}{M}\right) \otimes disk\left(\Delta v\, r\right) \right] + Ce^{j\left[\phi_0+\phi_m(t)\right]} \right|^2 \tag{17.10}$$

assuming linear phase variation in time, and calling the initial phase difference between imaging and reference beams as ϕ_o and the linear phase increment introduced in time between two subsequent intensity images as ϕ_m. Notice that the spherical phase factor, common to both terms has been removed due to the absolute value operation.

As the phase step ϕ_m is limited by the video frequency necessary for the storage of each intensity image, we can express the time dependence of the recorded intensity as a function of the intensity image number m multiplied by the phase step between two consecutive images—that is, $\phi_m = m\,\phi_k$. So, we can rewrite Equation 17.10 considering that different intensity distribution will be taken in time sequence:

$$I_{CCD}^m(x,y;t_m) = \left| \left[f\left(-\frac{x}{M}, -\frac{y}{M}\right) \otimes disk(\Delta vr) \right] + Ce^{j[\phi_0+m\phi_k]} \right|^2$$

$$= \left| f\left(-\frac{x}{M}, -\frac{y}{M}\right) \otimes disk(\Delta vr) \right|^2 + C^2 + C\left[f\left(-\frac{x}{M}, -\frac{y}{M}\right) \otimes disk(\Delta vr) \right] e^{-j[\phi_0+m\phi_k]}$$

$$+ C\left[f\left(-\frac{x}{M}, -\frac{y}{M}\right) \otimes disk(\Delta vr) \right]^* e^{j[\phi_0+m\phi_k]} = A(x,y) + B(x,y)\cos(\phi_0 + m\phi_k) \tag{17.11}$$

where $A(x,y)$ is representative of the object image intensity plus a constant given by the reference beam, and $B(x,y)$ is coming from the real part of the object amplitude distribution:

$$A(x,y) = \left| f\left(-\frac{x}{M}, -\frac{y}{M}\right) \otimes disk\left(\Delta v\, r\right) \right|^2 + C^2$$

$$B(x,y) = 2C\, \text{Re}\left[f\left(-\frac{x}{M}, -\frac{y}{M}\right) \otimes disk\left(\Delta v\, r\right) \right] \quad (17.12)$$

Then, we are able to recover the initial phase distribution ϕ_o between object and reference by applying a phase-shift algorithm for known phase shifts and computing the different stored intensity distributions. In particular, we have applied the method that takes into account m intensity images in one phase shift period and that permits the recovering of the initial phase distribution as

$$\phi_0(x,y) = \arctan \frac{-\sum_{i=1}^{m} I_i(x,y)\sin\left[\frac{2\pi}{m}(i-1)\right]}{\sum_{i=1}^{m} I_i(x,y)\cos\left[\frac{2\pi}{m}(i-1)\right]} \quad (17.13)$$

17.3.2 Experimental Results

We have tested the experimental setup presented in Figure 17.9. We used a Mitutoyo infinity corrected microscope objective with 0.55NA and 40× magnification. All the images of the input object correspond to a histological sample of a mouse small intestine, dissected and fixed in a 10% formol-saline solution. Figure 17.10 shows the images obtained using the microscope lens and without reference beam. We use a coherent illumination beam coming from a 532 nm laser source. For an initial configuration of the experimental setup, we have an on-axis interferogram recorded at the CCD. Then, we applied a sawtooth function to the PZT mirror using an external function generator to produce the continuous phase shifting between both interferometric beams. This allows us to store a sequence of intensity images that contains the information of the different phase shifts produced by the PZT mirror. The overall process takes nearly 2 seconds to record 18 intensity images (0.11 seconds per image). The phase shift between subsequent images is 0.11π.

After the image sequence is stored, we perform digitally the algorithm (Equation 17.13) to recover the phase information of the biological sample (see Figure 17.11). As can be clearly seen, the phase information for this sample is much more relevant than the amplitude one. It also should be noted that the phase extraction process, to a given extent, produces a cleaner image than the direct

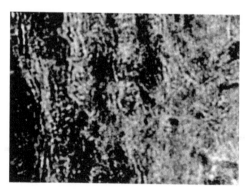

FIGURE 17.10 Conventional intensity image.

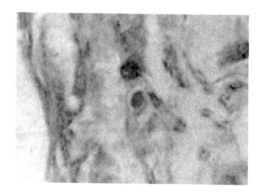

FIGURE 17.11 Phase extracted from the phase shifting interference microscope.

intensity image. This fact derives from the implicit removal of the reference inhomogeneities along the process.

Aside of the phase observation, the combination of amplitude and phase information includes the full information of the field of the sample. This means that from the combination of both images, any optical image processing can be performed digitally. A clear example is refocusing that can be done by any conventional Fresnel diffraction procedure applied to this image [33]. One more possibility is the digital reproduction of the phase visualization techniques that are usually performed optically, such as the phase contrast or differential interference contrast (DIC) techniques. The main advantage is that now the same system serves to provide all possible phase visualization techniques, from the obtained phase and amplitude image, without the need to change to special lenses or system.

Figure 17.12 depicts two examples of phase contrast obtained for the sample under examination. To stress the potential of the method, we have performed the phase contrast with two different phase plates (π and $\pi/2$). In an optical microscope this involves the physical change of the lens, or even a very complex objective with programmable phase delay. Figure 17.13 shows the output of digitally generated DIC images. This is very simply obtained by subtracting two displaced images. In contrast with the optical equivalent, the change of image shift as well as the image shearing orientation are immediately done in the computer.

As a final example, Figure 17.14 shows a dark field image, obtained by blocking the DC term of the image.

We can observe that each possible technique enhances the visibility of different features of the sample. It is noteworthy that the computing time is extremely low, once the phase information is

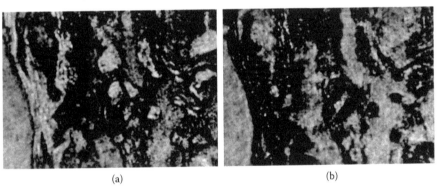

(a)   (b)

FIGURE 17.12 Digitally obtained phase contrast image. The phase plate introduces (a) π phase shift and (b) $\pi/2$ phase shift.

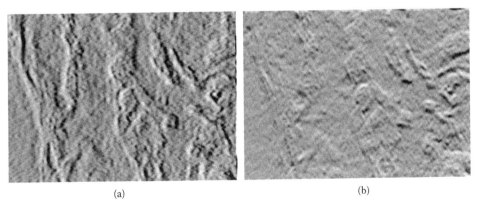

(a)   (b)

FIGURE 17.13 Digitally obtained differential interference contrast image. The shearing is introduced in the (a) horizontal and (b) vertical axes.

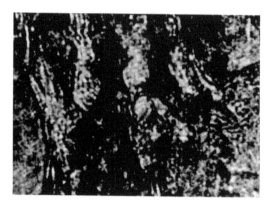

FIGURE 17.14   Digitally obtained dark field image.

provided by the optoelectronic system. The processing at most involves a couple of digital Fourier transforms, giving a clear chance for real-time processing.

ACKNOWLEDGMENTS

This work was supported by FEDER Funds and the Spanish "Ministerio de Educación y Ciencia" under the Project FIS2004-06947-C02-01.

REFERENCES

1. G. Toraldo di Francia, "Resolving power and information," *J. Opt. Soc. Am.* 45, 497–501 (1955).
2. W. Lukosz, "Optical sytems with resolving powers exceeding the classical limits," *J. Opt. Soc. Am.* 56, 1463–1472 (1967).
3. W. Lukosz, "Optical sytems with resolving powers exceeding the classical limits. II" *J. Opt. Soc. Am.* 57, 932–941 (1967).
4. I.J. Cox and C.J.R. Sheppard, "Information capacity and resolution in an optical system," *J. Opt. Soc. Am.* A 3, 1152–1158 (1986).
5. A.J. den Dekker, A. van den Bos, "Resolution: A survey," *J. Opt. Soc. Am.* A 14, 547–557 (1997).
6. M.A. Grimm and A.W. Lohmann, "Superresolution image for one-dimensional object," *J. Opt. Soc. Am.* 56, 1151–1156 (1966).
7. A.W. Lohmann and D.P. Paris, "Superresolution for nonbirefringent objects," *Appl. Opt.* 3, 1037–1043 (1964).
8. R.W. Gerchberg, "Superresolution through error energy reduction," *Optica Acta* 21, 709–720 (1974).
9. A.M. Tai, "Two-dimensional image transmission through a single optical fiber by wavelength-time multiplexing," *Appl. Opt.* 22, 3826–3832 (1983).
10. D. Mendlovic, J. Garcia, Z. Zalevsky, E. Marom, D. Mas, C. Ferreira, and A.W. Lohmann, "Wavelength multiplexing system for single-mode image transmission," *Appl. Opt.* 36 8474–8480 (1997).
11. P. Naulleau, E. Leith, "Imaging through optical fibers by spatial coherence encoding methods," *J. Opt. Soc. Am.* A 13, 2096 (1996).
12. Z. Zalevsky, J. García, P. García-Martínez, and C. Ferreira, "Spatial information transmission using orthogonal mutual coherence coding," *Opt. Let.* 20, 2837–2839 (2005).
13. Z. Zalevsky and D. Mendlovic, *Optical Super Resolution*, (Springer, New York, 2004).
14. Z. Zalevsky, D. Mendlovic, and A.W. Lohmann, "Optical systems with improved resolving power," in *Progress in Optics*, Vol. XL, E. Wolf Ed. (Elsevier, Amsterdam, 1999).
15. C.J. Schwarz, Y. Kuznetsova, and S.R.J. Brueck, "Imaging interferometric microscopy," *Opt. Lett.* 28, 1424–1426 (2003).
16. P.C. Sun, E.N. Leith, "Superresolution by spatial-temporal encoding methods," *Appl. Opt.*, 31, 4857 (1992).
17. V. Mico, Z. Zalevsky, P. Garcia-Martinez, and J. Garcia, "Single-step superresolution by interferometric imaging," *Opt. Exp.* 12, 2589–2596 (2004).

18. A. Shemer, D. Mendlovic, Z. Zalevsky, J. Garcia, and P. Garcia-Martinez, "Superresolving optical system with time multiplexing and computer decoding," *Appl. Opt.* 38, 7245–7251 (1999).

19. D. Granrath and J. Lersch, "Fusion of images on Affine sampling grids," *J. Opt. Soc. Am.* A 15, 791 (1998).

20. C. Gillette, T. Stadtmiller, and R. Hardie, "Aliasing reduction in staring infrared imagers utilizing sub-pixel techniques," *Opt. Eng.* 34, 3130–3137 (1995).

21. Z. Zalevsky, D. Mendlovic, and E. Marom, "Special sensor masking for exceeding system geometrical resolving power," *Opt. Eng.* 39, 1936–1942 (2000).

22. I. Yamaguchi and T. Zhang, "Phase shifting digital holography," *Opt. Lett.* 22, 1268–1270 (1997).

23. J.H. Bruning, D.R. Herriot, J.E. Gallagher, D.P. Rosenfeld, A.D. White, and D.J. Brangaccio, "Digital wavefront measuring interferometer for testing optical surfaces and lenses," *Appl. Opt.* 13, 2693–2703 (1974).

24. C. Iemmi, A. Moreno, and J. Campos, "Digital holography with a point diffraction interferometer," *Opt. Express* 13, 1885–1891 (2005).

25. W. Xu, M.H. Jericho, I.A. Meinertzhagen, and H.J. Kreuzer, "Digital in-line holography of micro-spheres," *Appl. Opt.* 41, 5367–5375 (2002).

26. J. Garcia-Sucerquia, W. Xu, S.K. Jericho, P. Klages, M.H. Jericho, and H.J. Kreuzer, "Digital in-line holographic microscopy," *Appl. Opt.* 45, 836–850 (2006).

27. I. Yamaguchi, J. Kato, S. Ohta, and J. Mizuno, "Image formation in phase-shifting digital holography and applications to microscopy," *Appl. Opt.* 40, 6177–6186 (2001).

28. V. Mico, Z. Zalevsky, P. Garcia-Martinez, and J. Garcia, "Superresolved imaging in digital holography by superposition of tilted wavefronts," *Appl. Opt.* 45, 822–828 (2006).

29. P. Marquet, B. Rappaz, P.J. Magistretti, E. Cuche, Y. Emery, T. Colomb, and C. Depeursinge, "Digital holographic microscopy: A non-invasive contrast imaging technique allowing quantitative visualization of living cells with subwavelength axial accuracy," *Opt. Lett.* 30, 468–470 (2005).

30. V. Mico, Z. Zalevsky, and J. Garcia, "Superresolution optical system by common-path interferometry," *Opt. Express* 14, 5168–5177 (2006).

31. D. Fixler, J. Garcia, Z. Zalevsky, A. Weiss, and M. Deutsch, "Pattern projection for subpixel resolved imaging in microscopy," *Micron* 38, 115–120.

32. T. Kreis, *Handbook of Holographic Interferometry* (Wiley-VCH, Weinheim, Germany, 2005).

33. D. Mas, J. Garcia, C. Ferreira, L.M. Bernardo, and F. Marinho, "Fast algorithms for free-space diffraction pattern calculation," *Opt. Comm.* 164, 233–245 (1999).

18 DNA as a Scaffold for Nanostructure Assembly

Michael Connolly
Integrated Nano-Technologies, Henrietta, New York, USA

CONTENTS

18.1   INTRODUCTION

Advances in technology have led to increasingly smaller levels of manipulation to achieve greater results. As this trend continues, we begin to approach the limits of conventional processes to the point where improvements go from small, to incremental, to significant. In microelectronics, the theoretical end of the silicon-based lithographic process is drawing near, while in material science the need to control characteristics beyond the macroscale is becoming necessary to continue innovating. There is an imminent need for controlled manipulation at the nanoscale, and until it is reliably achieved, progress in some arenas may plateau. Once it is achieved, however, a new chapter in technical innovation will be open.

Nanotechnology has been put forward as the approach to solve this fabrication problem. Nanoscale structures would be self-assembled from the "bottom-up." Over the past several years, new materials having unique properties have been identified. For example, carbon nanotubes have extraordinary strength and electrical properties [1–2], and numerous labs have demonstrated simple patterning of these materials. However, more complex structures require a material capable of directing multiple components to precise locations.

Molecular biotechnology is a promising point from which nanotechnology can evolve, because living systems are successful examples of atomic and molecular manipulation on the nanoscale. Although enzymes manipulate atoms and molecular fragments on the Angstrom scale, biological systems make their structural components on the nanometer scale, where weak intermolecular energies direct the self-assembly process. This latter approach is likely to be the simplest motif for building the first nanoscale devices whose synthesis and assembly are controlled, to a degree, with current technology. These objects can be used for scaffolding to orient and juxtapose other molecules to form devices, mechanisms, and structures [3].

Because the predominant examples of nanotechnology in nature derive from living systems, it is reasonable to look to those systems for the components of the first nanotechnological objects and devices. Because of this, nanotechnology, which can come from many routes, is likely to evolve in part from molecular biotechnology. The bottom-up approach entails making objects and devices on the nanoscale from molecular and macromolecular components. There is good reason to believe that this approach is practical to some extent, because living systems already exemplify its success: cells manipulate chemical structure on the Angstrom scale via their enzymatic proteins, and in addition, they contain self-assembling structural components. Self-assembly is spontaneous. Manipulation that involves the breaking or formation of bonds requires the control of processes in which large amounts of energy can be liberated or consumed [3].

Although processes for manipulating, enhancing, and modifying biological entities exist, they are predominantly effective on a sample en masse, rather than on an individual cell or molecule. By applying chemical, electrical, mechanical, and biological processes, a volume of a sample can be affected. This is useful in a serial manner, but what about when a multistep process is required to achieve a result? Batch processes quickly become either too cumbersome or ineffective, rendering the desired result either too complicated or functionally impractical. If there were a way to direct manipulation at the nanoscale, whereby molecules would behave in a predictable and reliable fashion, the goal of nanoscale achievement could begin to be realized.

Nature has been doing nanotechnology for millennia, and one of the most powerful nanotechnology mechanisms is present in every living thing. That mechanism is called Watson–Crick base pairing of DNA. DNA can store and transfer information, perform computations, and build structures. These things already occur in nature without human interaction; however, manipulating them for a specific purpose and connecting the activity to a useful human interface is what lies between theory and practical application.

18.2 DNA AS A SCAFFOLD FOR BUILDING STRUCTURES

Nature has designed DNA such that it is an ideal molecule for building nanoscale structures. Because of how the molecule is constructed, strands of DNA can be "programmed" to assemble themselves into complex arrangements in two and three dimensions. Self-assembly requires information to be carried on the substrates; therefore, the greater the information-carrying capacity, the greater is the complexity of the structures that are produced. This specific programmability stems from the four nucleic acids that make up the rungs of the DNA double helix. These four acids (adenine, thymine, guanine, and cytosine) are represented by the letters A, T, G, and C, respectively. The characteristics of these acids (also referred to as nucleotides or bases) dictate specific pairings of A and T, and G and C. Each nucleic acid makes up one half of each rung in the twisted ladder structure of a complete DNA strand. Because each base will only match up with an appropriate complement, both naturally occurring and synthesized DNA are predictable and stable.

An important property of DNA is its double-stranded nature. Because every DNA sequence has a natural complement (i.e., if a sequence is ATTACGTCG, its complement is TAATGCAGC), these two strands will come together (or hybridize) to form double-stranded DNA, as shown in Figure 18.1.

By producing strands with the appropriate combinations of complementary bases, DNA can be designed to automatically create a nearly infinite number of self-assembling structures. These structures can be as simple as a naturally occurring double helix or a highly complex, three-dimensional shape.

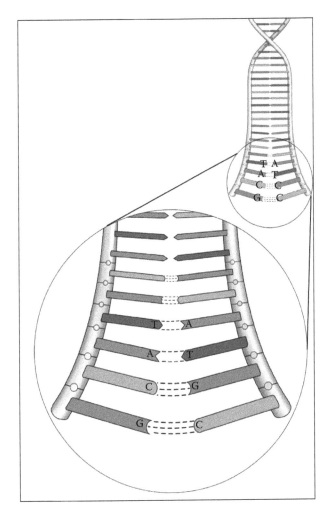

FIGURE 18.1   Structure of DNA.

Nadrian C. Seeman, Professor of Chemistry at New York University, has published extensively on this subject and has put forth several models for using DNA to create simple, compound, and complex structures. Using a combination of ligation, restriction, and hybridization steps, Seeman has been able to create three-dimensional structures like the cube in Figure 18.2 [4].

A group of scientists at The Scripps Research Institute has designed, constructed, and imaged a single strand of DNA that spontaneously folds into a highly rigid, nanoscale octahedron [5].

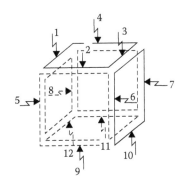

FIGURE 18.2   Seeman cube.

Beyond the specificity and predictability of DNA, nature has provided a comprehensive toolbox to manipulate DNA. In fact, nature provides virtually every tool an engineer would need to build. Restriction enzymes cut DNA at specific sequences of bases. Ligase fuses the ends of two molecules. More complex sets of enzymes can be used to make

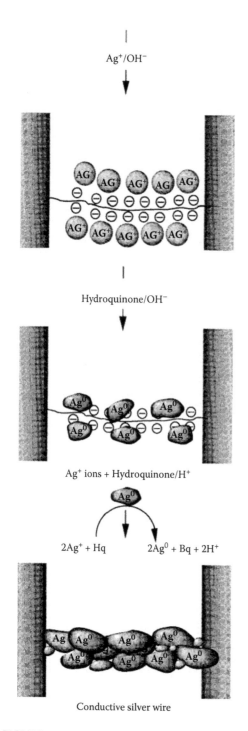

Ag^+/OH^-

Hydroquinone/OH^-

Ag^+ ions + Hydroquinone/H^+

$2Ag^+$ + Hq $2Ag^0$ + Bq + $2H^+$

Conductive silver wire

FIGURE 18.3 Silver DNA wire development. (From Braun, E., Eichen, Y., Sivan, U., and Ben-Joseph, G., *Nature*, 391, 775, 1998. With permission.)

virtually unlimited copies of a DNA molecule. Polymerase chain reaction technology uses DNA polymerases to exponentially copy DNA molecules. Enzymes facilitate restriction (cutting), ligation (fusing), and polymerization (copying) of DNA strands. Numerous site-specific DNA-binding proteins can be used to mask specific locations on DNA. Other enzymes can edit mistakes and twist or untwist DNA. Similar proteins also work with ribonucleic acid (RNA). When combined with the macro effects of temperature and pH, these naturally occurring substances provide for a nearly infinite number of creation and assembly combinations.

18.3   COATING DNA WITH METALS OR PLASTICS

As discussed earlier, DNA is uniquely suited for the formation of complex three-dimensional structures. However, to produce electronic circuits and nanoscale structures, it is necessary to be able to alter the properties of the DNA molecules. Fortunately, DNA reacts with a wide variety of materials. Charged molecules are attracted to the negatively charged phosphate backbone. Other reagents react with the active groups on the bases. Yet other compounds intercalate between the stacked bases of the single- or double-stranded molecules. Using these compounds, it is possible to alter the electronic and physical properties of the DNA molecules.

Over the last decade, several research efforts have concluded that a strand of DNA can act as an electrical conductor or semiconductor [6]. During the same time period, an almost equal number of studies concluded just the opposite, that DNA is either a poor conductor or a resistor [7]. Although it is true that under certain circumstances, DNA may appear to carry a current, no research has been able to put forth consistent circumstances under which DNA, in its natural form, is a reliable conductor.

In order to convert a DNA molecule into a highly conductive wire, researchers at the Technion (Israel Institute of Technology) began work on a process to coat DNA with metal. To instill electrical functionality, silver metal is deposited along the DNA molecule, as shown in Figure 18.3. The three-step chemical deposition process is based on selective localization of silver ions along the DNA through Ag$^+$/Na$^+$ ion-exchange 18 and formation of complexes between the silver and the DNA bases. The Ag$^+$/Na$^+$ ion-exchange process is monitored by following the almost instantaneous quenching of the fluorescence signal of the labeled DNA. The ion-exchange process, which is highly selective and restricted to the DNA template alone, is terminated when the fluorescence signal drops to 1 to 5% of its initial value (the quenching is much faster than normal bleaching of the fluorescent dye). The silver ion-exchanged DNA is then reduced to form nanometer-sized metallic silver aggregates bound to the DNA skeleton. These silver aggregates are subsequently further "developed," much as in the standard photographic procedure, using an acidic solution of hydroquinone and silver ions under low light conditions. Such solutions are metastable, and spontaneous metal deposition is normally very slow. However, the silver aggregates on the DNA act as catalysts and significantly acceler-

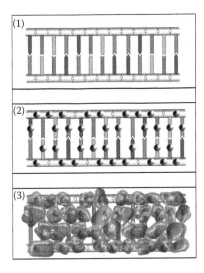

FIGURE 18.4 DNA coating (1) interdigitated wires DNA strand. (2) Oligonucleotide probes primary ions applied. (3) Hybridized target secondary material developed on the primary. (From Keren, K., Berman, R., Buchstab, E., Sivan, U., and Braun, E., *Science*, 302, 1380–1382, 2003. With permission.)

ate the process. Under the experimental conditions, metal deposition therefore occurs only along the DNA skeleton, leaving the passivated glass practically clean of silver. The silver deposition process is monitored *in situ* by differential interference contrast (DIC) microscopy and terminated when a trace of the metal wire is clearly observable under the microscope. The metal wire follows precisely the previous fluorescence image of the DNA skeleton. The structure, size, and conduction properties of the metal wire are reproducible and dictated by the "developing" conditions [8].

The coating process works equally well with compounds other than metal and has therefore been developed into a process for coating DNA for purposes beyond conductivity. The process in its most simplified form (as shown in Figure 18.4) is as follows:

1. Synthesized or naturally occurring DNA is isolated.
2. Positively charged (primary) ions are introduced and are attracted to the negatively charged phosphates (along the DNA backbone) and the negatively charged NH_2 on the bases.
3. Once the ions attach, the DNA is rinsed and the secondary substance is added. This substance develops (or grows) on the primary ions creating a continuum of material along the entirety of the DNA strand [9].

The coated DNA can now possess a wide range of characteristics depending on the type of material used in the coating process. The coated strand can now bear electrical characteristics of a conductor, semiconductor, or a resistor as shown in Figure 18.5. Structurally it can be rigid, semirigid, or with any level of flexibility. By running a wide range of materials through this process, DNA becomes an ideal foundation for both electrical and structural tasks. A nanoscale wire (or any other coated strand) by itself is not particularly useful; however, once a reliable process for coating DNA was established, a whole new world of possibility was opened.

Three DNA metallization chemistries have been used to convert DNA into a conducting wire [5,8–10]. These chemistries fall into two reaction categories: (1) ion-exchange of a metal (silver) ion for the positive sodium counterion associated with the phosphate groups of the DNA backbone and (2) formation of a covalent bond between a metal ion (palladium or platinum) and amine groups

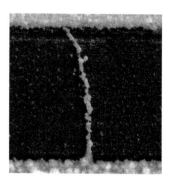

FIGURE 18.5 Scanning electron micrograph (SEM) of metal-coated DNA.

of the DNA bases. In both cases, the attached ions are reduced to form a metal that can act as a catalytic site for the deposition of an alternate metal on the surface of the DNA [11].

18.4 SODIUM–SILVER ION EXCHANGE

A silver-catalyzed gold chemistry has also been studied [8]. This chemistry also involves ion exchange of silver ions for sodium followed by reduction with hydroquinone. In this case, the particles of metallic silver are incubated in a three-part solution containing potassium tetra-chloroaurate to develop gold on the surface of the DNA. Keren et al. [9] performed electronic testing on such metalized DNA and reported ohmic behavior with a resistance of 25 Ω for a 2.5 μm-long wire at a voltage of 0 to 2 mV. This chemistry has also been shown to form 60 to 70 nanometer-sized gold particles along the surface of DNA with no background deposition elsewhere. However, a problem was encountered with a precipitate of gold thiocyanate produced during synthesis of the tetrachloroaurate solution. This precipitate is formed as particles with diameters ≤ 0.8 μm and then redissolved into a phosphate buffer. If these particles are not completely redissolved or removed before metallization, they can cause aberrant results.

18.5 PALLADIUM–AMINE COVALENT BINDING

Another metallization chemistry [9–11] involves the formation of a covalent bond between palladium and platinum ions and the amine groups of DNA bases. A solution of palladium acetate is mixed with a solution of DNA, and the palladium ions become associated with DNA by forming covalent bonds with the amine groups of the DNA bases. Subsequent reduction of the palladium ions allows them to form autocatalytic sites for the deposition of a palladium metal coating on the surface of the DNA. Reduction of the palladium bonded to the DNA results in very small metal deposits of palladium. However, there may not be enough metal to form a continuous conducting wire after a single treatment with the palladium acetate solution. The initial palladium deposits can serve as catalytic sites for the further deposition of palladium. Subsequent rounds of treatment with palladium acetate and the reducing agent enhance these deposits with additional metallic palladium.

Richter et al. [12,13] electrically characterized wires formed by the palladium-catalyzed deposition of palladium on lambda DNA immobilized between gold electrodes with an interelectrode gap of 5 to 10 μm. DNA was dried on the surface of a comb-shaped contact structure so that the DNA strands were perpendicular to the gold electrodes. Palladium metallization was then carried out on the microchips and voltages applied across the wires. Applied voltages in the range of tens of millivolts produced currents in the range of tens of microamps. Resistance measurements were reported for individual wires by comparing the system resistance before and after individual wires were broken. Resistance measurements were made for more than 100 wires and all showed ohmic behavior at room temperature. Data showed that a diameter of approximately 50 nm is sufficient to achieve continuous metallization of the DNA. Although the initial resistance of the wires was proportional to their length, none of the wires exhibited resistance less than 5 kΩ, even when wire diameter was increased to 200 nm. These higher resistances were attributed to contact resistances between the palladium wire and the gold electrodes when electron-beam–induced carbon lines were written over the ends of the wires where contact was made with the gold electrodes. The resistance of individual wires was less than 1 kΩ. For example, a 16.5 μm wire with an average diameter of 50 nm had a resistance of 743 Ω. The two-terminal *I–V* curve of this wire was recorded after cutting all other wires. Linear current–voltage dependence was observed for bias voltages down to 1 mV, and no evidence of a nonconducting region or diode-like behavior was found at room temperature.

18.6 PATTERNING MATERIALS ON DNA

To fully take advantage of DNA as a substrate, one would like to direct coatings to specific regions of the DNA molecule. Reaching into nature's toolbox, it is possible to mask regions of the DNA molecule using sequence-specific DNA-binding proteins. During the synthesis of the DNA molecules, binding sites for masking proteins are engineered into the molecules. With the engineered molecules, it is possible to mimic the lithography process on DNA to produce structures with more than one coating on a DNA molecule. Nanoscale electronic components are created through the multistep process outlined below [14]:

1. A specifically designed DNA strand is synthesized with single-stranded "tailed" ends. These ends will be used in the self-assembly/manipulation process once the component is created.
2. Blocking proteins are applied to specific locations along the strand, providing a mask from the coating process. Depending on the complexity of the component, several different proteins may be used.
3. Once the proteins are applied, the first coating step is applied to the unblocked section of DNA.
4. The masking proteins are removed with enzymes.
5. The now-exposed area is coated with the next coating material. Steps 4 and 5 may be repeated several times to achieve the desired characteristics.
6. Once the internal areas are coated, the blocking proteins on the ends are removed and the nanoscale component is created (see Figure 18.6).

   Proper design of the components is the key to assembly. The unique ends of each component will only bond to its intended counterpart. Designs can create circuits in two or three dimensions and can be independent (free floating) or be joined to a substrate like silicon for connection to more conventional circuitry. With complete circuits thousands of times smaller than their conventional counterparts, an entirely new (previously unthinkable) world of development is created.

   In 2002, Keren et al. [9] demonstrated a detailed masking process for creating DNA-based electronic components. A region of DNA was coated with RecA protein. The DNA was then exposed to silver nitrate, and then gold was deposited onto the regions of DNA unprotected

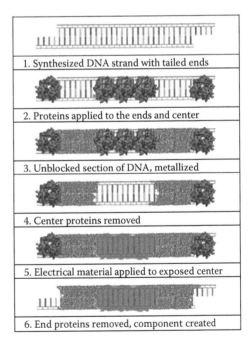

1. Synthesized DNA strand with tailed ends

2. Proteins applied to the ends and center

3. Unblocked section of DNA, metallized

4. Center proteins removed

5. Electrical material applied to exposed center

6. End proteins removed, component created

FIGURE 18.6 DNA patterning process. (From Thorstenson, Y.R., Hunicke-Smith, S.P., Oefner, P.J., and Davis, R.W., *Genome Res.*, 8, 848–855, 1998. With permission.)

by RecA. The RecA protein was then removed to expose an uncoated region of the DNA molecule. This demonstration of sequence-specific lithography on a single molecule was an important step toward DNA-templated electronics.

In November 2003, the same group took the approach further with the creation of a DNA-templated field effect transistor (FET). The details of the FET creation are summarized here.

Assembly of a DNA-templated FET:

1. RecA monomers polymerize on a ssDNA molecule to form a nucleoprotein filament.
2. A homologous recombination reaction leads to binding of the nucleoprotein filament at the desired address on an aldehyde-derivatized scaffold dsDNA molecule.
3. The DNA-bound RecA is used to localize a streptavidin-functionalized single-walled nanotubule (SWNT), utilizing a primary antibody to RecA and a biotin-conjugated secondary antibody.
4. Incubation in an AgNO₃ solution leads to the formation of silver clusters on the segments that are unprotected by RecA.
5. Electroless gold deposition, using the silver clusters as nucleation centers, results in the formation of two DNA-templated gold wires contacting the SWNT bound at the gap (Figure 18.7) [9].

These DNA-based components can be synthesized in solution and then combined to make electronic circuits. As indicated above, single-stranded DNA ends can be protected using single-stranded DNA binding protein (SSB) to leave the ends available for binding with other components. The single-stranded regions are designed to specifically bind to the end of another component. Using this approach, it is possible to create self-assembling electronic circuits or devices in solution (see Figure 18.8).

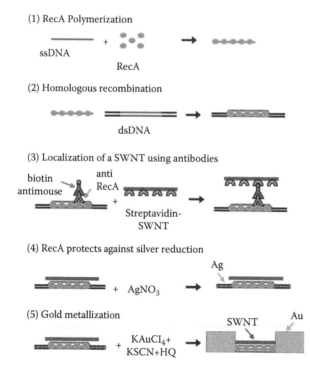

FIGURE 18.7   Assembly of a DNA-based field effect transistor. (From Keren, K., Berman, R., Buchstab, E., Sivan, U., and Braun, E., *Science*, 302, 1380–1382, 2003. With permission.)

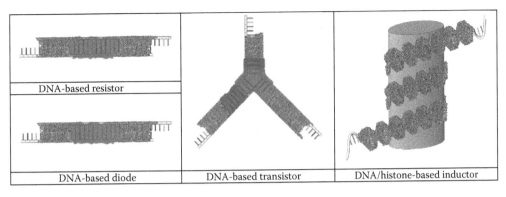

FIGURE 18.8 DNA-based, nanoscale, electronic components. (From Thorstenson, Y.R., Hunicke-Smith, S.P., Oefner, P.J., and Davis, R.W., *Genome Res.*, 8, 848–855, 1998. With permission.)

18.7 COATED DNA STRUCTURES IN PRACTICE—PCR-FREE, BIOLOGICAL DETECTION, AND IDENTIFICATION SYSTEMS

A growing percentage of the things that threaten health, safety, economy, and national security are nearly invisible. From bacteria on a piece of uncooked chicken, to a stranger with a contagious cough, to the looming threat of bioterrorism, pathogens can cause disease ranging from a simple inconvenience to a catastrophic pandemic. Because of their size, detection and accurate identification of biological pathogens is difficult through traditional means. Current technology has proven moderately accurate, but often too slow to be effective in many situations.

A technology that could read the DNA from a sample and rapidly and accurately identify the organisms therein would address this problem. Using the mechanisms discussed previously in this chapter, a system has been developed with these capabilities. Currently, polymerase chain reaction (PCR) amplification followed by fluorescent analysis is the most common method of DNA identification. PCR is a well-understood and reliable laboratory process, but it is highly susceptible to contamination, is labor intensive, and requires a skilled operator and specialized equipment. It is best suited for use within a laboratory or other controlled environment. Attempts to deploy PCR to field environments have proven largely ineffective.

This electronic approach to detection and identification of biological organisms actually uses DNA to identify DNA. This sensor technology does not require any PCR and is rapid and highly accurate. The sensor uniquely combines a biological event (DNA hybridization), a chemical event (metal coating), and microelectronics, to electronically produce a strong electrical signal that indicates the presence of an organism [16].

The sensor consists of oligonucleotide probes attached to multiple pairs of interdigitated electrodes on a microchip. Biological samples are processed to produce a solution of DNA fragments that are passed over the sensor's surface. Hybridization of a target DNA to the DNA capture probes bound to the electrodes forms a DNA bridge connecting the two electrodes. Coating this DNA bridge with metal converts it to a conductive wire (Figure 18.9). The sensor is then electrically analyzed to determine if any bridges have formed. When as little as one bridge is formed and metalized, the electrical resistance of the sensor is reduced more than 1000-fold.

Once prepared, the DNA is introduced to the chip surface containing capture probes complementary to a DNA target sequence. Hybridization occurs with a high degree of specificity because (1) two complementary binding events are required (one to each electrode) and (2) the DNA fragment must be of sufficient length to span the interelectrode gap.

Because DNA by itself is not a reliable conductor [7], the DNA must be made conductive using the coating techniques previously discussed. The decreased resistance of test structures with metalized (coated) DNA bridges indicates the presence of a target DNA. Several metallization chemistries

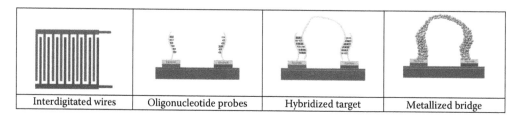

| Interdigitated wires | Oligonucleotide probes | Hybridized target | Metallized bridge |

FIGURE 18.9 DNA/electronic biosensor process.

have been developed for use with the biosensor focusing on the ideal balance of rapid reaction time, minimal background, and no adverse effect on hybridization.

The final step in the process is to measure the electrical resistance of each of the test structures. Voltage is applied to one of the two electrodes in each test structure, and the resistance is obtained by probing the opposite electrode. It has been observed that a single DNA bridge formation results in at least a 1000-fold reduction in resistance on the test structure.

Other groups have developed technologies that rely on electronic signals for the detection of modified DNA [17–19], but these systems are limited by a requirement for a high concentration of target DNA within the sample. Motorola used a gold electrode to form a complicated sandwich of target and reporter probes that contained ferrocene capable of donating electrons [19]. In the presence of a specific target DNA, ferrocene reporters donated electrons that were captured by the gold electrode. However, in this system, many molecules of target DNA were required to generate enough electrons to give a robust signal. Park et al. [10] reported an electronic DNA detection technology based on a gold sol hybridization technique. Gold sols were used to immobilize specific target DNAs from solution to gold surfaces in between two electrodes. Silver was then deposited on the gold sols to close the electrode gap and form a conductive bridge. This technique also required a high concentration of specific target DNA molecules to capture sufficient gold sols to produce an electrical signal.

The approach detailed above is not dependent on the presence of a high concentration of target DNA within the sample, and it has a high signal-to-noise ratio. In this approach, prior amplification of the target DNA is unnecessary. The target DNA forms the connection between the two electrodes and is converted to a conductive wire by direct metallization. Thus, in theory, with this technology it might be possible to detect even a single hybridization event.

18.8 COMPONENTS

The biosensor system is built around a two-component design: a self-contained disposable test cartridge and an analyzer into which the cartridge is loaded for testing. Within each test cartridge, there is a simple silicon chip with multiple independently addressable test structures arrayed upon it. Current chip architecture supports 14 and 64 test structures, each of which may test for the same or multiple-target organisms simultaneously. Each independently addressable test structure measures about 400 × 400 μm. The next chip in development will have more than 250 independently addressable test structures with embedded logic that will permit quantitation assays in addition to identification.

The scanning electron micrographs (SEMs) of the sensors after metallization reveal the conductive nanowires formed between electrodes when a target biological is present. Figure 18.10 shows a test structure of target DNA hybridized to capture probes, and Figure 18.11 shows a magnified view of a single wire (20 to 40 nm in diameter).

The system has successfully demonstrated electronic detection of gene targets from samples of genomic DNA from *Bacillus anthracis*. No amplification of the target sequences is performed before detection. Current research involves an expanding list of pathogens.

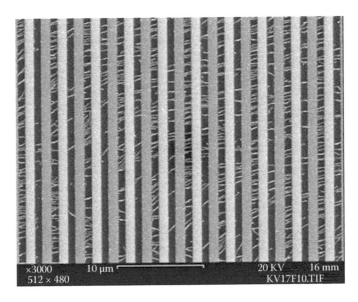

FIGURE 18.10 Thousands of DNA bridges on a sensor.

FIGURE 18.11 Single DNA bridge.

18.9 SAMPLE PREPARATION

For a sample to be read by the electronic biosensor, preparation requires efficient release and isolation of DNA and breaking the DNA down to an appropriate size. Because the system does not require amplification of the target nucleic acid molecules, sample preparation and processing requirements can be incorporated into a simple, automated procedure. Most inhibitors of the enzymes required for amplification will not have an effect on the process utilized. Additionally, detergents and organic solvents can be used to decrease nonspecific binding and inhibit degradation of target nucleic acid molecules, especially RNA targets.

There are several effective methods for releasing DNA from cells, including chemical lysis and sonication. Chemical lysis works well for human cells, common bacteria, and viruses. Sonication is more effective for disrupting bacterial spores. For most samples, chemical disruption is sufficient. After lysis, the sample is filtered to remove anything in the sample that could aberrantly short the sensors. Again, because the system does not use enzymes, it is not necessary to remove

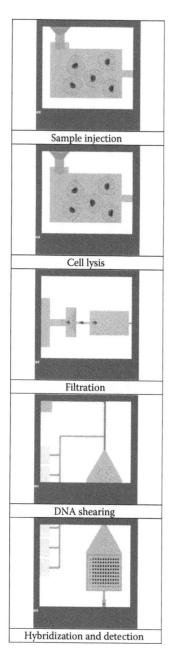

Sample injection

Cell lysis

Filtration

DNA shearing

Hybridization and detection

FIGURE 18.12 DNA sample preparation for electronic detection.

all chemical contaminants. Furthermore, the nucleic acid analog probes are not affected by salt concentrations.

The released DNA must be sheared to a length that works with the BioDetect sensor. Current DNA fragment size is between 1000 and 6000 base pairs in length. Future chip designs will lower the required target DNA length to several hundred base pairs. A mechanical shearing method provides fragments within the desired ranges [15,20]. This method involves pushing DNA through a small bore opening into a larger-bore vessel. The average length can be controlled by changes to flow rate and the size of the opening. The resulting fragments fall within a twofold size distribution.

The sheared DNA is then moved into the hybridization chamber where it can bind to the test sites on the sensor. The DNA is manipulated by mechanical mixing, electrical fields, and pulsing of the fluids. After hybridization, all unbound DNA is washed into a waste chamber.

In summary, the system requires minimal sample preparation. The process that involves cell lysis, DNA shearing, and filtering can be accomplished in a single pass through a cartridge that can be integrated with the detection sensor to produce a fully automated system (Figure 18.12).

18.10 FUTURE CAPABILITIES

The system can be highly multiplexed to test for numerous biological agents simultaneously, to provide confirmatory tests for different unique sequences from a target organism, and to provide highly accurate and quantitative results.

The new design under development will incorporate CMOS-based logic. This will allow the system to produce highly multiplexed results quickly and inexpensively using a combination of on-chip logic, statistics, and bioinformatics. The logic chips will have 256 separately addressable test sites. Future chips may have several thousands of test sites with each site possessing a unique set of probes. Additionally, each test site will be subdivided into thousands of subsensors, allowing for the collection of data for statistical and quantitative analysis. The data will be analyzed using algorithms that will weigh results from the various sensors and calculate the statistical level of certainty of a positive or negative result and provide quantitative results. Error-recognition software will be utilized to recognize patterns from handling damage to electrical signals from debris, further increasing the reliability of the system.

Due to the ability to multiplex, the system can provide for more information regarding an agent than simply a yes or no identification. Through the proper design of probe sets, the chips can identify genetically altered organisms, determine drug resistances, and even provide a taxonomic analysis of an unknown organism.

This detection system provides a glimpse into the possibilities of DNA-based nanostructures. Using a simple nanowire based on a naturally occurring strand, the system provides a significant advance over current technologies, matching or exceeding the sensitivity and accuracy of PCR-based assays in the field while delivering the speed, portability, and ease-of-use of much simpler assays.

18.11   CONCLUSION

DNA-directed assembly is making the promise of self-assembling nanosized devices a reality. The ability to realize the dream of self-assembly allows for low-cost fabrication of simple devices that, to date, could not be produced. DNA-based nanoelectronics and mechanisms will start appearing in products in the foreseeable future. Research and development in this arena hold the promise of great possibilities. Imagine materials that can communicate with the devices that they are made up of, nanoscale machines that can accurately perform medical tasks currently dependent on high-risk surgery, high-efficiency hydrogen fuel cells, radio-frequency identification (RFID) tags embedded into products at the material level, virtually eliminating time spent checking out in stores. These things and many more are not only possible but are likely in a world where DNA-based nanoelectronics are used. The example of the DNA sensor system is only the beginning of the kinds of things that are possible as this exciting new arena begins to take shape.

REFERENCES

1. Dürkop, T., Brintlinger, T., and Fuhrer, M.S., 2002, Nanotubes are high mobility semiconductors, in *Structural and Electronic Properties of Molecular Nanostructures*, H. Kuzmany, J. Fink, M. Mehring, and S. Roth (Eds.) (API Conference Proceedings, New York, 2002), pp. 242–246.
2. Tersoff, J. and Ruoff, R.S., 1994, Structural properties of a carbon-nanotube crystal, *Phys. Rev. Lett.*, 73, 676–679.
3. Seeman, N.C., 1989, Nanoscale assembly and manipulation of branched DNA: A biological starting point for nanotechnology, *NANOCON Proceedings*, J. Lewis and J.L. Quel, NANOCON, Bellevue, WA, pp. 101–123; transcript of oral presentation, pp. 30–36.
4. Seeman, N.C., 2003, DNA in a material world, *Nature*, 421, 427.
5. Shih, W.M., Quispe, J.D., and Joyce, G.F., 2004, A 1.7-kilobase single-stranded DNA that folds into a nanoscale octahedron, *J. Nature Lett.*, 427, 618–621.
6. Henderson, P.T., Jones, D., Hampikian, G., Kan, Y., and Schuster, G.B., 1999, Long-distance charge transport in duplex DNA: The phonon-assisted polaron-like hopping mechanism, *Proc. Natl Acad. Sci. USA*, 96, 8353–8358.
7. Zhang, Y., Austin, R.H., Kraeft, J., Cox, E.C., and Ong, N.P., 2002, Insulating behavior of Lambda-DNA on the micron scale, *Phys. Rev. Lett.*, 89, 198–102.
8. Braun, E., Eichen, Y., Sivan, U., and Ben-Joseph, G., 1998, DNA-templated assembly and electrode attachment of a conducting silver wire, *Nature*, 391, 775–778.
9. Keren, K., Berman, R., Buchstab, E., Sivan, U., and Braun, E., 2003, DNA-templated carbon nanotube field-effect transistor, *Science*, 302, 1380–1382.
10. Park, S.J., Taton, T.A., and Mirkin, C.A., 2002, Array-based electrical detection of DNA with nanoparticle probes, *Science*, 295, 1503–1506.
11. Onoa, G.B. and Moreno, V., 2002, Study of the modifications caused by cisplatin, transplatin, and Pd(II) and Pt(II) mepirizole derivatives on pBR322 DNA by atomic force microscopy, *Int. J. Pharm.*, 245, 55–65.
12. Richter, J., Seidel, R., Kirsch, R., Mertig, M., Pompe, W., Plaschke, J., and Schackert, K., 2000, Nanoscale palladium metallization of DNA, *Adv. Mater.*, 12, 507–510.
13. Richter, J., Mertig, M., Pompe, W., Monch, I., and Schackert, H.K., 2001, Construction of highly conductive nanowires on a DNA template, *Appl. Phys. Lett.*, 78, 536–538.

14. Connolly, D.M., Integrated Nano-Technologies, 2001, Method of Chemically Assembling Nano-Scale Devices, US Patent 6,248,529 B1.
15. Thorstenson, Y.R., Hunicke-Smith, S.P., Oefner, P.J., and Davis, R.W., 1998, An automated hydrodynamic process for controlled, unbiased DNA shearing, *Genome Res.*, 8, 848–855.
16. Connolly, D.M., Integrated Nano-Technologies, 2003, High Resolution DNA Detection Methods and Devices, US Patent 6,593,090.
17. LaBean, T.H., Yan, H., Kopatsch, J., Liu, F., Winfree, E., Reif, J.H., and Seeman, N.C., 2000, The construction, analysis, ligation and self-assembly of DNA triple crossover molecules, *J. Am. Chem. Soc.*, 122, 1848–1860.
18. Mao, C., Sun, W., and Seeman, N.C., 1999, Designed two-dimensional DNA holliday junction arrays visualized by atomic force microscopy, *J. Am. Chem. Soc.*, 121, 5437–5443.
19. Umek, R.M., Lin, S.W., Vielmetter, J., Terbrueggen, R.H., Irvine, B., Yu, C.J., Kayyem, J.F., Yowanto, H., Blackburn, G.F., Farkas, D.H., and Chen, Y.P., 2001, Electronic detection of nucleic acids: Versatile platform for molecular diagnostics, *J. Mol. Diagn.*, 3, 74–84.
20. Oefner, P.J., Hunicke-Smith, S.P., Chiang, L., Dietrich, F., Mulligan, J., and Davis, R.W., 1996, Efficient random subcloning of DNA sheared in a recirculating point-sink flow system, *Nucleic Acids Res.*, 24, 3879–3886.
21. Winfree, E., Sun, W., and Seeman, N.C., 1998, Design and self-assembly of two-dimensional DNA crystals, *Nature*, 394, 539–544.

19 Directed Evolution of Proteins for Device Applications

Jeremy F. Koscielecki, Jason R. Hillebrecht, and Robert R. Birge
University of Connecticut, Storrs, Connecticut, USA

CONTENTS

19.1 PROTEIN-BASED DEVICES

Protein-based photonic devices gain comparative advantage from the unique properties of proteins, and the fact that nature has optimized many proteins for the efficient conversion of light to structural changes. Additional advantages derive from the fact that many proteins produce a voltage, a current, or a change in polarizability in response to light absorption, and carry out this function with a high quantum efficiency and speed [1]. More recently, investigators have been approaching the use of proteins in device applications from the perspective that nature has provided a template for optimization rather than a material with optimal properties. This view is made possible by significant advances in genetic engineering and the use of techniques such as directed evolution. The combination of *in vitro* genetic diversification with tunable selective pressures has enabled investigators to tailor biological macromolecules for electronic and photonic device applications [2–5].

The topic of this brief chapter is bacteriorhodopsin (BR) and the use of genetic engineering to optimize the protein for devices that are based on the long-lived Q state. BR is grown by the halophile archaeon *Halobacterium salinarum*, which uses the protein as a solar energy converter [6]. *H. salinarum* has survived on Earth for more than 3 billion years and has evolved a light-transducing protein that has intrinsic properties appropriate to device applications. These properties include high quantum efficiency, thermal stability, and photochemical cyclicity that combine to make the native protein useful for making thin-film memories [7,8], photovoltaic converters [9], holographic processors [10], artificial retinas [11–13], associative memories [14], logic gates [15], and protein-semiconductor hybrid devices [16]. In all cases investigated, however, a genetically or chemically modified form of the protein outperforms the native protein when a systematic study is carried out to identify or create an optimized variant [17,18]. There are a number of techniques currently used to make modifications to the structure of proteins at different levels of variability. These techniques include site-directed, semirandom, and random mutagenesis. However, the most efficient technique in optimizing the structure or function of biological materials is known as directed evolution where

repeated mutagenesis experiments of screening and selection yield a material with a particular characteristic [3]. In this chapter, mutagenesis techniques are discussed as a method for the optimization
of BR for use in biomolecular devices.

19.2 BACTERIORHODOPSIN

Bacteriorhodopsin is a membrane-bound protein with seven transmembrane helices, 248 amino
acid residues, and a chromophore (retinal) covalently bound via a protonated Schiff base linkage
to Lys-216 near the center of the protein. BR is used by *H. Salinarum* as a photosynthetic protein.
Upon the absorption of light, this protein pumps protons across the cell membrane. The resulting
pH gradient is used to convert ADP and inorganic phosphate into ATP [6]. The primary photochemical event involves the photoisomerization of the chromophore from all-*trans* to 13-*cis*, which forms
the ground state species called K (Figure 19.1). The proton-pumping process then takes place in the
dark through a complex photocycle as shown in Figure 19.1. Figure 19.1 also shows the branched
photocycle involving the P and Q states [19]. The change in the absorption maximum of each intermediate is caused by three factors: the conformation and protonation state of the chromophore
(indicated under the absorption max in Figure 19.1), protonation changes of amino acids near the
chromophore, and other protein–chromophore interactions [20]. Note that the symbol bR is used to
reference the light-adapted resting state of bacteriorhodopsin and the symbol BR is used to reference the protein or a protein variant in an undefined state.

The two states of primary interest for photonic applications are the blue-shifted M and Q states.
In general, the M state is used for real-time holographic devices and memories [10,17,21,22], and
the Q state, the long-lived state within the "branched photocycle," is used for both long-term holographic storage and three-dimensional memories [11,19]. The M state has significant advantages for
holography because it is produced with high quantum efficiency (0.65) and generates a significant
change in refractive index (Figure 19.2). A thin film of BR adjusted to have an absorptivity of 5 at
280 nm produces a film with a 6.4% holographic efficiency at 670 nm. The only disadvantage is that
the holographic image is relatively short-lived (milliseconds to hours), and M-state holograms find
primary application in real-time holographic processing [10,17,21,22]. A single mutation involving the replacement of Asp96 with Asn (D96N) has generated one of the most useful holographic
materials known [17]. Long-term holography and data storage are carried out using the branched
photocycle to form the Q state (Figure 19.1). The branching reaction from O to P is the gateway to
the Q state and involves an all-*trans* to 9-*cis* photochemical event. The Q state is unique because
the chromophore separates from the covalent bond to the protein to form an isolated 9-*cis* retinal
chromophore, which remains caged in the binding site. The Q state is a very long-lived blue-shifted
intermediate that has a lifetime of many years at ambient temperature (Figure 19.2). Because this
state has a lower oscillator strength and is blue shifted relative to the M state, the holographic

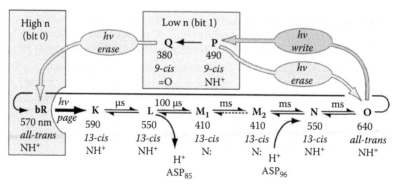

FIGURE 19.1   The main and branched photocycles of bacteriorhodopsin.

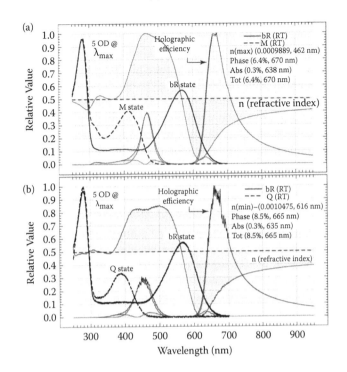

FIGURE 19.2   The diffractive and refractive properties of M-state (a) and Q state (b) films of BR.

efficiency of comparable BR films based on Q-state formation is higher (8.5%) than M-state films (see Figure 19.2). Furthermore, the Q state has additional uses as a binary storage component in three-dimensional memories. Space constraints prevent a detailed discussion of the three-dimensional memory, and the interested reader is directed to the literature [3,11] for details.

Holographic systems based on the M state and the D96N variant are near optimal for real-time holography [10,17,21,22]. But the native protein can also be used for real-time holography with adequate results. In contrast, competitive devices based on the Q state cannot be generated by using the native protein. Some form of chemical or genetic optimization of the protein is required to create a viable system [3,11]. The problem can be explained by reference to Figure 19.1. Note that the only method of generating Q is via a branching reaction involving photoconversion of the red-shifted O intermediate. In the native protein, the O-state concentration rarely exceeds 3 to 5% of the activated protein concentration, and thus the O $\rightarrow$ P photoreaction is compromised by a lack of available O states. One of the first characteristics that needs to be optimized is the O-state concentration, and the remaining portion of this chapter will focus on methods to manipulate the formation and decay kinetics of this red-shifted intermediate. In addition, the quantum efficiency of the O $\rightarrow$ P all-*trans* to 9-*cis* photochemistry is a problem, but one that can be addressed by using site-directed mutagenesis.

19.3   PROTEIN OPTIMIZATION VIA MUTAGENESIS

Genetic modifications can be created by using site-directed (SDM) or semirandom mutagenesis (SRM) to insert, delete, or more often, replace one or more residues. SDM is a method that carries out the substitution of a given residue with a specific replacement residue, normally involving a single site. Using a commercially available mutagenesis kit, protein variants are easily engineered once an expression system for protein production is in place. SDM requires mutation of the protein coding sequence, then expression of this gene in a host (*Escherichia coli* or *H. salinarum*) for protein production. Because of the large number of mutations that are possible, SDM methods are not useful tools for protein optimization unless a good structure–function model of the protein is available to guide mutation. In most cases, theory is incapable of accurately predicting the impact of a given mutation, and hence SDM plays a minor role in protein optimization.

Semirandom mutagenesis or saturation mutagenesis allows each amino acid in a specified region to be mutated with an equal probability, leaving the rest of the protein unchanged. SRM generates a large number of mutant proteins and therefore requires an effective screening method to analyze the photokinetic properties of a library of mutations. In some cases, it is more efficient to generate mutations throughout the entire protein. Error-prone polymerase chain reaction (PCR) is one such technique and introduces mutations at a frequency from 1 to 20 mutations per 1 kb by enhancing the natural error rate of polymerase (usually *Taq*) by the modification of standard PCR methods [23]. In the case of BR, there is enough structural and prior mutagenesis work to preempt the need for a purely random search. For example, we know that residues in the region from 190 to 210 are involved in the exit channel of the protein. Thus, it is not surprising that the mutation of residues in this region has a profound impact on the lifetime of the O-state, which must transfer a proton from Asp85 into this region prior to reformation of the bR resting state. Carrying out random mutagenesis in this region has a much higher probability of generating long-lived O-state mutants than random mutations throughout the entire protein. As we explore in the next section, directed evolution may offer the best combination of efficiency and speed in finding an optimal mutation or set of mutations.

19.4 DIRECTED EVOLUTION

Biological systems have evolved over the past 3 billion years via an algorithm of mutation and natural selection [23]. Present-day mutagenesis techniques and *in vitro* recombination methods are aimed to mimic the genetic diversification that occurs in natural ecosystems. The major advantage to the artificial creation of genetic diversity is its attenuated timescale, relative to natural mutation rates. Countless mutations can be inserted into the genetic code of nearly any biological macromolecule in a matter of hours. Combining this technology with a selection method that is specific to a parameter of interest is the fundamental essence of directed evolution.

Although mutagenesis and *in vitro* recombination methods can be applied to any genetically amenable system, the limiting factor in most directed evolution investigations is the screening system used to select for optimized variants. The architectural design and stringency of the system must be tailored to the biochemical and biophysical parameters that are being optimized. Selection of optimized variants is followed by genetic and phenotypic characterization of the molecule. Select variants are then used as parental templates in subsequent rounds of directed evolution. The selective pressure in ensuing rounds of directed evolution is typically increased, in order to drive and direct the evolution of the macromolecule.

For BR to serve as a biomaterial in photochromic and optoelectronic device applications, the branched photochemistry of the protein must be optimized. Optimization of the O-state concentration and the quantum efficiency of the O → P transition are key to the architecture of BR-based optical memories. Although traditional mutagenesis techniques (SDM and SRM) are useful for probing localized regions of a protein, global modifications account for more of the complicated and oftentimes distant molecular interactions that contribute to the photochemistry of a protein. Random mutagenesis, DNA shuffling, and *in vitro* recombination are just a few of the methods available for introducing global diversity into a biological macromolecule [23].

Photochemical selection of BR variants requires a high-throughput screening method for mutants with long O-state concentrations and efficient O → P transitions. Two types of selection methods currently exist for BR-based libraries. Type 1 selection involves the *in vitro* characterization of isolated purple membrane fragments, and type 2 involves screening the photochemistry of whole cell cultures. The photokinetic properties of each variant protein are directly measured in type 1 screening. The drawback to this method is the time and cost involved with isolating each variant protein from whole cell cultures. Type 2 screening probes the photocycle of whole cell cultures and indirectly selects for variants with optimized branched photochemistry (long Q-state concentrations). Selecting for variants with long Q-state concentrations is automated by using a modified, bioflow cell reactor. These apparatuses sort out variants with undesirable photochemistry while cataloging the cells with high Q-state yields. Candidate mutants are then grown in large quantities and isolated for more detailed photochemical analysis. Mutants with favorable photochemistry are used as starting points for subsequent rounds of diversification and differential selection.

The ability to tailor a biomaterial to the demands of protein-based applications is a testament to the flexibility of directed evolution and its usefulness to the field of biomolecular electronics.

19.5   CONCLUSIONS

By using directed evolution and other mutagenesis techniques, we greatly increase the possibility for discovering a genetic variant of BR with optimal performance in a biomolecular device. Starting with a protein that nature has provided as a template will allow researchers to optimize proteins for device applications by achieving a high level of optimization in a short period of time. BR variants with altered photochemistry have already been produced that are functional and stable; directed evolution could enhance these properties even further.

REFERENCES

1. Xu, J. et al., Direct measurement of the photoelectric response time of bacteriorhodospin via electro-optic sampling. *Biophysic. J.*, 2003, 85: 1128–1134.
2. Arnold, F.H., Design by directed evolution. *Acc. Chem. Res.*, 1998, 31: 125–131.
3. Wise, K.J. et al., Optimization of bacteriorhodopsin for bioelectronic devices. *Trends Biotechnol.*, 2002, 20: 387–394.
4. Arnold, F. and J.C. Moore, Optimizing industrial enzymes by directed evolution. *Adv. Biochem. Eng.*, 1997, 58: 1–14.
5. Dalby, P.A., Optimising enzyme function by directed evolution. *Curr. Opin. Struct. Biol.*, 2003, 13: 500–505.
6. Oesterhelt, D. and W. Stoeckenius, Rhodopsin-like protein from the purple membrane of *Halobacterium halobium. Nature (London), New Biol.*, 1971, 233: 149–152.
7. Lawrence, A.F. and R.R. Birge, Communication with submicron structures. Perspectives in the application of biomolecules to computer technology. In *Nonlinear Electrodynamics in Biological Systems*, H.R. Adey and A.F. Lawrence, Eds. 1984, Plenum: New York, pp. 207–218.
8. Schick, G.A., A.F. Lawrence, and R.R. Birge, Biotechnology and molecular computing. *Trends Biotechnol.*, 1988, 6: 159–163.
9. Marwan, W., P. Hegemann, and D. Oesterhelt, Single photon detection by an archaebacterium. *J. Mol. Biol.*, 1988, 663–664.
10. Hampp, N. and T. Juchem, Fringemaker—The first technical system based on bacteriorhodopsin. In L. Keszthelyi (ed.), *Bioelectronic Applications of Photochromic Pigments*, 2000, IOS Press: Szeged, Hungary, pp. 44–53.
11. Birge, R.R. et al., Biomolecular electronics: Protein-based associative processors and volumetric memories. *J. Phys. Chem. B*, 1999, 103: 10746–10766.
12. Miyasaka, T., K. Koyama, and I. Itoh, Quantum conversion and image detection by a bacteriorhodopsin-based artifical photoreceptor. *Science*, 1992, 255: 342–344.
13. Chen, Z. and R.R. Birge, Protein based artificial retinas. *Trends Biotechnol.*, 1993, 11: 292–300.
14. Birge, R.R., Photophysics and molecular electronic applications of the rhodopsins. *Annu. Rev. Phys. Chem.*, 1990, 41: 683–733.
15. Mobarry, C. and A. Lewis, Implementations of neural networks using photoactivated biological molecules. *Proc. SPIE*, 1986, 700: 304.
16. Bhattacharya, P. et al., Monolithically integrated bacteriorhodopsin-GaAs field-effect transistor photoreceiver. *Opt. Lett.*, 2002, 27: 839–841.
17. Hampp, N., Bacteriorhodopsin: Mutating a biomaterial into an optoelectronic material. *Appl. Microbiol. Biotechnol.*, 2000, 53: 633–639.

18. Wise, K.J. and R.R. Birge, Biomolecular photonics based on bacteriorhodopsin. In *CRC Organic Handbook of Photochemistry and Photobiology*, W. Horspool and F. Lenci, Eds. 2003, Boca Raton, FL, CRC Press, Chapter 135.
19. Gillespie, N.B. et al., Characterization of the branched-photocycle intermediates P and Q of bacteriorhodopsin. *J. Phys. Chem. B*, 2002, 106: 13352–13361.
20. Ebrey, T.G., Light energy transduction in bacteriorhodopsin. In *Thermodynamics of Membrane Receptors and Channels*, M.B. Jackson, Ed. 1993, CRC Press: Boca Raton, FL, pp. 353–387.
21. Juchem, T. and N. Hampp, Interferometric system for non-destructive testing based on large diameter bacteriorhodopsin films. *Optics Lasers Eng.*, 2000, 34: 87–100.
22. Oesterhelt, D., C. Bräuchle, and N. Hampp, Bacteriorhodopsin: A biological material for information processing. *Quart. Rev. Biophys.*, 1991, 24: 425–478.
23. Arnold, F.H. and G. Georgiou, Eds. Directed evolution library creation. *Meth. Mol. Biol.*, 2003, 231: 3.

20 Semiconductor Quantum Dots for Molecular and Cellular Imaging

Andrew Michael Smith and Shuming Nie
Emory University, Georgia Institute of Technology,
Atlanta, Georgia, USA

CONTENTS

20.1 INTRODUCTION

Biological probes are indispensable tools for studying biological samples, cells in culture, and animal models. Exogenous probes are frequently multifunctional, having one component that can detect a biological molecule or event, and another component that reports the presence of the probe. A fundamental example of this functionality is a fluorescently labeled antibody: when administered to a monolayer of fixed cells, the antibody binds to its target molecule, and the fluorophore emits light to signal its presence. Of the many available reporters (e.g., radioactive isotopes, chromophores, and fluorophores), fluorescent molecules have been found to be invaluable due to their inherently high sensitivity of detection, low cost, ease of conjugation to biological molecules, and lack of ionizing radiation. Indeed, organic fluorophores and fluorescent proteins have been used in nearly all avenues of biological sensing, from *in vitro* assays to living animal imaging. Recently, quantum dots (QDs) have been developed as a new class of biological fluorophore. With easily tunable properties and significant spectral advantages over conventional fluorophores, QDs have already been used for ultrasensitive biological detection.

Semiconductor QDs have captivated scientists and engineers over the past two decades due to their fascinating optical and electronic properties that are not available from isolated molecules

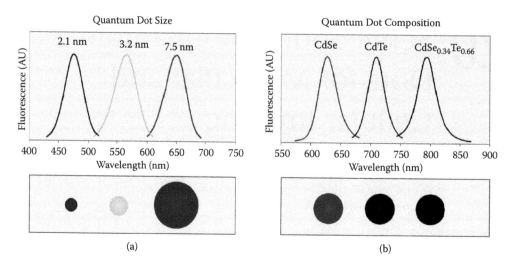

FIGURE 20.1 Size and composition tuning of optical emission for binary CdSe and ternary CdSeTe quantum dots. (a) CdSe QDs with various sizes (given as diameter) may be tuned to emit throughout the visible region by changing the nanoparticle size while keeping the composition constant. (b) The size of QDs may also be held constant, and the composition may be used to alter the emission wavelength. In the above example, 5 nm diameter quantum dots of the ternary alloy $CdSe_xTe_{1-x}$ may be tuned to emit at longer wavelengths than either of the binary compounds CdSe and CdTe due to a nonlinear relationship between the alloy bandgap energy and composition. (From Bailey, R.E. and Nie, S.M., *J. Am. Chem. Soc.* 125, 7100–7106, 2003. With permission.)

or from bulk solids. QDs are nanocrystals of inorganic semiconductors that are restricted in three dimensions to a somewhat spherical shape, typically with a diameter of 2 to 8 nm (on the order of 200 to 10,000 atoms). Bulk-phase semiconductors are characterized by valence electrons that can be excited to a higher-energy conduction band. The energy difference between the valence band and the conduction band is the bandgap energy of the semiconductor. The excited electron may then relax to its ground state through the emission of a photon with energy equal to that of the bandgap. When a semiconductor is of nanoscale dimensions, the bandgap is dependent on the size of the nanocrystal. As the size of a semiconductor nanocrystal decreases, the bandgap increases, resulting in shorter wavelengths of light emission. This quantum confinement effect is analogous to the quantum mechanical "particle in a box," in which the energy of the particle increases as the size of the box decreases. Cadmium selenide (CdSe) is the prototypical QD, and its size-tunable fluorescence throughout the visible light spectrum is depicted in Figure 20.1a. Other semiconductor materials display fluorescence in different spectral ranges, so that QDs can be synthesized to emit at wavelengths between 400 and 2000 nm by changing their composition and size [1–3]. An important consequence of this quantum confinement effect for biologists is that these size-tunable properties occur at the same size regime as biological macromolecules like proteins and nucleic acids.

20.2 QUANTUM DOTS VERSUS ORGANIC FLUOROPHORES

Fluorescent dyes have been valuable in the study of biological phenomena due to their inherent high sensitivity of detection and ease of use. QDs may provide a new class of biological labels that could overcome the limitations of organic dyes and fluorescent proteins. With size-tunable fluorescence emission, QDs can be generated for any specific wavelength, from the UV through the near-infrared [4]. QD emission peaks are narrow (FWHM typically 25 to 35 nm) and symmetric compared to organic fluorophores, making them ideal for applications involving the simultaneous detection of multiple fluorophores [5]. In addition, the broad absorption spectra of QDs allow the excitation of multiple fluorophores with a single light source, at any wavelength shorter than the emission peak wavelength [5]. QDs are highly resistant to photobleaching, a commonly occurring problem for organic fluorophores, thus making them useful for continuous monitoring of fluorescence [6]. QDs have very large molar extinction coefficients and high quantum yields, resulting in bright fluorescent probes in aqueous solution [7]. Moreover, QDs have long fluorescence lifetimes on the order of 20 to 50 nsec, which may allow them to be distinguished from background and other fluorophores for increased sensitivity of detection [4].

It should be noted, however, that QDs are unlikely to replace organic dyes. Although QDs are commercially available, they are currently expensive compared to organic dyes, and changing an already established biological detection system from dyes to QDs will require time and optimization. Also, QDs are an order of magnitude larger than organic dyes. Therefore, applications such as real-time monitoring of biomolecular interactions (in which steric hindrance is of concern) may require the use of organic dyes, as QDs smaller than 1 nm are inherently unstable. In addition, most organic dyes are of similar sizes, so that fluorophores of different emissions are similar sterically, compared to the large difference in QD size required to tune their wavelength. However, it has been shown that the emission wavelengths of alloy QDs may be tuned by altering the alloy composition, while keeping the size constant (Figure 20.1b) [1].

20.3 SYNTHESIS AND BIOCONJUGATION

20.3.1 SYNTHESIS AND CAPPING

The prototypical QD is CdSe because colloidal syntheses for monodisperse nanocrystals of this semiconductor are well established. CdSe is most often synthesized through the combination of cadmium and selenium precursors in the presence of a QD-binding ligand that stabilizes the growing QD particles and prevents their aggregation into bulk semiconductor. Among various synthetic methods reported in the literature, high-temperature synthesis in coordinating solvents has yielded the best size monodispersity and fluorescence efficiencies. A coordinating solvent serves as a solvent and as a ligand, and is most commonly a mixture of trioctylphosphine (TOP), trioctylphosphine oxide (TOPO), and hexadecylamine (HDA). The basic functional groups of these ligands (phosphines, phosphine oxides, and amines) attach to the QD surface during synthesis, leaving the ligand alkyl chains directed away from the surface. The resulting QDs are highly hydrophobic, and only soluble in nonpolar solvents such as chloroform and hexane. The CdSe core is often capped with a thin layer of a higher bandgap material, such as ZnS or CdS, which removes surface defects, significantly improving fluorescence quantum yields.

20.3.2 WATER SOLUBILIZATION AND BIOCONJUGATION

For use in biological labeling, QDs must be rendered hydrophilic so that they are soluble in aqueous buffers. Two general strategies have been developed for phase transfer of QDs to aqueous solution (Figure 20.2). In the first approach, hydrophobic surface ligands are replaced with bifunctional ligands such as mercaptoacetic acid, which contains a thiol group that binds strongly to the QD surface as well as a carboxylic acid group that is hydrophilic [7]. Other functional groups may also be used; for example, silane groups can be polymerized into a silica shell around the QD after ligand exchange [4]. In the second method, coordinating ligands (e.g., TOPO) on the QD surface are used to interact with amphiphilic polymers or lipids [6,8], resulting in micelle-like encapsulation of the QD. This latter method is more effective than ligand exchange at maintaining the QD optical properties and storage stability in aqueous buffer, but it increases the overall size of QD probes. Water-soluble QDs may be rendered biologically active through conjugation to biomolecules, such as nucleic acids, proteins, or small molecules. Attachment of these biomolecules has been demonstrated using a variety of intermolecular interactions, including covalent coupling [4,7], ionic attraction [9], and streptavidin–biotin bridging [6].

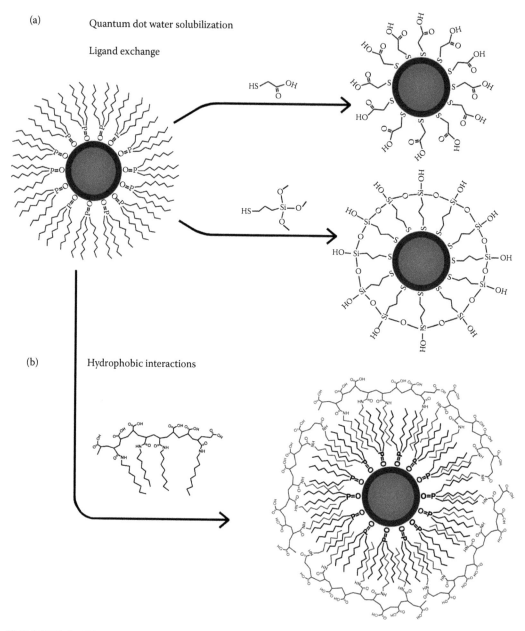

FIGURE 20.2 Diagram of two general strategies for phase transfer of trioctylphosphine oxide (TOPO)-coated quantum dots (QDs) into aqueous solution. Ligands are drawn disproportionately large for detail. (a) TOPO ligands may be exchanged for heterobifunctional ligands for dispersion in aqueous solution. This scheme can be used to generate a hydrophilic QD with carboxylic acids or a shell of silica on the QD surface. (b) The hydrophobic ligands may be retained on the QD surface and rendered water soluble through micelle-like interactions with an amphiphilic polymer-like octylamine-modified polyacrylic acid.

20.4 BIOLOGICAL APPLICATIONS

Fluorescence is a sensitive and routine means for monitoring biological events using fluorescent dyes and fluorescent proteins. Since 1998, QDs have also been used as biological labels in a variety of bio-assays, some of which would not have been possible with conventional fluorophores. *In vitro* bioana-lytic assays were developed by using QD-tagged antibodies, fluorescence resonance energy transfer quantum dot (FRET-QD) biosensors, as well as by using QD-encoded microbeads. In addition to solution-based assays, the spectroscopic advantages of QDs have allowed for sensitive optical imag-ing in living cells and animal models. Many reports have concentrated on simply replacing organic dyes with QDs, without utilizing their unique properties. This analysis will focus on the publications that have exploited their resistance to photobleaching and potential for multiplexed detection.

20.4.1 BIOANALYTIC ASSAYS

Organic fluorophores are commonly used as reporters in a large number of *in vitro* bioassays, such as quantitative immunoassays and fluorescence quenching assays for macromolecular interactions. High sensitivity has been realized with the use of organic dyes, but the spectral properties of QDs could lead to further improvements. Research in the application of QDs for *in vitro* bioanalysis has been advanced primarily by Mattoussi and his coworkers at the U.S. Naval Research Laboratory [9,10], and can be divided into two areas: immunoassays and biosensors.

Immunoassays typically involve the specific binding of a labeled antibody to an analyte, fol-lowed by physical removal of unbound antibody to allow the quantification of the bound label. QDs have been conjugated to antibodies for use in an assortment of these fluoroimmunoassays for detec-tion of proteins and small molecules [10]. The results of these studies proved that QDs may be used as "generalized" reporters in immunoassays, but did not demonstrate an advantage over organic fluorophores, in that their sensitivity was comparable to that of commercial assays (protein con-centrations down to 2 ng/ml, or 100 pM) [11]. The main advantages of QDs for immunoassays are their narrow, symmetric emission profiles and the excitability of many different QDs with a single light source, allowing the detection of multiple analytes simultaneously. Taking advantage of these spectral properties, Goldman et al. [10] simultaneously detected four toxins using four different QDs, emitting between 510 and 610, in a sandwich immunoassay configuration. Although there was spectral overlap of the emission peaks, deconvolution of the spectra revealed fluorescence contribu-tions from all four toxins. This assay was far from quantitative, however, and it is apparent that fine-tuning of antibody cross-reactivity will be required to make multiplexed immunoassays useful.

Whereas immunoassays require the physical separation of unbound QD conjugates prior to analysis, biosensors can be developed to detect biomolecular targets on a real-time or continuous basis. QDs are ideal for biosensor applications due to their resistance to photobleaching, allowing for continuous monitoring of a signal. FRET has been the major proposed mechanism to render QDs switchable from a quenched "off" state to a fluorescent "on" state. FRET is the nonradiative energy transfer from an excited donor fluorophore to an acceptor. The acceptor can be any mol-ecule (another nanoparticle, a nonemissive organic dye, or fluorophore) that absorbs radiation at the wavelength of donor emission. QDs are promising donors for FRET-based applications due to their continuously tunable emissions that can be matched to any desired acceptor, and their broadband absorption, allowing excitation at a short wavelength that does not directly excite the acceptor.

It has been confirmed that QDs can be FRET donors, quenchable with efficiencies up to 99%, using organic fluorophores, nonemissive dyes, gold nanoparticles, or other QDs as acceptors. Med-intz et al. [9] used QDs conjugated to maltose binding proteins as an *in situ* biosensor for carbo-hydrate detection. Adding a maltose derivative covalently bound to a FRET acceptor dye caused QD quenching, and fluorescence was restored upon addition of native maltose, which displaced the sugar–dye compound. A key element of this work was that the physical orientation and stoichiometry of the maltose receptors on the QDs were controlled so that the restoration of QD fluorescence upon maltose addition could be directly related to maltose concentration. Although the FRET quenching efficiency was low, this work demonstrates the potential of QD-based *in situ* biosensing.

20.4.2 QD-Encoding

Rather than using single QDs for biological detection schemes, it has been proposed that different colors of QDs can be combined into a larger structure, such as a microbead, to yield an "optical barcode" [5]. With the combination of six QD emission colors and ten QD intensity levels for each color, one million different codes are theoretically possible. Biological molecules may be optically encoded by conjugation to these beads, opening the door to the multiplexed identification of many biomolecules for high-throughput screening of biological samples. Pioneering work was reported by Han et al. [5] in 2001, in which 1.2 µm polystyrene beads were encoded with three colors of QDs (red, green, and blue) and different intensity levels. The beads were then conjugated to DNA, resulting in different nucleic acids being distinguished by their spectrally distinct optical codes. These encoded probes were incubated with their complementary DNA sequences, which were also labeled with a fluorescent dye as a target signal. The hybridized DNA was detected through colocalization of the target signal and the probe optical code, via single-bead spectroscopy, using only one excitation source. The bead code identified the sequence, and the intensity of the target signal corresponded to the presence and abundance of the target DNA sequence.

The high-throughput potential of this seminal report was realized in 2003 with the use of a similar system to detect DNA sequences that differed by only one nucleotide (single nucleotide polymorphisms) [12]. In this work, 194 samples of ten different DNA sequences from specific alleles of the human cytochrome P450 gene family were correctly identified by hybridization to encoded probes. High-throughput analysis was achieved by the use of flow cytometry to identify spectral codes, rather than single-bead spectroscopy. This identification would have been considerably more difficult with organic fluorophores due to the fact that their emission peaks overlap, obscuring the distinct codes, and the fact that multiple excitation sources would be required.

Once encoded libraries have been developed for identification of nucleic acid sequences and proteins, solution-based multiplexing of QD-encoded beads could quickly produce a vast amount of genomic and protein expression data. Another approach to gene multiplexing has been the use of planar chips, but bead-based multiplexing has advantages of greater statistical analysis, faster assaying time, and the flexibility to add additional probes at lower costs.

20.4.3 Imaging of Cells and Tissues

Fluorescent dyes are used routinely for determining the presence and location of biological molecules in cultured cells and tissue sections. Two original papers in 1998 demonstrated the feasibility of using QDs for cell labeling, displaying distinct advantages over organic dyes (Figure 20.3). Bruchez et al. [4] demonstrated dual-color labeling of fixed mouse fibroblasts, staining the nucleus with green QDs, and labeling the F-actin filaments in the cytoplasm with red QDs. Chan et al. [7] showed that QDs maintained their bright fluorescence in live cells, by imaging the uptake of transferrin-conjugated QDs by HeLa cells. These studies showed that QDs were brighter and more photostable than organic fluorophores, a claim that has been verified by independent reports [6].

In 2003, QDs were used for the first time to visualize cellular structures at high resolution, as Wu et al. [6] illustrated immunocytochemical stains of membrane, cytoplasmic, and nuclear antigens in fixed cells. Although imaging of fixed cells is useful and sufficient for many applications, live cell microscopy is ideal for visualizing cellular processes but is considerably more difficult. It has been shown that many cell types naturally engulf QDs through a nonspecific uptake mechanism [13]. This mechanism was used to track the migration of breast tumor cells on a substrate coated with red QDs; the fluorescence inside the cells increased as the cells transversed and engulfed the QDs, leaving behind a dark path [13]. This and other studies demonstrated that QDs can be imaged

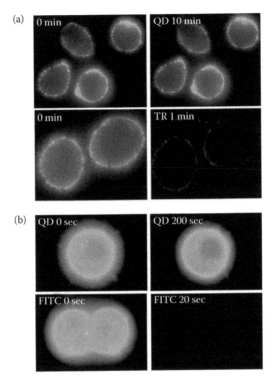

FIGURE 20.3 Immunofluorescent labeling of human breast tumor cells with antibody-conjugated quantum dots (QDs), and comparison of signal brightness and photostability with organic dyes. (a) Cancer cells labeled with antibody-conjugated QD or Texas Red (TR) targeting cell surface antigen uPAR. (b) Cancer cells labeled with antibody-conjugated QD or fluorescein isothiocyanate (FITC) targeting cell surface antigen Her-2/neu. Excitation from a 100 W mercury lamp caused negligible photobleaching of QDs, compared to the two organic fluorophores. (Courtesy of Dr. Xiaohu Gao, Emory University.)

inside living cells for long periods of time (over a week), a task that is not possible with organic fluorophores due to photobleaching. Indeed, QDs have opened up a new avenue for studying biomolecular processes inside living cells.

Two true marvels of real-time live cell imaging have recently been demonstrated using QDs. Dahan et al. [14] labeled glycine receptors on neuronal membranes with QDs. Imaging of the cells revealed the ability to observe the motion of single QDs, in real time (single, isolated QDs can be identified visually because they "blink"). This first example of single molecule detection using QDs in living cells produced remarkable movies of glycine receptor diffusion. In a second report, Lidke et al. [15] used QDs conjugated to epidermal growth factor (EGF) to monitor the interactions between EGF and the erbB/HER receptor on living cell membranes. Single molecules of EGF were visualized in real time as they bound to receptors and were endocytosed. This allowed the study of receptor interactions and revealed a new cellular filopodial transport phenomenon.

Because the use of QDs as reporters in living cells may soon become conventional, the possibility of QD cytotoxicity is of interest and concern. Almost all of the reports of QDs in living cells have revealed little or no obvious cytotoxicity or changes in cellular differentiation [8,13]. Although QDs contain toxic elements, most importantly divalent cadmium, cytotoxicity issues may only become relevant for truly long-term (months to years) visualization of QDs in cells, a time period in which QD degradation could become significant.

QDs have been used as labels for studying single molecules that interact with the membranes of living cells. Performing the same task with intracellular targets will be much more difficult but is essential for visualizing processes in living cells. Advanced delivery methods are needed to deliver

QD probes into living cells, and the delivered probes must be available for binding to intracellular targets, and not trapped in endosomes, lysosomes, or other organelles. Until this becomes possible, intracellular processes can only be modeled by using isolated macromolecules under *in vitro* conditions. For example, QD-actin bioconjugates have been used to observe single-molecule motorized motion of actin filaments sliding across myosin proteins in an ATP-driven reaction [16]. These model systems should be visualizable intracellularly once a protocol for translocation across the cellular membrane is established.

20.4.4   IN VIVO ANIMAL IMAGING

The progression from optical microscopy of cells *in vitro* to optical imaging of entire organisms has mainly been inhibited by poor penetration of visible light through tissue. Due to this attenuation problem, QDs were initially used as optical imaging contrast agents only in simple model systems. In 2002, Akerman et al. [17] conjugated QDs to peptides for targeting endothelial cell receptors in specific tissues (lung, tumor blood vessels, or tumor lymphatic vessels). Intravenous injection of these bioconjugated nanoparticles into a mouse revealed accumulation of QDs in the targeted tissue, visualized histologically. Whole-organism imaging was not performed in this work but was achieved by Dubertret et al. [8] on small *Xenopus* embryos containing intracellular QDs. Micro-injection of more than a billion QDs into single cells allowed cell lineage tracking and real-time imaging of stably fluorescent QDs.

To solve the attenuation problem for optical imaging in larger organisms and in deeper tissue, it has been shown that the far-red and near-infrared (NIR) spectral regions are characterized by less scattering and absorption by biological tissue. For sensitive detection, wavelengths must be chosen so that excitation light can penetrate tissue to the desired depth, and the emitted light must be able to travel back to a photodetector. Several semiconductor materials have been used to generate bright QDs that emit between 650 and 2000 nm [1,3]. There are no conventional dyes that are bright and photostable that can emit fluorescence light beyond ~850 nm, which is why QDs are expected to provide substantial advantages for NIR optical imaging. NIR QDs with emission maxima between 750 and 860 nm were used to image coronary vasculature in a rat model [18], and to visualize sentinel lymph nodes in a pig, in 1-cm deep tissue [19].

Most imaging systems generate image contrast based on attenuation of radiation through tissue. Imaging with contrast based on molecular differences in tissue is called molecular imaging. Organic fluorescent dyes and fluorescent proteins have already been used as contrast agents for fluorescent molecular imaging in animal models. NIR QDs will be powerful tools for molecular imaging because they can be imaged in real time with multiplexed detection to monitor biomolecular phenomena *in vivo*. In our own lab we have recently been able to perform molecular imaging for the detection of subdermal tumors. Targeting of antibody-conjugated QDs to tumors has allowed generation of whole-body fluorescence imaging of mice with contrast based on biomolecular differences between normal and cancerous tissue [20].

20.5   FUTURE DIRECTIONS

Quantum dots have already fulfilled some of their promise as groundbreaking biological labels. The tremendous amount of interest in QDs is sure to quickly improve the previous applications and inspire new ones. Organic fluorophores may never be completely supplanted due to their inherently small size, but research in the past 5 years has shown that QDs offer remarkable advantages.

In the near future, the development of efficient QD biosensors may make QDs into powerful tools not just for *in vitro* biosensing, but also for living cell studies and *in vivo* imaging. Biosensors based on organic fluorophores have already shown promise *in vivo*, as tumors in mice were detected with "stealth" quenched probes, activatable upon exposure to proteases in the tumor microenvironment. The ability of a biochemical signal to switch a QD from an "off" to an "on" state would be an important and powerful tool for studying intracellular signaling. The development of a "quantum dot beacon" would be a monumental advance. For this to become a reality, however, biosensing QDs must be translocated across the cell membrane. Although microinjection and nonspecific uptake have been used, widespread applications must wait for a generalizable methodology for efficient delivery of QD probes into living cells. Recent research in our lab has shown that delivery peptides can be used for rapid intracellular QD translocation, and other groups have already demonstrated their efficacy for the delivery of other types of nanoparticles. The QD combination of real-time imaging, multiplexing capabilities, single-molecule detection, and biological sensing on the nanoscale should allow scientists to address a broad array of analytical problems and biological questions.

A major goal in nanotechnology research is to develop smart multifunctional devices with nanometer dimensions. Although this is a lofty goal seemingly for the distant future, many multifunctional devices have already been created, and many tools have been developed for the assembly of QDs into complex, ordered structures. One proposed multifunctional device is a nanoscale contrast agent for multimodality imaging. QDs are fluorescent contrast agents, but they can also be used as markers for electron microscopy due to their high electron density. Multimodal imaging has already been performed in cells to correlate fluorescence staining with electron micrographs using QDs [14]. QDs may also be combined with magnetic resonance imaging (MRI) contrast agents like Fe_2O_3 and FePt nanoparticles. By correlating the deep imaging capabilities of MRI with ultrasensitive optical fluorescence, a surgeon could visually identify tiny tumors or other small lesions during an operation and remove the diseased cells and tissue completely. Medical imaging modalities such as MRI and PET (positron emission tomography) can identify diseases noninvasively, but they do not provide a visual guide during surgery. The development of magnetic or radioactive QD probes could solve this problem.

Another desired multifunctional device would be the combination of a QD imaging agent with a therapeutic agent. Not only would this allow tracking of pharmacokinetics, but diseased tissue could be treated and monitored simultaneously and in real time. Surprisingly, QDs may be innately multimodal in this fashion, as they have been shown to have potential activity as photodynamic therapy agents. These combinations are only a few possible achievements for the future. Practical applications of these multifunctional nanodevices will not come without careful research, but the multidisciplinary nature of nanotechnology may expedite these goals by combining the great minds of many different fields. The success seen so far with QDs points toward the success of QDs in biological systems, and also predicts the success of other avenues of bionanotechnology.

ACKNOWLEDGMENTS

This work was supported by a grant from the National Institutes of Health (R01 GM60562), the Georgia Cancer Coalition (Distinguished Cancer Scholar Award to S.N.), and the Coulter Translational Research Program at Georgia Tech and Emory University. We are grateful to Dr. Xiaohu Gao for stimulating discussions and for providing Figure 20.3 for this chapter. A.M.S. acknowledges the Whitaker Foundation for generous fellowship support.

REFERENCES

1. R.E. Bailey and S.M. Nie (2003) Alloyed semiconductor quantum dots: Tuning the optical properties without changing the particle size. *J. Am. Chem. Soc.* 125, 7100–7106.
2. X.H. Zhong, Y.Y. Feng, W. Knoll, and M.Y. Han (2003) Alloyed $Zn_xCd_{1-x}S$ nanocrystals with highly narrow luminescence spectral width. *J. Am. Chem. Soc.* 125, 13559–13563.
3. B.L. Wehrenberg, C.J. Wang, and P. Guyot-Sionnest (2002) Interband and intraband optical studies of PbSe colloidal quantum dots. *J. Phys. Chem. B* 106, 10634–10640.
4. M. Bruchez, M. Moronne, P. Gin, S. Weiss, and A.P. Alivisatos (1998) Semiconductor nanocrystals as fluorescent biological labels. *Science* 281, 2013–2016.

5. M.Y. Han, X.H. Gao, J.Z. Su, and S. Nie (2001) Quantum-dot-tagged microbeads for multiplexed optical coding of biomolecules. *Nat. Biotechnol.* 19, 631–635.

6. X.Y. Wu, H.J. Liu, J.Q. Liu, K.N. Haley, J.A. Treadway, J.P. Larson, N.F. Ge, F. Peale, and M.P. Bruchez (2003) Immunofluorescent labeling of cancer marker Her2 and other cellular targets with semiconductor quantum dots. *Nat. Biotechnol.* 21, 41–46.

7. W.C.W. Chan and S.M. Nie (1998) Quantum dot bioconjugates for ultrasensitive nonisotopic detection. *Science* 281, 2016–2018.

8. B. Dubertret, P. Skourides, D.J. Norris, V. Noireaux, A.H. Brivanlou, and A. Libchaber (2002) *In vivo* imaging of quantum dots encapsulated in phospholipid micelles. *Science* 298, 1759–1762.

9. I.L. Medintz, A.R. Clapp, H. Mattoussi, E.R. Goldman, B. Fisher, and J.M. Mauro (2003) Self-assembled nanoscale biosensors based on quantum dot FRET donors. *Nat. Mater.* 2, 630–638.

10. E.R. Goldman, A.R. Clapp, G.P. Anderson, H.T. Uyeda, J.M. Mauro, I.L. Medintz, and H. Mattoussi (2004) Multiplexed toxin analysis using four colors of quantum dot fluororeagents. *Anal. Chem.* 76, 684–688.

11. E.R. Goldman, G.P. Anderson, P.T. Tran, H. Mattoussi, P.T. Charles, and J.M. Mauro (2002) Conjugation of luminescent quantum dots with antibodies using an engineered adaptor protein to provide new reagents for fluoroimmunoassays. *Anal. Chem.* 74, 841–847.

12. H.X. Xu, M.Y. Sha, E.Y. Wong, J. Uphoff, Y.H. Xu, J.A. Treadway, A. Truong, E. O'Brien, S. Asquith, M. Stubbins, N.K. Spurr, E.H. Lai, and W. Mahoney (2003) Multiplexed SNP genotyping using the Qbead™ system: A quantum dot-encoded microsphere-based assay. *Nucleic Acids Res.* 31, e43.

13. W.J. Parak, R. Boudreau, M. Le Gros, D. Gerion, D. Zanchet, C.M. Micheel, S.C. Williams, A.P. Alivisatos, and C. Larabell (2002) Cell motility and metastatic potential studies based on quantum dot imaging of phagokinetic tracks. *Adv. Mater.* 14, 882–885.

14. M. Dahan, S. Levi, C. Luccardini, P. Rostaing, B. Riveau, and A. Triller (2003) Diffusion dynamics of glycine receptors revealed by single-quantum dot tracking. *Science* 302, 442–445.

15. D.S. Lidke, P. Nagy, R. Heintzmann, D.J. Arndt-Jovin, J.N. Post, H.E. Grecco, E.A. Jares-Erijman, and T.M. Jovin (2004) Quantum dot ligands provide new insights into erbB/HER receptor-mediated signal transduction. *Nat. Biotechnol.* 22, 198–203.

16. A. Mansson, M. Sundberg, M. Balaz, R. Bunk, I.A. Nicholls, P. Omling, S. Tagerud, and L. Montelius (2004) *In vitro* sliding of actin filaments labelled with single quantum dots. *Biochem. Biophys. Res. Commun.* 314, 529–534.

17. M.E. Akerman, W.C.W. Chan, P. Laakkonen, S.N. Bhatia, and E. Ruoslahti (2002) Nanocrystal targeting *in vivo. Proc. Natl Acad. Sci. USA* 99, 12617–12621.

18. Y.T. Lim, S. Kim, A. Nakayama, N.E. Stott, M.G. Bawendi, and J.V. Frangioni (2003) Selection of quantum dot wavelengths for biomedical assays and imaging. *Mol. Imag.* 2, 50–64.

19. S. Kim, Y.T. Lim, E.G. Soltesz, A.M. De Grand, J. Lee, A. Nakayama, J.A. Parker, T. Mihaljevic, R.G. Laurence, D.M. Dor, L.H. Cohn, M.G. Bawendi, and J.V. Frangioni (2004) Near-infrared fluorescent type II quantum dots for sentinel lymph node mapping. *Nat. Biotechnol.* 22, 93–97.

20. X.H. Gao, Y.Y. Cui, R.M. Levenson, L.W.K. Chung, and S.M. Nie (2004) *In vivo* cancer targeting and imaging with semiconductor quantum dots. *Nat. Biotechnol.* 22, 969–976.

21 Bionanotechnology for Bioanalysis

Lin Wang and Weihong Tan
University of Florida, Gainesville, Florida, USA

CONTENTS

21.1 OVERVIEW

Bionanotechnology is defined by science's growing ability to work at the molecular level, atom by atom, combining biological materials and the rules of physics, chemistry, and genetics to create tiny synthetic structures. The end result of bionanotechnology is to create a highly functional system of biosensors, electronic circuits, nanosized microchips, molecular "switches," and even tissue analogs for growing skin, bones, muscle, and other organs of the body—all accomplished in ways that allow these structures to assemble themselves, molecule by molecule. On the other side, medical and biotechnological advances in the area of disease diagnosis and treatment are dependent on an in-depth understanding of biochemical processes. Diseases can be identified based on anomalies at the molecular level, and treatments are designed based on activities in such low dimensions. Although a multitude of methods for disease identification as well as treatment already exists, it would be ideal to use research tools with dimensions close to the molecular level to better understand the mechanisms involved in the processes. These tools can be nanoparticles (NPs), nanoprobes, or other nanomaterials, all of which exist in ultrasmall dimensions and can be designed to interrogate a biochemical process of interest.

Nanomaterials are at the leading edge of the rapidly evolving field of nanotechnology. NPs usually form the core of nanobiomaterials [1]. The unique size-dependent physical and chemical properties of NPs make them superior and indispensable in many areas of human activity. Typical size dimensions of biomolecular components are in the range of 5 to 200 nm, which is comparable with the dimensions of man-made NPs. Using NPs as biomolecular probes allows us to probe biological processes without interfering with them [2].

The representative NP probes include semiconductor NPs (quantum dots), gold NPs, polystyrene latex NPs, magnetic NPs, and dye-doped NPs. In our laboratory, dye-doped silica NPs have been developed, which possess unique properties of high signal amplification [3], excellent photostability, and easy surface modification.

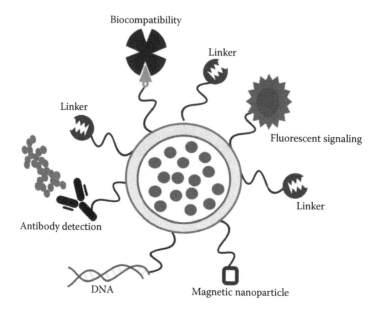

FIGURE 21.1 Typical configurations utilized in functionalized nanoparticles applied to bioanalysis.

21.2 NANOPARTICLE SURFACE MODIFICATION

To employ NPs as biological tags, a biological or molecular coating or layer acting as a bioinorganic interface should be attached to the NPs. The approaches used in constructing nanobiomaterials are schematically presented in Figure 21.1. To prepare such conjugates from NPs and biomolecules, the surface chemistry of the NPs must be such that the ligands are fixed to the NPs and possess terminal functional groups that are available for biochemical coupling reactions. A variety of surface modification and immobilization procedures have been utilized in our laboratory [4–8]. Recently we developed new methods by cohydrolysis of organosilanes with TEOS (tetra ethyl orthosilicate) [9–13] for NP surface modification, which facilitates NP bioconjugation as well as NP dispersion.

Dye-doped silica NPs are first prepared using a water-in-oil microemulsion system. After a 24 hour polymerization process, organosilanes with a range of terminal functional groups (Figure 21.2) are introduced into the microemulsion together with TEOS. Thiol groups (Figure 21.2a) are immobilized onto NPs by cohydrolysis of TEOS with MPTS (3-mercaptopropyltrimethoxy-silane). Amino groups can be introduced onto NPs with the addition of APTS (3-aminopropyltriethoxysilane) (Figure 21.2b), and carboxyl group modified NPs can be obtained by cohydrolyzing CTES (carboxyethylsilanetriol, sodium salt) (Figure 21.2d) with TEOS. To produce an overall negative surface charge, inert phosphonate groups (Figure 21.2c) can also be introduced onto NP surfaces.

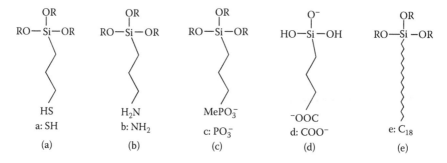

FIGURE 21.2 Structure of representative organosilanes for nanoparticle surface modification.

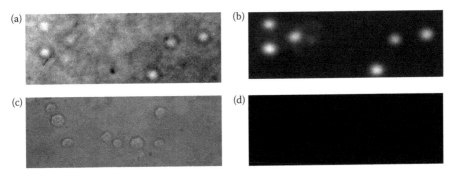

FIGURE 21.3 Nanoparticles (NPs) as biomarkers for cell labeling. (a) Optical and (b) fluorescence images of leukemia cells incubated with antibody-immobilized dye-doped NPs, (c) optical, and (d) fluorescence images of leukemia cells incubated with unmodified dye-doped NPs as a control.

Silica NPs are hydrophilic in nature and can be easily dispersed in water. For reactions in a nonpolar medium, it is essential to coat the NPs with hydrophobic alkyl groups. These hydrophobic silica NPs can be prepared during a postcoating process by cocondensation of alkyl functionalized triethoxy silane, such as octadecyl triethoxysilane (Figure 21.2e) and TEOS. The surface-modified NPs thus act as a scaffold for the grafting of biological moieties (DNA oligonucleotides or aptamers, enzymes, proteins, etc.) to the functional groups by means of standard covalent bioconjugation schemes or electrostatic interactions between NPs and charged adapter molecules.

21.3 NANOPARTICLES FOR CELLULAR IMAGING

For effective cellular labeling techniques, biomarkers need to have excellent specificity toward biomolecules of interest and also have optically stable signal transducers. Dye-doped silica NPs are ideal candidates for cellular membrane labeling and imaging. An example was demonstrated for the biomarking of leukemia cells. Mouse antihuman CD10 antibody was used as the cell recognition element and labeled with NPs pretreated with CNBr [7]. The mononuclear lymphoid cells were incubated with CD10 labeled NPs. After incubation, unbound NPs were washed away with phosphate-buffered saline (PBS) buffer (pH 6.8). The cell suspension was then imaged with both optical and fluorescence microscopy. As shown in Figure 21.3, all of the cells in the field of view of the microscope were labeled, indicated by the bright emission of the dye-doped NPs. The optical image (Figure 21.3a) correlated well with the fluorescence image (Figure 21.3b). The control experiments with bare dye-doped NPs (no antibody attached) did not show labeling of the cells as shown in Figure 21.3c and Figure 21.3d (optical image and fluorescence image). This clearly shows that the NPs conjugated with antibody are able to perform as biomarkers for cells via antibody–antigen recognition. With further development of this system, the NPs can serve as an efficient biomolecular analysis tool.

21.4 NANOPARTICLES FOR MICROARRAY TECHNOLOGY

Dye-doped NPs have distinct advantages over conventional dye molecules in terms of their excellent photostability and extremely high signal amplification, which allow them to be favorably used as luminescent probes for bioassays. For every binding event, one NP provides thousands of dye molecules rather than only a few, resulting in an increased sensitivity for most bioanalytical applications such as ultrasensitive DNA detection.

Dye-doped NPs can be potentially applied as staining probes for DNA/protein microarray-based technology. Current imaging and detection of microarrays suffer from weak signal intensities and low photobleaching threshold of the staining probes. To overcome these problems, NP-based microarray detection has been proposed as an alternative for microarray technology. Metal NPs [14–16], magnetic

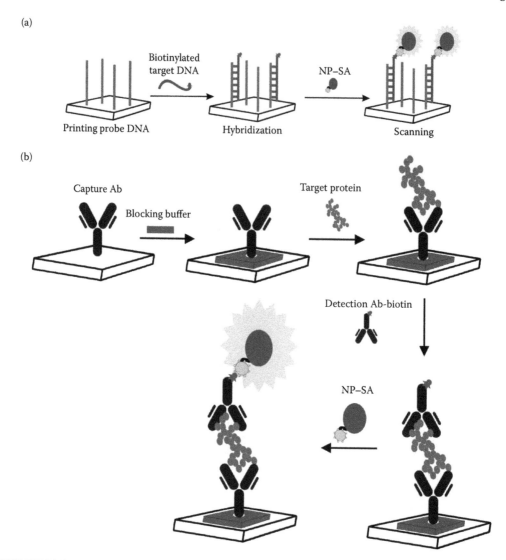

FIGURE 21.4   Strategies of nanoparticle-based labeling for (a) DNA microarray and (b) protein microarray technology.

NPs [17–18], and semiconductor nanocrystals [19] have been employed as labels for chip-based DNA detection. To further increase the sensitivity to lower molecular concentrations, dye-doped silica NPs can be employed as fluorescent labels for DNA and protein microarray detection. The strategies are shown in Figure 21.4a and Figure 21.4b for DNA and protein microarray applications. Basically, streptavidin-labeled NPs bind to biotinylated target DNA (DNA microarray) or biotinylated detection antibody (protein microarray). The highly fluorescent NPs provide amplified signal for trace amounts of samples, solve the major sensitivity limitation of microarray technology, and push the boundaries of discovery. This advance is of significant importance when microarray analysis is applied in areas such as genetic screening, proteomics, safety assessment, and medical diagnosis.

21.5   FUTURE PERSPECTIVES

Although NPs have been successfully utilized as biomolecular probes, they have not yet been exploited to their full potential. Some key advances include making NPs for drug-delivery regimes and targeting biologically relevant diseases, using NPs for whole-cell labeling and cytoplasmic or nuclear target labeling, and developing NP detection probes for single molecule separation and detection techniques. All of these promising techniques, designed with nanometer dimensions, show that NPs will have a far-reaching impact on the ultrasensitive detection and monitoring of biological events.

ACKNOWLEDGMENTS

This work is partially supported by NIH and NSF grants and by a Packard Foundation Science and Technology Award. We thank our colleagues at the University of Florida for their contributions.

REFERENCES

1. Feynman, R., There's plenty of room at the bottom. *Science*, 1991, 254, 1300–1301.
2. Taton, T.A., Nanostructures as tailored biological probes. *Trends Biotechnol.*, 2002, 20, 277–279.
3. Zhao, X., Bagwe, R.P., and Tan, W., Development of organic-dye-doped silica nanoparticles in a reverse microemulsion. *Adv. Mater.*, 2004, 16, 173–176.
4. Qhobosheane, M., Santra, S., Zhang, P., and Tan, W., Biochemically functionalized silica nanoparticles. *Analyst*, 2001, 126, 1274–1278.
5. Zhao, X., Tapec, R., and Tan, W., Ultrasensitive DNA detection using bioconjugated nanoparticles. *J. Am. Chem. Soc.*, 2003, 125, 11474–11475.
6. Tapec, R., Zhao, X., and Tan, W., Development of organic dye-doped silica nanoparticle for bioanalysis and biosensors. *J. Nanosci. Nanotechnol.*, 2002, 2, 405–409.
7. Hilliard, L., Zhao, X., and Tan, W., Immobilization of oligonucleotides onto silica nanoparticles for DNA hybridization studies. *Anal. Chim. Acta*, 2002, 470, 51–56.
8. Santra, S., Zhang, P., Wang, K., Tapec, R., and Tan, W., Conjugation of biomolecules with luminophore-doped silica nanoparticles for photostable biomarkers. *Anal. Chem.*, 2001, 73, 4988–4993.
9. Kriesel, J.W. and Tilley, T.D., Synthesis and chemical functionalization of high surface area dendrimer-based xerogels and their use as new catalyst supports. *Chem. Mater.*, 2000, 12, 1171–1179.
10. Epinard, P., Mark, J.E., and Guyot, A., A novel technique for preparing organophilic silica by water-in-oil microemulsions. *Polym. Bull.*, 1990, 24, 173–179.
11. Izutsu, H., Mizukami, F., Sashida, T., Maeda, K., Kiyozumi, Y., and Akiyama, Y., Effect of malic acid on structure of silicon alkoxide derived silica. *J. Non-Cryst. Solids*, 1997, 212, 40–48.
12. van Blaaderen, A. and Vrij, A., In Bergna, H.E., Ed., *The Colloid Chemistry of Silica,* American Chemical Society, Washington, DC, 1994, p. 83.
13. Markowitz, M.A., Schoen, P.E., Kust, P., and Gaber, B.P., Surface acidity and basicity of functionalized silica particles. *Colloids Surfaces A*, 1999, 150, 85–94.
14. Fritzsche, W., Craki, A., and Moller, R., Nanoparticle-based optical detection of molecular interactions for DNA-chip technology. *Proc. SPIE*, 2002, 4626, 17–22.
15. Caski, A., Maubach, G., Born, D., Reichert, J., and Fritzsche, W., DNA-based molecular nanotechnology. *Single Mol.*, 2002, 3, 275–280.
16. Fritzsche, W. and Taton, T.A., Metal nanoparticles as labels for heterogeneous, chip-based DNA detection. *Nanotechnology*, 2003, 14, R63–R73.
17. Schotter, J., Kamp, P.B., Beckere, A., Puhler, A., Reiss, G., and Bruckl, H., Comparison of a prototype magnetoresistive biosensor to standard fluorescent DNA detection. *Biosensors Bioelectron.*, 2004, 19, 1149–1156.
18. Zhao, X., Tapec, R., Wang, K., and Tan, W., Efficient collection of trace amounts of DNA/mRNA molecules using genomagnetic nano-capturers. *Anal. Chem.*, 2003, 75, 11474–11475.
19. Gerion, D., Chen, F., Kannan, B., Fu, A., Parak, W.J., Chen, D.J., Majurndar, A., and Alivisatos, A.P., Room-temperature single-nucleotide polymorphism and multiallele DNA detection using fluorescent nanocrystals and microarrays. *Anal. Chem.*, 2003, 75, 4766–4772.

22 Nanohydroxyapatite for Biomedical Applications

Zongtao Zhang, Yunzhi Yang, and Joo L. Ong
Inframat Corporation, Farmington, Connecticut, USA

CONTENTS

22.1 INTRODUCTION

Pure crystalline hydroxyapatite (HA) has the composition of $Ca_{10}(PO_4)_6(OH)_2$ with the calcium to phosphor mole ratio Ca/P = 1.67. The main composition of the biological bone is nanograined hydroxyapatite (HA) with the grain size of about 5 to 50 nm (see Figure 22.1). In a physiological environment, bone is a nonstoichemical HA of Ca/P mole ratio = 1.5 to 1.67, dependent on the age and bone site. Some ions such as HPO_4^{2-}, CO_3^{2-}, and F^{-1} replace partial PO_4^{3-} and OH^{-1}. Some other earth elements such as Mg^{2+} and Sr^{2+} can replace Ca^{2+}, too. The common formula should be $Ca_{10-x+y}(PO_4)_{6-x}(OH)_{2-x-2y}$, where $0 < x < 2$ and $0 < y < x/2$. For example, the bone composition can be represented by the formula of

$$Ca_{8.3}(PO_4)_{4.3}(HPO_4, CO_3)_{1.7}(OH, CO_3)_{0.3}$$

Bone is a living tissue and undergoes constant change in composition by either dissolving or deposition of bone minerals through osteoclast and osteoblast cells, respectively. The nano-HA has a nanocrystalline feature similar to the bone, thus being used as the bone substitute material [1–3].

Synthetic nano-HA has been used in medical application since the 1970s. The major products are coatings on metallic dental, hip, and spine implants for the acceleration of early stage healing and decreasing the pain. Other products such as nano-HA powders or porous blocks are used as bone fillers. From the 1980s to 1990s, a calcium phosphate cement (nano-HA formed after cementation)

FIGURE 22.1 Biological hydroxyapatite from dental enamel etched 30 seconds with 35% phosphoric acid. (From Buddy D. Ratner, Allan S. Hoffman, Frederick J. Schosen, and Jack E. Lemons, *Biomaterials Science*, Academic Press, New York, 1996, p. 321. With permission.)

has been used for cosmetic surgery and spine fusion. From 1990s to the present, nano-HA has been used for tissue engineering and drug delivery. In this chapter, we will briefly introduce the basic science of HA material, manufacturing process, application, and try to reflect on the latest progress of applications.

22.2 BASIC SCIENCE OF NANOHYDROXYAPATITE

Nanostructured hydroxyapatite is defined as the HA material with the grain size of less than 100 nm. The nanostructured materials exhibited some unique properties that normal microstructured materials do not have, such as high hardness and low wear rate for engineering materials. For hydroxyapatite, the nanomaterial will have extremely high surface area. Because the atoms in the surface layer have unsaturated atomic bond, nano-HA exhibit a high bioactivity, which accelerates the early stage bone growth and tissue healing [3–10]. Supposing the 0.8 nm thick surface layer (about one crystal lattice parameter, $a = 9.418$ Å, $c = 6.884$ Å) on the spherical ball, the volume fraction of surface bioactive atoms can be calculated to be $1 - (1 - 0.8 / D)^3$, where D is the diameter of the particle. The calculated result is shown in Figure 22.2. For 10 nm grain size, the surfaced atoms account for about 22%, and the 50 nm grains have only 5% atoms on the surface. If the grain size is larger than 100 nm, the surface atoms account for less than 2.5%. The smaller the grain size is, the higher the number of surface atoms, resulting in quicker bone growth and faster dissolution rate. Experimental results of magnetic nano-$NiZnFe_2O_4$ particles are demonstrated in Figure 22.2 [11–13].

22.3 NANO-HA CHEMISTRY

Hydroxyapatite is composed of CaO, P_2O_5, and H_2O, and thus its stability is dependent on both the temperature and water-vapor pressure. Figure 22.3 shows the phase diagram of CaO and P_2O_5 at 500 mm Hg partial pressure of water. At 500 mm Hg water vapor, HA is the stable phase below

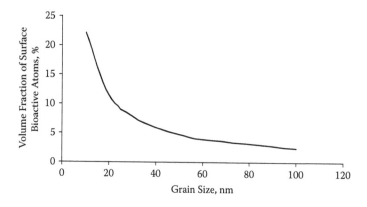

FIGURE 22.2 Surface atoms fraction as a function of grain size for a spherical particle, supposing the surface layer is 0.8 nm.

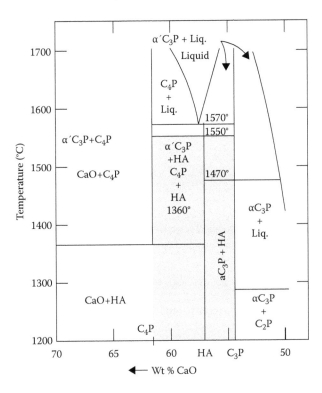

FIGURE 22.3 Phase diagram of CaO–P_2O_5. (From Buddy D. Ratner, Allan S. Hoffman, Frederick J. Schosen, and Jack E. Lemons, *Biomaterials Science*, Academic Press, New York, 1996, p. 82. With permission.)

1360°C, and decomposed into tetracalcium phosphate (C_4P or TTCP), ($Ca_4(PO_4)_2O$) and alfa-tricalcium phosphate (α-C_3P), $Ca_3(PO_4)_2$ at >1360°C. The complete decomposition occurs at 1550 to 1570°C, and HA becomes the mixture of liquid, C_4P, and α-C_3P at >1570°C. Without water vapor, HA starts losing OH at even as low as 900°C, and the decomposition is significant at >1000°C in air, N_2, or Ar. For sintering, the preferred temperature is at <1360°C under at least 500 mm Hg water-vapor pressure. For plasma thermal spray at temperature >5000°C, the decomposition is inevitable. The final phase composition of HA coating contains crystalline HA, amorphous phase, C_4P, α-C_3P, β-C_3P, and CaO.

Hydroxyapatite stability in water is shown in Figure 22.4 [15]. HA has the lowest calcium ion concentration and is the stable phase at the

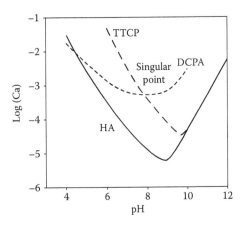

FIGURE 22.4 Phase diagram of Ca(OH)$_2$-H$_3$PO$_4$-H$_2$O at 25°C. (From L.C. Chow and S. Takagi, *J. Res. Natl Inst. Standard Technol.* 2001; 106: 1029–1033. With permission.)

same pH when pH > 4.2, and all other calcium phosphate compounds, such as $Ca_4(PO_4)_2O$ (TTCP) and $CaHPO_4$ (DCPA), transfer to HA through reaction with water. At pH < 4.2, $CaHPO_4·2H_2O$ is the stable compound, and HA decomposes into $CaHPO_4·2H_2O$. Figure 22.4 is the base of calcium phosphate cement. For powder synthesis, the chemical precipitation process should be controlled at pH 8.5 to 9.5, where the Ca ion concentration in the solution is at minimum. In a body fluid system, the pH is 7.3 at 37°C, therefore any kind of calcium phosphate fillers or coatings will finally form

TABLE 22.1

Mechanical Properties of Hydroxyapatite

	Sintered HA (99% density)	Coralline HA	Cemented HA	Bone
Grain size (nm)	>500	50–500	50–500	5–50
Tensile strength (MPa)	80–180	2.8	2.1	3 (Cancellous)
				151 (Cortical)
Compression strength (MPa)	470	9.3	55	5.5 (Cancellous)
				162 (Cortical)

HA, then the HA is gradually transformed into bone through balancing with the Ca^{2+} and PO_4^{3-} as well as other ions such as Mg^{2+}, F^- in body fluids.

22.4 NANO-HA MECHANICS

The HA mechanical properties are listed in Table 22.1 for different forms of products as compared with bone. It is found that the sintered HA has the grain size >500 nm, which has the highest compressive and tensile strength. Researchers have demonstrated that the mechanical properties are dominated by composition, porosity, and microstructure. For sintered HA bulk material, the compressive strength (σ_c) and tensile strength (σ_t) exponentially decrease with volume porosity (V_p)—that is, $\sigma_c = 700\exp(-5V_p)$ and $\sigma_t = 220\exp(-20V_p)$ [1]. Using hot pressing or sintering in water vapor at 1150 to 1250°C, the HA relative density reached 99%. The CO_3^{2-} or F^- replaced HA has different values. The sintered HA is a hard and brittle material, too. Its Young's modulus is 110 GPa, Vicker's hardness is 500 kg/mm^2 at 200 g load, and fracture toughness is about 1.0 to 1.2 MPa $m^{1/2}$ [16–19]. Because the sintered HA has a very low Weibull factor ($n = 12$) in physiological solution, the bulk HA cannot be used in load-bearing applications [1]. Most applications of the bulk HA are under the nonload conditions such as in the middle ear.

The majority of HA products are powder blocks, cements, and coatings, with grain size about 50 to 100 nm, and they are highly porous (30 to 80% pores). These pores make the HA have extremely low tensile strength of 2 to 3 MPa and compressive 9 to 60 MPa, which are similar to cancellous bone of 3 and 5.5 MPa, but far lower than the cortical bone. For nonload applications, the porous coralline and cemented HA have little foreign body reaction and finally merge and transform into physiological bone 3 to 5 years after surgery; this is the real advantage of HA materials [15,20].

22.5 NANO-HA BIOLOGY *IN VITRO* AND *IN VIVO*

Because the biological apatite of bone mineral is a nanoapatite, the synthesized nano-HA is expected to be recognized as belonging to the body [21]. In other words, the synthesized nano-HA could be directly involved in the natural bone remodeling process, rather than being phagocytosized [21]. On the other hand, the biocompatibility of HA is being governed by a thin layer of apatitic mineral matrix as a result of the dissolution and reprecipitation of HA [22,23]. The biological responses to HA surfaces are known to be influenced by the size, morphology, and structure of HA particles [3,4,24,25]. For example, as-received HA with smaller crystal particles have been reported to stimulate greater inflammatory cytokine release as compared to well-sintered HA with bigger crystal particles [5]. In addition, well-sintered HA was observed to significantly enhance osteoblast differentiation as compared to as-received HA [6]. In contrast, nano-HA is expected to lead to different biological responses because nano-HA may have higher solubility.

In the *in vitro* study, human monocyte-derived macrophages and human osteoblast-like (HOB) cell models have been used to study the biocompatibility of nano-HA coatings [5]. The nano-HA coatings were prepared by means of electrospraying nano-HA particles onto glass substrates. The cells were seeded onto the nano-HA coatings and cultured for a week. The release of lactate dehydrogenase (LDH) and tumor necrosis factor alpha (TNF-α) from cells were used to evaluate the cytotoxicity and inflammatory responses, respectively. Although there was some evidence of LDH release from macrophages in the presence of high concentrations of nano-HA particles, there was no significant release of TNF-α. In addition, nano-HA was observed to support the attachment and the spread of HOB cells [5]. In another study using porous nano-HA scaffolds, periosteal-derived osteoblast (POB) was isolated from the periosteum of 4-month human embryos aborting and seeded on porous nano-HA scaffolds. The attachment and the growth of POB on the scaffolds were evaluated by measuring the POB morphology, proliferative abilities, and osteogenic activity. POB could fully attach to and extend on HA scaffolds, and form extracellular matrix. Moreover, the presence of nano-HA scaffolds in cell culture could promote cell proliferation as compared to tissue culture plate ($p < 0.5$) [7]. In another separate study, a porous nano-HA/collagen composite was produced in sheet form and convolved into a three-dimensional scaffold. Bone-derived mesenchymal cells from neonatal Wistar rats were seeded on the scaffolds and cultured up to 21 days. Spindle-shaped cells were found to continuously proliferate and migrate throughout the network of the coil. Eventually, three-dimensional polygonal cells and new bone matrix were observed within the composite scaffolds [8].

In a comparison study between nanosize HA filler and micronsize HA filler using a rat calvarial defect model, histological analysis and mechanical evaluation showed a more advanced bone formation and a more rapid increase in stiffness in the defects with the nanosize HA augmented poly(propylene glycol-co-fumaric acid), suggesting an improved biological response to the nano-HA particles [9]. Other studies investigating the tissue response to a nano-HA/collagen composite implanted in the marrow cavity of New Zealand rabbit femur reported implant degradation and bone substitution during bone remodeling [10]. The process of implant degradation and bone substitution during bone remodeling suggested that the composite can be involved in bone metabolism. In addition to the process of degradation and bone substitution, the composite exhibited an isotropic mechanical behavior and similar microhardness when compared to the femur compacta [10].

22.6 HA PRODUCTS AND THEIR APPLICATIONS

As indicated in HA biology, the HA as bone substitute has the ability to perform osteointegration and osteoconduction. The clinical applications are basically divided into two categories: nano-HA granular and calcium phosphate cement for bone repair and nano-HA coating for bone replacement. In research areas, nano-HA combined with polymer and bone growth factors (BMPs) have been used for drug delivery and tissue engineering. The details are described below.

22.6.1 Porous Nano-HA Granules or Blocks

Porous nano-HA can be prepared by two methods: sintering and hydrothermal reaction. The sintering method is first mixing naphthalene with HA particles, then compact the particles into a composite, and finally sintering. Naphthalene particles would sublime during the sintering, thus leaving a pore-filled structure. This method can make spherical porous HA particles, but pores are isolated. Hydrothermal reaction is using natural coral as raw material. This method was developed in 1971, marked as ProOsteon (Interpore International, Irvine, California). The coral is first treated in boiling water with about 5% sodium hypochrorite (NaClO) to remove the organics and left calcium carbonate ($CaCO_3$) and pores. Then the $CaCO_3$ is reacted with ammonium monohydrogen phosphate ((NH_4)$_2$$HPO_4$) to form a coralline HA under temperature around 250°C at 3.8 MPa for 24 hours. The coralline HA has both needle-like grains with a length of 1 to 2 μm and a width of 100 to 200

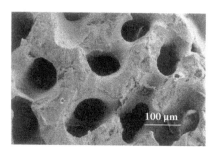

FIGURE 22.5 Scanning electron micrograph (SEM) of coralline hydroxyapatite. (From B. Ben-Nissan, A. Milev, and R. Vago, *Biomaterials* 2004; 25: 4971–4975. With permission.)

FIGURE 22.6 Nanohydroxyapatite formed by calcium phosphate cementation. (From L.C. Chow and S. Takagi, *J. Res. Natl Inst. Standard Technol.* 2001; 106: 1029–1033. With permission.)

nm. The most important is that the coralline HA has the microconnective pores of 100 to 150 μm (Figure 22.5), which provide bone in the growth channel. Fifty to eighty percent of the voids are filled within 3 months. When the fibro-osteous tissue ingrowth is complete, the implant consists of about 17% bone, 43% soft tissue, and 40% residual HA. Therefore, the coralline HA becomes as strong and tough as the cancellous bone [26–28].

22.6.2 NANO-HA CEMENT

Calcium phosphate cements are another kind of porous HA product used as a bone filler. Different from coral HA, the cements are paste-like slurries or gels that can be directly injected to the void site and molded to shape and set *in vivo*. This flow ability is unique and especially suitable for cosmetic surgery. There are two kinds of calcium cements. One was invented by Chow et al. and made by equal mole percent of $Ca_4(PO_4)_2O$ and $CaHPO_4$ mixed with water to form a paste [15]. This cement sets via isothermal reaction in 15 minutes and then fully hardens over 4 hours. Another calcium cement uses $CaHPO_4H_2O$ and $CaCO_3$ as raw materials and H_3PO_4 as the liquid [26,27]. At 37°C, the cement has 2 minutes working time (mixing and injection while flowing ability is kept) and sets in about 8 minutes. In all the calcium phosphate cements, the raw materials' particle size, powder-to-liquid ratio, temperature, pH, and seeding are all variables to influence the microstructure, working and setting times, and mechanical properties [31–37]. There are many manufacturers in the world to commercially supply these two types of cements. In the U.S. market, Bone Source, made by Leibinger (Dallas, Texas) is a trade name of the first type of cement. Norrian SRS Cement (Norian Corp., Cupertino, California) is another commercial product for the second type of cement [26,27].

All these products have similar mechanical properties as mentioned before. A porous needle-like structure $Ca_{10}(PO_4)_6(OH)_2$ crystals are their typical microstructure (Figure 22.6). These cements have been successfully used in sinus repair, cranioplasty, and distal radial and calcaneal fracture. It has also been reported that the calcium phosphate cement has been used for percutanous vertebroplasty [15,26–37].

22.6.3 NANO-HA COATING

Hydroxyapatite has been coated on metallic dental and orthopedic implants by high-temperature plasma thermal spray since the 1980s [24,25]. In this process, HA powders are fed into a plasma flame (temperature 5000 to 15,000°C). The powders are quickly melted and quenched on the metallic implant substrate to form a thick film coating. Because the temperature is high, the coating contains melted and crystallized HA, unmelted HA, amorphous phase, and some decomposed phases

 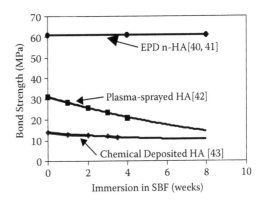

FIGURE 22.7 Scanning electron microscope (SEM) picture (left) and tensile bond strength of electrophoresis deposited (EPD) nanohydroxyapatite coating in SBF at 37°C, as compared to the commercial products. (See also Zhang Z., Xiao T.D., and Tongsan D., U.S. Patent No. 2003099762 A1 2003; Zhang Z., Dunn M.F., Xiao T.D., Tomsia A.P., and Saiz E., *Proc. Mater. Res. Soc. Symp. 703 Nanophase and Nanocomposite Mater. IV*, 2002, pp. 291–296; Yang C.Y., Wang B.C., Chang E., and Wu B.C., *J. Mater. Sci. Mater. Med.* 1995; 6: 258–265; and Han Y., Fu T., Lu J., and Xu K., *J. Biomed. Mater. Res.* 2000; 54: 96–101.)

such as C_4P, α-C_3P, β-C_3P, and CaO, as shown in Figure 22.2. Clinically, plasma thermal-sprayed HA coating has been successfully used in dental implants and femoral stems for hip replacement, but the HA coating on cups has a high failure rate. The special advantage of plasma-sprayed HA coating on the stems is the acceleration of the cup's early stage healing, no sigh-pain, and the enhancement of bone growth across a gap of 1 mm between the bone and the implants under stable and unstable mechanical conditions. Filling the gap is very important for the revision of hips where patients had bone loss during the primary hip replacement.

However, the plasma thermal-sprayed HA has the disadvantage of low bond strength at the coating–implant interface, and the strength decreases over time in simulated fluid (SBF) (Figure 22.7). This is attributed to its low crystallinity (typically 70%) and microcracks formed due to high-temperature quenching, decomposition, and thermal expansion mismatch between the HA and metallic substrates ($\alpha_{HA} = 13 \times 10^{-6}/°C$, $\alpha_{Ti} = 9.8 \times 10^{-6}/°C$, $\alpha_{CoCr} = 16 \times 10^{-6}/°C$). There was a controversy about the amorphous HA. Theoretically, the amorphous HA is quickly dissolved *in vivo*, which generates high local Ca^{2+} and PO_4^{3-} ion concentration and forms supersaturated conditions. Based on thermodynamics (Figure 22.4), the supersaturated Ca^{2+} and PO_4^{3-} will be deposited into the biological bone, thus proving beneficial to the early stage gap healing, about 30% amorphous phase is necessary [24,25,38,39]. Experimentally, the beneficial effect of the amorphous phase did not happen [40–44]. In a recent specific study, HA-coated titanium of 50% (low), 70% (medium), and 90% (high) crystallinity were inserted into the canine femur for 1, 4, 12, and 26 weeks. No significant differences could be found in the percentage of bone contact and interfacial attachment strength between the three types of HA-coated implants throughout the four implantation periods [44]. In another study, after 3 months of implantation, the high crystallinity (98%) coating showed the higher shear bond strength and integrated bone/coating interface, whereas the separation of the coating fragments was clearly observed in the coating that had low crystallinity (56%) [45].

In order to avoid the disadvantages of the plasma thermal spray, a chemical precipitation was introduced in the 1990s [46,47]. This method is a room-temperature process, deposing nano-HA thin film (grain size 60 to 100 nm and thickness <5 μm) on porous beaded implants in the supersaturated Ca^{2+} and PO_4^{3-} solution. This process has been successfully used for coating porous cups and porous stems. Because the chemical precipitation is a liquid-based room-temperature process, the coating is uniform on the beads and in the inside of pores, which helps the tissue and bone ingrowth in the pores. Therefore, this is the fixation of combined mechanical interlocking and osteointegration. Some companies even coat only β-C_3P on the porous implant and let β-C_3P transform into HA *in vivo*. The basic function of the thin-film nano-HA coating is stimulation of early bone growth into the macropores. Unfortunately, the chemically precipitated HA is only mechanically bonded on the implant surface with an extremely low bond strength, even one's finger can scratch the coating off. This requires a careful handling of the coated implants during surgery. Any chips of the coating might migrate in the articulating area and result in accelerated wear of the bearing surface.

Because both plasma thermal-sprayed and chemically precipitated HA coatings have disadvantages in low bond strength or the strength decrease over time, many other processes have been used for coating HA, including sol-gel, dip coating, electrophoretic deposition, ion implantation, chemical vapora deposition, and sputtering [48–55]. In all of these new processes, Inframat Corporation has developed an innovative electrophoretic deposition process to coat a special nano-HA coating. This nano-HA coating has the combined advantages of nanograin structure, high crystallinity (>90%), and the capability to coat the beaded porous surfaces. Figure 22.7 shows the typical surface microstructure and bond strength comparison [40,41].

In addition to the dental, hip, knee, and maxillofacial applications, nano-HA is also used for drug delivery and tissue engineering. These usually make a composite containing HA and other ingredients, such as anti-infection drugs, bone growth factors, and biodegradable polymers [56–65].

ACKNOWLEDGMENT

The authors gratefully acknowledge support from NIH SBIR programs under grants 1 R43 DE 015881-01 and 2-R44-AR047278-02A1.

REFERENCES

1. Ratner B.D., Hoffman A.S., Schosen F.J., and Lemons, J.E. *Biomaterials Science.* Academic Press, New York, 1996, pp. 82, 321.
2. Pasteris J.D., Wopenka B., Freeman J.J., Rogers K., Valsami-Jones E., Van der Houwen J.A.M., and Silva M.J. Lack of OH in nanocrystalline apatite as a function of degree of atomic order: Implications for bone and biomaterials. *Biomaterials* 2004; 2: 229–238.
3. Sun J., Tsuang Y., Chang W.H., Li H., Liu H., and Lin F. Effect of hydroxyapatite particle size on myoblasts and fibroblasts. *Biomaterials* 1997; 18: 683–690.
4. Ninomiya J.T., Struve J.A., Stelloh C.T., Toth J.M., and Crosby K.E. Effects of hydroxyapatite particulate debris on the production of cytokines and proteases in human fibroblasts. *J. Ortho. Res.* 2001; 19: 621–628.
5. Ong J.L., Hoppe C.A., Cardenas H.L., Cavin R., Carnes D.L., Sogal A., and Raikar G.N. Osteoblast precursor cell activity on HA surfaces of different treatments. *J. Biomed. Mater. Res.* 1998; 39: 176–183.
6. Huang J., Best S.M., Bonfield W., Brooks R.A., Rushton N., Jayasinghe S.N., and Edirisinghe M.J. *In vitro* assessment of the biological response to nano-sized hydroxyapatite. *J. Mater. Sci. Mater. Med.* 2004; 15: 441–445.
7. Zhang Q., Zhao S., Guo Z., Dong Y., Lin P., and Pu Y. Research of biocompatibility of nano-bioceramics using human periosteum *in vitro*. *J. Southeast Univ. Natural Science Edition* 2004; 34: 219–223.
8. Du C., Cui F.Z., Zhu X.D., and de Groot K. Three-dimensional nano-Hap/collagen matrix loading with osteogenic cells in organ culture. *J. Biomed. Mater. Res.* 1999; 44: 407–415.
9. Doherty S.A., Hile D.D., Wise D.L., Ying J.Y., Sonis S.T., and Trantolo D.J. Nanoparticulate hydroxyapatite enhances the bioactivity of a resorbable bone graft. *Proc. Mater. Res. Soc. Symp.* 2003; 735: 75–79.
10. Du C., Cui F.Z., Zhu X.D., and de Groot K. Tissue response to nano-hydroxyapatite/collagen composite implants in marrow cavity. *J. Biomed. Mater. Res.* 1998; 42: 540–548.
11. Zhang Z. Non-magnetic surface for magnetic nano-materials, a limit of exchange coupling. Inframat Corporation, Inner report, February 27, 2002.
12. Yang D-P., Lavoie L.K., Zhang Y., Zhang Z., Ge S. Mossbauer spectroscopic and x-ray diffraction studies of structural and magnetic properties of heat-treated $Ni0.5Zn0.5Fe_2O_4$ nanoparticles. *J. Appl. Phys.* 2003; 93: 7492–7494.

13. Zhang Z., Zhang Y.D., Xiao T.D., Ge S., and Wu M. Nanostructured NiFe$_2$O$_4$ ferrite. *Proc. Mater. Res. Soc. Symp. Nanophase Nanocomposite Mater. IV*, 2002; 703: 111–116.

14. Hines W.A., Budnick J.I., Gromek J.M., Yacaman M.J., Troiani H.E., Chen Q.Z., Wong C.T., Lu W.W., Cheung K.M.C., Leong J.C.Y., and Luk K.D.K. Strengthening mechanisms of bone bonding to crystalline hydroxyapatite *in vivo*. *Biomaterials* 2004; 25: 4243–4254.

15. Chow, L.C. and Takagi, S. A natural bone cement—A laboratory novelty led to the development of reutional new biomaterials. *J. Res. Natl Inst. Standard Technol.* 2001; 106: 1029–1033.

16. Raynaud S., Champion E., and Bernache-Assollant D. Calcium phosphate apatites with variable Ca/P atomic ratio II. Calcination and sintering. *Biomaterials* 2002; 23: 1073–1080.

17. Landi E., Tampieri A., Celotti G., Vichi L., and Sandri M. Influence of synthesis and sintering parameters on the characteristics of carbonate apatite. *Biomaterials* 2004; 23: 1763–1770.

18. Raynaud S., Champion E., Lafon J., and Bernache-Assollant D. Calcium phosphate apatites with variable Ca/P atomic ratio III. Mechanical properties and degradation in solution of hot presses ceramics. *Biomaterials* 2002; 23: 1081–1089.

19. Gross K.A., and Bhadang K.A. Sintered hydroxyfluorapatite. Part II: Sintering and resultant mechanical properties of sintered blends of hydroxyapatite and fluorapatite. *Biomaterials* 2004; 25: 1395.

20. Hing K.A., Best S.M., Tanner K.E., and Bonfield W. Quantification of bone ingrowth within bone-derived porous hydroxyapatite implants of varying density. *J. Mater. Sci.* 1999; 10: 663–670.

21. Driessens F.C.M., Boltong M.G., Khairoun I., De Maeyer E.A.P., Ginebra M.P., Wenz R., Planell J.A., and Verbeeck R.M.H. Applied aspects of calcium phosphate bone cement application. In Wise D.L., Trantolo D.J., Lewandrowski K.U., Gresser M.V., and Yaszemski M.J. (Eds). *Biomaterials Engineering and Devices. Human Applications*. Volume 2. Humana Press, Totowa, New Jersey. 2000, pp. 253–260.

22. Hench L.L. Bioceramics: From concept to clinic. *J. Am. Ceram. Soc.* 1991; 74: 1487–1510.

23. Kokubo T. Recent progress in glass based materials for biomedical applications. *J. Ceram. Soc. Japan* 1991; 99: 965–973.

24. Yang Y., Kim K.-H., and Ong J.L. A review on calcium phosphate coatings produced using a sputtering process—An alternative to plasma spraying. *Biomaterials* 2005; 26: 327–337.

25. Yang Y., Bessho K., and Ong J.L. Plasma-sprayed hydroxyapatite-coated and plasma-sprayed titanium-coated implants. In Yaszemski M.J., Trantolo D.J., Lewandrowski K.U., Hasirci V., Altobelli D.E., and Wise D.L. (Eds). *Biomaterials in Orthopedics*. Marcel Dekker, New York, 2004, pp. 401–423.

26. Ben-Nissan B., Milev A., and Vago R. Morphology of sol–gel derived nano-coated Coralline hydroxyapatite. *Biomaterials* 2004; 25: 4971–4975.

27. Moore W.R., Graves S.E., and Bain G.I. Synthetic bone graft substitutes. *ANZ J. Surg.* 2001; 71: 354–361.

28. Lu W.W., Cheung K.M.C., Li Y.W., Luk D.K., Holmes A.D., Zhu Q.A., and Leong J.C.Y. Bioactive bone cement as a principal fixture for spinal fracture, An *in vitro* biomechanical and morphological study. *Spine* 2001; 26: 2684–2691.

29. Yussa T., Miyamoto Y., Ishikawa K., Takechi M., Momota Y., Tatehara S., and Nagayama M. Effects of apatite cements on proliferation and differentiation of human osteoblasts *in vitro*. *Biomaterials* 2004; 25: 1159–1166.

30. Liu C., Shen W., and Chen J. Solution properties of calcium phosphate cement hardening body. *Mater. Chem. Phys.* 1999; 58: 78–83.

31. Ginebra M., Driessens F.C.M., and Planell J.A. Effect of the particle size on the micro and nanostructural features of a calcium phosphate cement: A kinetic analysis. *Biomaterials* 2004; 25: 3453–3462.

32. Liu C., Huang Y., and Zheng H. Study of the hydration process of calcium phosphate cement by AC Impedance Spectroscopy. *J. Am. Ceram. Soc.* 1999; 82: 1052–1057.

33. Liu C., Gai W., Pan S., and Liu Z. The exothermal behaviour in the hydration process of calcium phosphate cement. *Biomaterials* 2003; 24: 2995–3003.

34. Bigi A., Bracci B., and Panzata S. Effect of added gelatin on the properties of calcium phosphate cement. *Biomaterials* 2004; 25: 2893–2999.

35. Savarino L., Breschi L., Tedaldi M., Ciapetti G., Tarabusi C., Greco M., Giunti A., and Prati C. Ability of restorative and fluoride releasing materials to prevent marginal dentine demineralization. *Biomaterials* 2004; 25: 1011–1017.

36. Liu C., Huang Y., and Chen J. The physicochemical properties of the solidification of calcium phosphate cement. *J. Biomed. Mater. Res. Appl. Biomater.* 2004; 69: B:73–78.

37. Apelt A., Theiss F., El-Warrak A.O., Zilinazky K., Bettschart-Wolfisberger R., Bohner M., Matter S., Auer J.A., and von Rechenberg B. *In vivo* behaviour of three injectable hydraulic calcium phosphate cements. *Biomaterials* 2004; 25: 1439–1451.

38. Souto R.M., Laz M.M., and Reis R.L. Degradation characteristics of hydroxyapatite coatings on orthopedic TiAlV in simulated physiological media investigated by electrochemical impedance spectroscopy. *Biomaterials* 2003; 24: 4213–4221.

39. Lu Y-P., Li M-S., Li S-T., Wang Z-G., and Zhu R-F. Plasma-sprayed hydroxyapatite+titania composite bond coat for hydroxyapatite coating on titanium substrate. *Biomaterials* 2004; 25: 4393–4403.

40. Zhang Z., Xiao T.D., and Tongsan D. Multi-layer coating useful for the coating of implants. U.S. Patent No. 2003099762 A1 2003.

41. Zhang Z., Dunn M.F., Xiao T.D., Tomsia A.P., and Saiz E. Nanostructured hydroxyapatite coatings for improved adhesion and corrosion resistance for medical implants. *Proc. Mater. Res. Soc. Symp. 703 Nanophase and Nanocomposite Mater. IV*, 2002, pp. 291–296.

42. Yang C.Y., Wang B.C., Chang E., and Wu B.C. Bond degradation at the plasma-sprayed HA coating/Ti-6Al-4V alloy interface: an *in vitro* study. *J. Mater. Sci. Mater. Med.* 1995; 6: 258–265.

43. Han Y., Fu T., Lu J., and Xu K. Characterization and stability of hydroxyapatite coatings prepared by an electrodeposition and alkaline-treated process. *J. Biomed. Mater. Res.* 2000; 54: 96–101.

44. Chang Y.L., Lew D., Park J.B., and Keller J.C., Biomechanical and morphometric analysis of hydroxyapatite-coated implants with varying crystallinity. *J. Oral Maxillofac. Surg.* 1999; 579: 1096–1108.

45. Xue W., Tao S., Liu X., Zheng X., and Ding C. *In vivo* evaluation of plasma sprayed hydroxyapatite coatings having different crystallinity. *Biomaterials* 2004; 253: 415–421.

46. Leitao Eugenia Ribeiro de Sousa Fidalgo, Implant material. US patent No. 6,069,295, 2000.

47. Barrere F., Snel M.M.E., Van Blitterswijk C.A., de Groot K., and Layrolle P. Nano-scale study of the nucleation and growth of calcium phosphate coating on titanium implants. *Biomaterials* 2004; 25: 2901–2910.

48. Kim H-W., Kong Y-M., Bae C-J., Noh Y-J., and Kim H-E. Sol–gel derived fluro-hydroxyapatite biocoatings on zircornia substrate. *Biomaterials* 2004; 25: 2919–2926.

49. Kusakabe H., Sakamaki T., Nihei K., Oyama Y., Yanagimoto S., Ichimiya M., Kimura J., and Toyama Y., Osseointegration of a hydroxyapatite-coated multilayered mesh stem. 2004; 25: 2957–2969.

50. Kim H-W., Kim H-E., and Knowles J.C. Fluro-hydroxyapatite sol–gel coating on titanium substrate for hardtissue implants. *Biomaterials* 2004; 25: 3351–3358.

51. Rocca M., Fini M., Giavaresi G., Nicoli Aldini N., and Giardino R. Osseointegration of hydroxyapatite-coated and uncoated titanium screw in long-term ovariectomized shee. *Biomaterials* 2002; 23: 1017–1023.

52. Fulmer M.T., Ison I.C., Hankermayer C.R., Constantz B.R., and Ross J. Measurements of the solubilities and dissolution rates of several hydroxyapatites. *Biomaterials* 2004; 23: 751–755.

53. Rößler S., Sewing A., Stölzel M., Born R., Scharnweber D., Dard M., and Worch H. Electrochemically assisted deposition of thin calcium phosphate coatings at near-physiological pH and temperature. *J. Biomed. Mater. Res.* 2002; 64A: 655–663.

54. Leeuwenburgh S.C.G., Wolke J.G.C., Schoonman J., and Jansen J.A. Influence of precursor solution parameters on chemical properties of calcium phosphate coatings prepared using electrostatic spray deposition (ESD). *Biomaterials* 2004; 641–649.

55. Nguyen H.Q., Deporter D.A., Pilliar R.M., Valiquette N., and Yakubovich R. The effect of sol–gel-formed calcium phosphate coatings on bone ingrowth and osteoconductivity of porous-surfaced Ti alloy implants. *Biomaterials* 2004; 25: 865–876.

56. Paul W. and Sharma C.P. Development of porous spherical hydroxyapatite granules: Application towards protein delivery. *J. Mater. Sci.* 1999; 10: 383–388.

57. Rotter N., Aigner J., and Naumann A. Behaviour of tissue-engineered human cartilage after transplantation in nude mice. *J. Mater. Sci.* 1999; 10: 689–693.

58. Hu Q., Li B., Wang M., and Shen J. Preparation and characterization of biodegradable chitosan/hydroxyapatite nanocomposite rods via *in situ* hybridization: a potential material as internal fixation of bone fracture. *Biomaterials* 2004; 25: 779–785.

59. Ramay Hassna R.R. and Zhang M. Biophasic calcium phosphate nanocomposite porous scaffolds for load-bearing bone tissue engineering. *Biomaterials* 2004; 25: 5171–5180.

60. Skrtic D., Antonucci J.M., Evance E.D., and Eidelman N. Dental composites based on hybrid and surface-modified amorphous calcium phosphates. *Biomaterials* 2004; 25: 1141–1150.

61. Ge Z., Baguenard S., Lim L-Y., Wee A., and Khor E. Hydroxyapatite-chitin materials as potential tissue engineered bone substitutes. *Biomaterials* 2004; 25: 1049–1058.

62. Ota Y., Iwashita T., Kasuga T., and Abe Y. Process for producing hydroxyapatite fibers. US Patent No. 6,228,339 B1, 2001.

63. Xu H.H.K., Quinn J.B., Takagi S., and Chow L.C. Synergistic reinforcement of *in situ* hardening of calcium phosphate composite scaffold for bone tissue engineering. *Biomaterials* 2004; 25: 1029–1037.

64. Ribeiro de Sousa Fildago Leitao E. Implant material and process for using it. US Patent No. 6,146,686, 2000.

65. Gu Y.W., Khor K.A., Pan D., and Cheang P. Activity of plasma sprayed yttria stabilized zirconia reinforced hydroxyl apatite/Ti–6Al–4V composite coatings in simulated body fluid. *Biomaterials* 2004; 25: 3177–3185.

23 Nanotechnology Provides New Tools for Biomedical Optics

Jennifer L. West, Rebekah A. Drezek, and Naomi J. Halas
Rice University, Houston, Texas, USA

CONTENTS

23.1 INTRODUCTION

Continuing advances in nanotechnology are generating a variety of nanostructured materials with highly controlled and interesting properties—from exceptionally high strength to the ability to carry and target drugs to unique optical properties. By controlling structure at the nanoscale dimensions, one can control and tailor the properties of nanostructures, such as semiconductor nanocrystals and metal nanoshells, in a very accurate manner to meet the needs of a specific application. These materials can provide new and unique capabilities for a variety of biomedical applications ranging from diagnosis of diseases to novel therapies. In particular, nanotechnology may greatly expand the impact of biophotonics by providing more robust contrast agents, fluorescent probes, and sensing substrates.

In addition, the size scale of nanomaterials is very interesting for many biomedical applications. Nanoparticles are similar in size scale to many common biomolecules, making them interesting for applications such as intracellular tagging and for bioconjugate applications such as antibody-targeting of imaging contrast agents. In many cases, one can make modifications to nanostructures to better suit their integration with biological systems (e.g., modifying their surface layer for enhanced aqueous solubility, biocompatibility, or biorecognition). Nanostructures can also be embedded within other biocompatible materials to provide nanocomposites with unique properties.

Many of the biomedical applications of nanotechnology will involve bioconjugates. The idea of merging biological and nonbiological systems at the nanoscale has actually been investigated for many years. The broad field of bioconjugate chemistry is based on combining the functionalities of biomolecules and nonbiologically derived molecular species for applications including markers for research in cellular and molecular biology, biosensing, and imaging [1]. Many current applications of nanotechnology, particularly in the area of biophotonics, are a natural evolution of this approach. In fact, several of the "breakthrough" applications recently demonstrated using nanostructure bioconjugates are in fact traditional applications originally addressed by standard molecular bioconjugate techniques that have been revisited with these newly designed nanostructure hybrids, often with far superior results. Typically, nanostructured materials possess optical properties far superior to the molecular species they replace—higher quantum efficiencies, greater scattering or absorbance cross-sections, optical activity over more biocompatible wavelength regimes, and substantially greater chemical or photochemical stability. Additionally, some nanostructures provide optical properties that are highly dependent on particle size or dimension. The ability to systematically vary the optical properties via structure modification not only improves traditional applications but also may lead to applications well beyond the scope of conventional molecular bioconjugates. In this chapter, we introduce several successful examples of nanostructures that have been applied to relevant problems in biotechnology and medicine.

23.2 QUANTUM DOTS AS FLUORESCENT BIOLOGICAL LABELS

Semiconductor nanocrystals, also referred to as "quantum dots," are highly light-absorbing, luminescent nanoparticles whose absorbance onset and emission maximum shift to higher energy with decreasing particle size due to quantum confinement effects [2,3]. These nanocrystals are typically in the size range of 2 to 8 nm in diameter. Unlike molecular fluorophores, which typically have very narrow excitation spectra, semiconductor nanocrystals absorb light over a very broad spectral range (Figure 23.1). This makes it possible to optically excite a broad spectrum of quantum dot "colors" using a single excitation laser wavelength, which may enable one to simultaneously probe several markers in imaging, biosensing, and assay applications. Although the luminescence properties of semiconductor nanocrystals have historically been sensitive to their local environment and nanocrystal surface preparation, recent core-shell geometries where the nanocrystal is encased in a shell of a wider bandgap semiconductor have resulted in increased fluorescence quantum efficiencies (>50%) and greatly improved photochemical stability. In the visible region, CdSe–CdS core-shell nanocrystals have been shown to span the visible region from approximately 550 nm (green) to

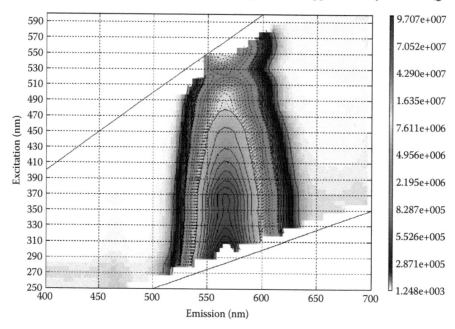

FIGURE 23.1 Semiconductor nanocrystals, also referred to as "quantum dots," are highly light-absorbing, luminescent nanoparticles whose absorbance onset and emission maximum shift to higher energy with decreasing particle size due to quantum confinement effects. As seen in this typical excitation–emission plot for a type of quantum dot, strong emission is observed over a broad range of excitation wavelengths.

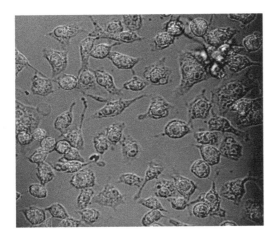

FIGURE 23.2 Quantum dots can be used for live cell imaging. They are advantageous due to their bright fluorescence over a broad range of excitation wavelengths, resistance to photobleaching, and stability. Surface modifications are required for biocompatibility.

630 nm (red). Other material systems, such as InP and InAs, provide quantum dot fluorophores in the near-infrared region of the optical spectrum, a region where transmission of light through tissues and blood is maximal [4]. Although neither II–VI nor III–V semiconductor nanocrystals are water soluble, let alone biocompatible, surface functionalization with molecular species such as mercaptoacetic acid or the growth of a thin silica layer on the nanoparticle surface facilitate aqueous solubility [5]. Both the silica layer and the covalent attachment of proteins to the mercaptoacetic acid coating permit the nanoparticles to be at least relatively biocompatible. Quantum dots have also been modified with dihydrolipoic acid to facilitate conjugation of avidin and subsequent binding of biotinylated targeting molecules [6]. Quantum dots can also be embedded within polymer nano- or microparticles to improve biocompatibility while maintaining the unique fluorescence.

Specific binding of quantum dots to cell surfaces, cellular uptake, and nuclear localization have all been demonstrated following conjugation of semiconductor nanocrystals to appropriate targeting proteins such as transferring, growth factors, peptides, or antibodies [2,4,7,8]. This could be useful in a variety of microscopy and imaging applications (Figure 23.2). Several preliminary reports of *in vivo* imaging using quantum dots show considerable promise. For example, cancerous cells labeled with quantum dots *ex vivo* were injected intravenously to track extravasation and metastasis [9]. Five different populations of cells, each labeled with a different size of quantum dot, could be simultaneously tracked *in vivo* using multiphoton laser excitation. Another study evaluated *in vivo* imaging with quantum dots following direct injection [10]. A surface modification with polyethylene glycol of at least 5000 Da was required for sustained (>15 min) circulation of the particles in the bloodstream.

Quantum dots may also be useful in a variety of *in vitro* diagnostic applications, particularly because concerns about semiconductor nanocrystal biocompatibility can be neglected in such uses. One example is the development of a fluorescent immunoassay using antibody-conjugated quantum dots [6]; several protein toxins have been successfully detected using this system. In another example, quantum dots embedded in polymer microbeads have been used for DNA hybridization studies [11]. Encasing the nanocrystals in the polymer beads allows for simultaneous reading of a huge number of optical signals. The emission of different nanocrystal species can be tuned by varying the particle size. Microbeads can then be prepared with varying colors and intensities of quantum dots. Using ten intensity levels and six colors, one could theoretically code 1 million optically differentiated signals to mark different nucleic acid or protein sequences for high-throughput screening and diagnostics. Similarly, quantum dots bound to oligonucleotides have been used as probes for fluorescent *in situ* hybridization (FISH) [12]. This could enable more detailed analysis of gene expression profiles with localization within tissue than is currently possible with conventional molecular fluorophores.

23.3   GOLD NANOPARTICLE BIOCONJUGATE-BASED COLORIMETRIC ASSAYS

The use of gold colloid in biological applications began in 1971 when Faulk and Taylor invented the immunogold staining procedure. Since that time, the labeling of targeting molecules, such as antibodies, with gold nanoparticles has revolutionized the visualization of cellular components by electron microscopy [13]. The optical and electron beam contrast properties of gold colloid have provided excellent detection capabilities for applications including immunoblotting, flow cytometry, and hybridization assays. Furthermore, conjugation protocols to attach proteins to gold nanoparticles are robust and simple [1], and gold nanoparticles were shown to have excellent biocompatibility [13].

Gold nanoparticle bioconjugates were recently applied to polynucleotide detection in a manner that exploited the change in optical properties resulting from plasmon–plasmon interactions between locally adjacent gold nanoparticles [14]. The characteristic red color of gold colloid has long been known to change to a bluish-purple color upon colloid aggregation due to this effect. In the case of polynucleotide detection, mercaptoalkyloligonucleotide-modified gold nanoparticle probes were prepared. When a single-stranded target oligonucleotide was introduced to the preparation, the nanoparticles aggregated due to the binding between the probe and target oligonucleotides, bringing the nanoparticles close enough to each other to induce a dramatic red-to-blue color change as depicted in Figure 23.3. Because of the extremely strong optical absorption of gold colloid, this colorimetric method can be used to detect ~10 fmol of an oligonucleotide, which is ~50 times more sensitive than the sandwich hybridization detection methods based on molecular fluorophores.

A similar approach has been used to develop a rapid immunoassay that can be performed in whole blood without sample preparation steps. This assay utilizes a relatively new type of gold nanoparticle called a gold nanoshell. Gold nanoshells are concentric sphere nanoparticles consisting of a dielectric core nanoparticle (typically gold sulfide or silica) surrounded by a thin gold shell [15]. By varying the relative dimensions of the core and shell layers, the plasmon-derived optical resonance of gold can be dramatically shifted in wavelength from the visible region into the mid-infrared as depicted in Figure 23.4 [16]. By varying the absolute size of the gold nanoshells, they may be designed to either strongly absorb (for particles <~75 nm) or scatter (for particles >~150 nm) the incident light [16]. The gold shell layer is formed using the same chemical methods that are

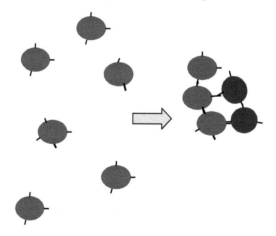

FIGURE 23.3   When gold nanoparticles come into close proximity, plasmon–plasmon interactions cause dramatic changes in optical properties. Using appropriately conjugated nanoparticles, this behavior can be exploited for DNA hybridization assays and immunoassays.

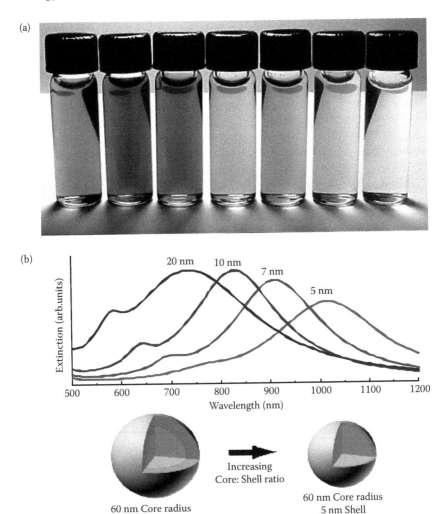

FIGURE 23.4 Gold nanoshells consist of a dielectric core nanoparticle surrounded by a thin metal shell. By varying the relative dimensions of the core and shell constituents, one can design particles to either absorb or scatter light over the visible and much of the infrared regions of the electromagnetic spectrum. (a) These vials contain suspensions of either gold colloid (far left with its characteristic red color) or gold nanoshells with varying core:shell dimensions. (b) The optical properties of nanoshells are predicted by Mie scattering theory. For a core of a given size, forming thinner shells pushes the optical resonance to longer wavelengths.

employed to form gold colloid; thus, the surface properties of gold nanoshells are virtually identical to gold colloid, providing the same ease of bioconjugation and excellent biocompatibility. To develop a whole blood immunoassay, gold nanoshells were designed and fabricated for near-infrared resonance, and antibodies against target antigens were conjugated to the nanoshell surfaces [17]. When introduced into samples containing the appropriate antigen, the antibody–antigen linkages caused the gold nanoshells to aggregate, shifting the resonant wavelength further into the infrared. This assay system was shown to have sub-ng/ml sensitivity. More importantly, this assay can be performed in whole-blood samples because the wavelengths utilized are in the near infrared, above the absorption of hemoglobin yet below the water absorption band, where penetration of light through blood is relatively high [4]. Additionally, because gold nanoshells have highly tunable optical properties, it may be possible to probe for several antigens simultaneously using nanoshells with varying optical resonances. Nanoshells are also effective substrates for surface-enhanced Raman scattering [18], which may enable alternative methods for near-infrared biosensing.

23.4 PHOTOTHERMAL THERAPIES

Gold nanoshells, described above, can be designed to strongly absorb light at desired wavelengths, in particular in the near infrared between 700 and 1100 nm where the tissue is relatively transparent [4]. Very few molecular chromophores are available in this region of the electromagnetic spectrum, let alone ones with low toxicity. When optically absorbing gold nanoshells are embedded in a matrix material, illuminating them at their resonance wavelength causes the nanoshells to transfer heat to their local environment. This photothermal effect can be used to optically modulate drug release from a nanoshell–polymer composite drug delivery system [19]. To accomplish photothermally modulated drug release, the matrix polymer material must be thermally responsive. Copolymers of N-isopropylacrylamide (NIPAAm) and acrylamide (AAm) exhibit a lower critical solution temperature (LCST) that is slightly above body temperature [20]. When the temperature of the copolymer exceeds its LCST, the resultant phase change in the polymer material causes the matrix to collapse, resulting in a burst release of any soluble material (i.e., drug) held within the polymer matrix [20]. As demonstrated in Figure 23.5, when gold nanoshells that were designed to strongly absorb near-infrared light were embedded in NIPAAm-co-AAm hydrogels, pulsatile release of insulin and other proteins could be achieved in response to near-infrared irradiation.

In another application, nanoshells are being used for photothermal tumor ablation [21,22]. When tumor cells are treated with nanoshells and then exposed to near-infrared light, cells are efficiently killed, while neither the nanoshells alone nor the near-infrared light had any effect on cell viability, as shown in Figure 23.6 [21]. Furthermore, particles in the size range of 60 to 400 nm will extravasate and accumulate in tumors due to the enhanced permeability and retention effect that results from the leakiness of tumor vasculature [23]. Near-infrared absorbing nanoshells are in the appropriate size range for this phenomenon. When polyethylene glycol-modified nanoshells were injected intravenously in tumor-bearing mice, the nanoshells accumulated in the tumor [22]. Subsequent exposure of the tissue region to near-infrared light led to complete tumor regression and

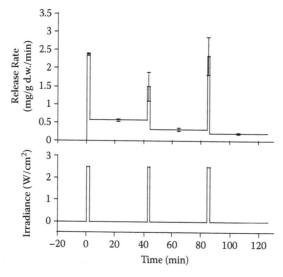

FIGURE 23.5 Release of insulin from NIPAAm-coAAm hydrogels with nanoshells embedded in their structure can be modulated by exposure to near-infrared light (832 nm, 1.5 W/cm²). The top panel is the release rate of insulin from the nanocomposite hydrogel materials versus time, and the bottom panel indicates the pattern of laser illumination.

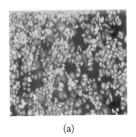 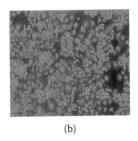 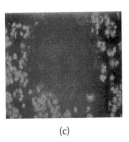

(a) (b) (c)

FIGURE 23.6 Breast carcinoma cells were exposed to either nanoshells (a), near-infrared light (b), or the combination of nanoshells and near-infrared light (c). As demonstrated by staining with the fluorescent viability marker calcein AM, the carcinoma cells in the circular region corresponding to the laser spot were completely destroyed, but neither the nanoshells nor the light treatment alone compromised viability.

survival of the nanoshell-treated mice. The light alone had no effect. Nanoshells can be conjugated to antibodies against oncoproteins to potentially have cellular-level specificity of therapy [17,21].

23.5 SILVER PLASMON RESONANT PARTICLES FOR BIOASSAY APPLICATIONS

Silver plasmon resonant particles have been used as reporter labels in microarray-based DNA hybridization studies [24] and sandwich immunoassays [25]. Silver plasmon resonant particles consist of a gold nanoparticle core onto which a silver shell is grown. Particles of this type in the size range of 40 to 100 nm scatter light very strongly, as many as 10^7 photons/sec to the detector [24], allowing them to act as diffraction-limited point sources that can be observed using a standard dark field microscope with white light illumination. In the bioassay applications that have been developed, bioconjugates are prepared with antibodies either against a target antigen for an immunoassay or against biotin for subsequent attachment of biotinylated DNA. In both the immunoassay and the hybridization assay, the results are determined by counting the number of particles bound to the substrate via microscopy. In the DNA hybridization assay, the sensitivity obtained was approximately 60× greater than what is typically achieved using conventional fluorescent labels.

REFERENCES

1. Hermanson, G.T. 1996. *Bioconjugate Techniques*. Academic Press, San Diego, CA.
2. Bruchez, M., Moronne, M., Gin, P., Weiss, S., and Alivisatos, A.P. 1998. Semiconductor nanocrystals as fluorescent biological labels. *Science* 281: 2013–2016.
3. Chan, W.C., Maxwell, D.J., Gao, X., Bailey, R.E., Han, M., and Nie, S. 2002. *Curr. Opin. Biotechnol.* 13: 40–46.
4. Weissleder, R. 2001. A clearer vision for *in vivo* imaging. *Nat. Biotechnol.* 19: 316–317.
5. Chan, W.C. and Nie, S. 1998. Quantum dot bioconjugates for ultrasensitive nonisotopic detection. *Science* 281: 2016–2018.
6. Goldman, E.R., Balighian, E.D., Mattoussi, H., Kuno, M.L., Mauro, J.M., Tran, P.T., and Anderson, G.P. 2002. Avidin: A natural bridge for quantum dot-antibody conjugates. *J. Am. Chem. Soc.* 124: 6378–6382.
7. Lidke, D.S., Nagy, P., Heintzmann, R., Arndt-Jovin, D.J., Post, J.N., Grecco, H.E., Jares-Erijman, E.A., and Jovin, T.M. 2004. Quantum dot ligands provide new insights into erbB/HER receptor-mediated signal transduction. *Nat. Biotechnol.* 22: 169–170.
8. Lagerholm, B.C., Wang, M., Ernst, L.A., Ly, D.H., Liu, H., Bruchez, M.P., and Waggoner, A.S. 2004. Multicolor coding of cells with cationic peptide coated quantum dots. *NanoLetters* 10: 2019–2022.
9. Voura, E.B., Jaiswal, J.K., Mattoussi, H., and Simon, S.M. 2004. Tracking metastatic tumor cell extravasation with quantum dot nanocrystals and fluorescence emission scanning microscopy. *Nat. Med.* 10: 993–998.
10. Ballou, B., Lagerholm, B.C., Ernst, L.A., Bruchez, M.P., and Waggoner, A.S. Noninvasive imaging of quantum dots in mice. *Bioconjugate Chem.* 15: 79–86.

11. Han, M., Gao, X., Su, J.Z., and Nie, S. 2001. Quantum-dot-tagged microbeads for multiplexed optical coding of biomolecules. *Nat. Biotechnol.* 19: 631–635.
12. Pathak, S., Choi, S.K., Arnheim, N., and Thompson, M.E. 2001. Hydroxylated quantum dots as luminescent probes for *in situ* hybridization. *J. Am. Chem. Soc.* 123: 4103–4104.
13. Hayat, M. 1989. *Colloidal Gold: Principles, Methods and Applications.* Academic Press, San Diego, CA.
14. Elghanian, R., Storhoff, J.J., Mucic, R.C., Letsinger, R.L., and Mirkin, C.A. 1997. Selective colorimetric detection of polynucleotides based on the distance-dependent optical properties of gold nanoparticles. *Science* 277: 1078–1081.
15. Averitt, R.D., Sarkar, D., and Halas, N.J. 1997. Plasmon resonance shifts of Au-coated Au_2S nanoshells: Insight into multicomponent nanoparticle growth. *Phys. Rev. Lett.* 78: 4217–4220.
16. Oldenburg, S.J., Averitt, R.D., Westcott, S.L., and Halas, N.J. 1998. Nanoengineering of optical resonances. *Chem. Phys. Lett.* 288: 243–247.
17. Hirsch, L.R., Jackson, J.B., Lee, A., Halas, N.J., and West, J.L. 2003. A whole blood immunoassay using gold nanoshells. *Anal. Chem.* 75: 2377–2381.
18. Jackson, J.B., Westcott, S.L., Hirsch, L.R., West, J.L., and Halas, N.J. 2003. Controlling the surface enhanced Raman effect via the nanoshell geometry. *Appl. Phys. Lett.* 82: 257–259.
19. Sershen, S.R., Westcott, S.L., Halas, N.J., and West, J.L. 2000. Temperature-sensitive polymer-nanoshell composites for photothermally modulated drug delivery. *J. Biomed. Mater. Res.* 51: 293–298.
20. Okano, T., Bae, Y.H., Jacobs, H., and Kim, S.W. 1990. Thermally on–off switching polymers for drug permeation and release. *J. Control. Release* 11: 255–265.
21. Hirsch, L.R., Stafford, R.J., Bankson, J.A., Sershen, S.R., Rivera, B., Price, R.E., Hazle, J.D., Halas, N.J., and West, J.L. 2003. Nanoshell-mediated near infrared thermal therapy of tumors under magnetic resonance guidance. *Proc. Natl Acad. Sci. USA* 100: 13549–13554.
22. O'Neal, D.P., Hirsch, L.R., Halas, N.J., Payne, J.D., and West, J.L. 2004. Photo-thermal tumor ablation in mice using near infrared-absorbing nanoparticles. *Cancer Lett.* 209: 171–176.
23. Maeda, H. 2001. The enhanced permeability and retention (EPR) effect in tumor vasculature. *Adv. Enzyme Regul.* 41: 189–207.
24. Oldenburg, S.J., Genick, C.C., Clark, K.A., and Schultz, D.A. 2002. Base pair mismatch recognition using plasmon resonant particle labels. *Anal. Biochem.* 309: 109–116.
25. Schultz, S., Smith, D.R., Mock, J.J., and Schultz, D.A. 2000. Single-target molecule detection with nonbleaching multicolor optical immunolabels. *Proc. Natl Acad. Sci. USA* 97: 996–1001.

24 Nanomaterials: Perspectives and Possibilities in Nanomedicine

Kimberly L. Douglas, Shawn D. Carrigan, and Maryam Tabrizian
McGill University, Montreal, Quebec, Canada

CONTENTS

24.1   INTRODUCTION

Rapid advances in the field of nanotechnology have resulted in the development of methods to prepare, modify, and study materials and mechanisms at the molecular and atomic levels, providing tools to probe biological structures and processes on a scale not previously possible [1]. This greater understanding of biology has, in turn, fuelled nanotechnology by directing research in medical materials, devices, and treatments [2]. Divisible into complementary branches of biological discovery and biological mimicry, research in this field forms a nascent multifaceted domain requiring collaborative expertise from biologists, physicists, chemists, and engineers, which is collectively defined as nanobiotechnology [3]. Specific application of nanobiotechnology to nano- and molecular-scale design of devices for the prevention, treatment, and cure of illness and disease is called *nanomedicine* [1].

Nanotechnology, the parent research domain of nanobiotechnology, represents one of the few fields in which government funding has continued to increase. Between 1997 and 2003, government organizations globally increased funding in the nanotechnology sector from $432 million to almost $3 billion [3]. A report by the National Nanotechnology Initiative (NNI, United States) indicates that funding in nanobiotechnology accounts for less than 10% of this worldwide government support. In sharp contrast, over 50% of nanotechnology venture capital is invested in nanobiotechnology, indicative of the potential benefit to be reaped from this burgeoning field [3].

Still in its infancy, nanomedicine has the potential to revolutionize the future practice of medicine. The majority of work in the area can be classified in one of the following three areas: (1) therapeutic delivery systems with the potential to deliver genes and pharmaceuticals through specific cellular pathways; (2) novel biomaterials and tissue engineering for active tissue regeneration; and (3) biosensors, biochips, and novel imaging techniques for the purposes of diagnostic monitoring and imaging. In the sections that follow, each of these areas is discussed and supplemented by examples of current research.

24.2   PARTICLE-BASED THERAPEUTIC SYSTEMS

Current pharmaceutical treatments for illnesses ranging from cystic fibrosis to cardiovascular intervention and the myriad of known cancers suffer a variety of drawbacks related to their methods of administration. Systemic delivery requires high concentrations, which can lead to adverse toxic side effects, unsustainable drug levels, and developed drug resistance. Additionally, oral medications are subject to individual patient compliance, further complicating the challenge of sustained therapeutic levels. Though illness-specific treatments each have unique associated detriments, the application of nanofabrication techniques promises to offer a range of delivery systems designed to remedy these obstacles by providing methods of controlled therapeutic delivery and release to specific tissues and tumors over a desired timeline.

Therapeutic delivery systems are designed to deliver a range of therapeutic agents, including poorly soluble drugs, proteins, vaccine adjuvants, and plasmid DNA (pDNA) for gene therapy, by exposing target cells to their payload. This requires the carrier to enter cells through endocytic or phagocytic pathways where, once internalized, the therapeutic agent is released through vehicle degradation and diffusion mechanisms. Accomplishing these tasks while addressing other issues, such as biocompatibility, biodegradability, and an ability to avoid capture and clearance by the reticuloendothelial system (RES), has proved challenging; systems excelling at certain aspects often fail to incorporate all required characteristics for *in vivo* application.

Current nanoscale delivery systems are divisible into two major categories: surface modification systems designed to prevent immune response or promote cell growth, and particle-based systems designed to deliver therapeutics to cells and tissues. The major research focus of surface modifications is antiproliferative drug-eluting stent coatings designed to prevent restenosis, a reocclusion affecting 30 to 50% of angioplasty patients [4–7].

Alternatively, particle-based systems include viral carriers, organic and inorganic nanoparticles, and peptides. Current trends indicate that particle-based systems will likely replace surface modifications as treatments move toward less-invasive interventions, as evidenced by recent research attempting to ameliorate the shortfalls in drug-eluting stent coatings with nanoparticle systems [8–10].

Within the field of particle-based delivery, the major focus of research now lies in biocompatible polymers for gene therapy. Although the efficient targeted delivery of therapeutic drugs remains a significant challenge with enormous potential benefit, the unfolding nature of proteomics has led more researchers to focus on the potential for gene therapy as the future of nanomedicine treatments. Accordingly, the majority of the studies discussed refer to gene transfection. However, the considerations necessary for efficient transfection are equally applicable to drug delivery systems.

Viruses, particularly retroviruses and adenoviruses, continue to be used for transfection, though their DNA carrying capacity is limited [11]. Despite their highly efficient transfection capabilities and the ability to achieve permanent insertion of the therapeutic gene into the cell genome, their immunogenic and mutagenic impacts have led research toward less-hazardous vectors. The present design conundrum is in developing vehicles that transfect as efficiently as viral vectors while avoiding the mutagenic and carcinogenic risks. Nonviral vectors include liposomes, and nanoparticles of peptides and synthetic and natural polymers, with vector selection governed by a myriad of factors including the therapeutic agent, desired pharmacokinetics, and the target cells.

Liposomes are the primary choice for plasmid transfection, favorable for their ability to condense pDNA, protecting it against degradation by serum nucleases. Cationic lipids electrostatically compact DNA up to 2.3 Mb [12], forming complexes having a positive surface potential and diameter less than 200 nm; such particles are capable of being internalized through endocytic pathways. Complexes may consist of lipid cores with adsorbed plasmids [13] or lipid shells with internalized plasmids [14–16]. Due to the associated high surface charge, circulatory proteins are easily adsorbed and allow for rapid clearance of these vehicles from the circulatory system by the RES. This has led to the development of "stealth" coatings with hydrophilic polyethylene glycol (PEG) or longer-chain polyethylene oxide (PEO), allowing circulation times in the range of hours rather than minutes [17]. Added through physical adsorption or as block copolymers, the flexible hydrophilic region of PEG chains form a "conformational cloud," preventing adhesion of opsonizing proteins [18] (Figure 24.1).

The incorporation of PEG is also commonplace for synthetic polymeric nanoparticles, generally used as block copolymers with the complementary polymer selected based on the properties of the drug to be delivered [19,20]. Cationic polymers commonly selected for their ability to condense negative plasmids demonstrate a positive surface charge, or zeta (ξ) potential, which permits binding of opsonizing proteins in the blood; as with liposomes, the integration of PEG in nanoparticles prevents opsonization. Although the list of polymers employed is lengthy, the most common is U.S. Food and Drug Administration (FDA)-approved poly(d,l-lactide-co-glycolide) (PLGA) [5]. It should be noted that polyethyleneimine (PEI), whose proton sponge behavior is thought to cause endosome disruption, is generally the comparison standard for synthetic polymer transfection, though its cytotoxicity inhibits clinical application. In addition to plasmid transfection, synthetic polymers are

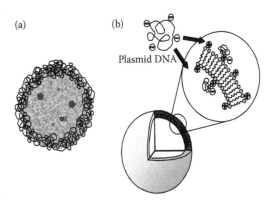

FIGURE 24.1 (a) The "conformational cloud" prevents opsonization of a polyethylene glycol (PEG)-coated polymeric nanoparticle loaded with a pharmaceutical. (b) Liposome section depicts how plasmid DNA is compacted by charge and encapsulated or adsorbed depending on fabrication method.

also being developed for delivery of poorly soluble hydrophobic drugs such as cisplatin [21], clonazepam [22], and paclitaxel [9,18,23].

A common concern with synthetic polymers is the inability of cells to adequately metabolize the polymer vehicles and constituents that may be used in their fabrication, such as poly(vinyl alcohol) (PVA), which may comprise nearly 10% w/w of PLGA vehicles [5,24]. Though synthetic polymers provide better sustained release and gene expression profiles than their natural counterparts [5], recognition that *in vivo* safety is as important as therapeutic success has pushed research efforts toward natural biopolymers in the hopes of achieving true immune transparency [5].

Natural polymers such as chitosan and alginate have received recent attention due to their desirable biodegradability characteristics [25–27]. Research in this field, to date limited in contrast to liposomes and synthetic polymers, indicates that these natural vectors have inferior transfection capabilities when compared to viral and synthetic vectors. As nanoparticles are producible with similar size and surface charge, it remains to be ascertained where the breakdown in natural polymer transfection occurs. Though biopolymeric nanoparticles have yet to reach the transfection efficiency of commercial liposome formulations such as Lipofectamine™ (Invitrogen Corporation, Carlsbad, California), or that of synthetic polymers, their slow biodegradation, ensuring a burst-free release of therapeutic plasmid, and the promise of transfection without immune response continues to fuel research in this field.

In addition to natural biopolymers, recent attention has focused on biocompatible peptides as delivery vectors. By condensing pDNA in a simple complexation fashion similar to polycations, protein transduction domain peptides demonstrate carrier uptake of 80 to 90% within 30 minutes, significantly more rapid than PEI or commercial liposomes [28]. These vectors are able to bypass traditional endocytic pathways to reach the nucleus within 1 hour. Similarly, nuclear localization sequence (NLS) peptides allow superior transfection over synthetic polyplexes, and may be incorporated in traditional liposome complexes to assist in plasmid nuclear penetration [29,30].

It is evident that particle-based vectors have yet to reach their envisioned capabilities. Research focus has shifted from viral vectors, which continue to offer the highest transfection efficiency, through synthetic polymer and liposome systems, commercially available and suitable for *in vitro* transfection, to natural compounds in search of transfection using a biodegradable vector. The enhanced biocompatibility of peptides and natural biopolymers will certainly drive research as the quest for suitable *in vivo* vectors continues, though the balance between attaining biocompatibility while preserving transfection efficiency has yet to be found.

24.2.1 PRACTICAL CONSIDERATIONS FOR NANOSCALE VECTORS

Evident within the research to date is the superiority of certain vectors for particular aspects of the delivery process. The initial stage in the transfection process is endocytosis, which generally requires vectors to be less than 200 nm in diameter, though size limit is dependent on the target cells [24,28,31]. Size trials of a nonphagocytic cell line indicate that smaller particles are much more rapidly internalized, with 100 and 200 nm particles being internalized 3 to 4 and 8 to 10 times more slowly, respectively, than 50 nm particles [31]. Particle sizing methods must also be considered, as dynamic light scattering (DLS) frequently gives larger measurement than electron microscopy and is particularly dependent on the presence of aggregate-inducing ions and proteins.

Cellular internalization is also strongly impacted by particle ξ potential. Though a positive ξ potential is required for binding to negative cell membranes, excessive surface charge leads to rapid systemic clearance and accumulation in the liver and spleen [27,32,33]. To control ξ potential, PEG and PEO are incorporated either as block copolymers for synthetic particles or as coatings for other vectors to produce a near neutral surface charge [17,21,34,35]. These elements also serve to prevent leeching of the therapeutic cargo by forming hydrogen bonds with the aqueous surroundings [20]. Unfortunately, the neutral ξ potential of these particles impairs the vector's ability to disrupt the lysosomal membrane for release to the cytosol, thought to occur upon a charge reversal as protons accumulate in nanoparticles within acidic lysosomes, limiting their transfection efficiency [5,33]. To exploit environmental pH change, polymers having degradable cross-linkers in acidic surroundings have been developed [36,37].

The increased circulation times provided by PEG coatings certainly increase the likelihood of vectors reaching the desired cells and demonstrate enduring *in vivo* protein production following transfection [38,39]. To further improve delivery characteristics, a broad range of targeting moieties, including transferrin, folate, peptides, vitamins, and antibodies, have been incorporated into particle surfaces to actively target cells [10,33,40,41]. Although transferrin and folate increase endocytosis in general, antibodies have the potential to target cell-specific membrane proteins. This enables targeting of particular cell types for specific therapies, such as localized delivery of costly and toxic chemotherapeutics to tumor growths. For example, antibody targeting of endothelial surface receptors results in a tenfold increase in cell binding and a doubling in transfection *in vitro* [10].

Vector targeting of specific tissues may also be achieved through magnetic, heat, and light-affected vehicles [17,19,36,42,43]. External fields retain magnetic vectors in the vasculature of a target region, giving cells greater contact time to endocytose the vector. Alternatively, heat-affected vectors can induce structural changes in polymeric vectors, leading to local pharmaceutical release in stimulated regions [17], and light-stimulated vectors provide localized heating of metallic vectors. By active oncoprotein antibody targeting, or passive enhanced permeation and retention (EPR) targeting, which allows 50 to 100 nm vectors to extravasate from leaky tumor vasculature, gold nanoshells demonstrate tumor ablation when stimulated with near-infrared lasers [43,44]. Even though such external targeting is effective, the biodegradability of the metallic and synthetic vectors remains an obstacle.

These noted considerations are further compounded by such factors as route of administration, target cells, vector stability, therapeutic cargo, and desired pharmacokinetics, as well as entrapment efficiency, loading rate, and release kinetics. Of primary importance is vector stability relative to the route of administration. Many synthetic polymer systems are rapidly cleared when injected intravenously as protein adsorption leads to agglomeration or RES removal, eliminating any potential therapeutic benefit. Similarly, a synergy is required between the target cell and the vector's pharmacokinetic characteristics. For example, slowly dividing cells require vectors capable of entering the nucleus or retarded system clearance and release profiles. Balance between the entrapment efficiency and the total mass loading of individual vectors must also be considered. Vectors with high encapsulation efficiency avoid waste of expensive therapeutic agents, and vectors with high mass loading require fewer particles to be delivered to achieve therapeutic benefit. One study found that encapsulation efficiency was reduced by more than half when loading was increased from 1 to 3% w/w [18]. The vector must also be capable of releasing sufficient quantities with controllable kinetics. Finally, the dosage must ensure a continued presence of vector in the vicinity of the target cells while simultaneously assuring that overdosing does not lead to tumor development or toxic effects [5]. Ultimately, the current findings and the corollaries within indicate that suitable therapeutic delivery systems need to be developed for specific *in vivo* or *in vitro* applications.

24.2.2 EXAMPLE DELIVERY NANOSYSTEMS

24.2.2.1 Vaccine Adjuvants

Plasmid vaccines that can successfully elicit humoral and cellular immune response have the potential to provide safer alternatives than live viral or heat-killed bacterial vaccines. Therapeutically relevant immune response is attainable with adjuvants such as cholera toxin and lipid A delivered using cationic nanoparticles by topical and subcutaneous administration, respectively [45]. Additionally, protein-based vaccines may be encapsulated in pH-sensitive polymers, where low pH in the phagosomes of antigen-presenting cells disrupts the vectors to release vaccines [37].

24.2.2.2 Drug Delivery

Delivery of paclitaxel, a poorly soluble antitumoral, benefits from vector delivery by eliminating the hypersensitivity found in 25 to 30% of patients administered commercial Cremophor™ (BASF Corporation, Mount Olive, New Jersey) [23]. Relying on EPR, PEG-coated liposomes and synthetic particles such as polycaprolactone [46], polylactide [18], and PLGA [47] allow lower doses over prolonged periods. Similarly, hepatoma cells can be targeted through galactosylated ligands using 10 to 30 nm PEG–PLA (poly(lactic acid)) nanoparticles to deliver clonazepam [22], while cisplatin-loaded PEG–polyglutamic acid block micelles demonstrate advanced tumor regression [21]. In an example of innovative drug delivery, temperature-sensitive PEG-coated liposomes, whose drug diffusivity greatly increases in the 40 to 45°C range, demonstrate prolonged circulation and increased accumulation of a model drug in tumor tissue [17]. (See LaVan for a drug delivery review [36].)

24.2.2.3 Antimicrobial Therapy

Bacterial detection using vancomycin-modified magnetic FePt particles demonstrates an ability to detect and entrap Gram-positive and Gram-negative strains, including highly lethal vancomycin-resistant *Enterococci* (VRE) [42]. Self-assembling peptides forming cylindrical nanotubes also display antimicrobial activity by penetrating bacterial cell membranes and increasing permeability, with demonstrated *in vivo* effectiveness against methicillin-resistant *Staphylococcus aureus* (MRSA) [48].

24.2.2.4 Antisense Therapy

Aimed at inhibiting the production of specific proteins, antisense oligonucleotides entrapped in PLGA nanospheres disrupt growth regulation of vascular smooth muscle cells and prevent restenosis following angioplasty [8]. Potential applications for this nascent field are innumerable and will become increasingly prominent as proteomics unfolds the nature of protein interactions and identifies the function of individual proteins.

24.2.2.5 Gene Therapy

Contrary to antisense therapy, the ability to induce production of specific proteins identified by proteomics has similar potential for future clinical therapeutics. *In vitro* application is widespread, and researchers continue to develop new liposome [10,49], synthetic [33,34,50–52] and natural polymer [25,27], and peptide [28] vectors in pursuit of more efficient and compatible formulations. Although human *in vivo* application of nonviral gene therapy remains limited, the wealth of research focused on developing suitable vectors cannot sufficiently emphasize the potential for successful application-specific transfection vehicles.

24.2.3 Summary

Given the range of therapeutic delivery systems presently available and the extent of continuing research, the above discussion is intended as a general background. Most evident in the majority of findings to date is the inability of individual vectors to simultaneously satisfy biocompatibility and delivery efficiency requirements. Further, results are nontransferable across applications, emphasizing the necessity of developing systems for specific therapeutic agents, specific cells, and even specific cellular compartments [5]. Novel screening methods for transfection efficiency of new vectors will certainly accelerate advancement in this process [50]. Though delivery systems have yet to attain their promise, they are certainly well poised to revolutionize nanomedicine, therapeutic delivery, and most importantly, the range and acuteness of illnesses that are considered treatable.

24.3 TISSUE ENGINEERING

One of the main goals of biomedical engineering is the repair or replacement of defective or damaged organs and tissues [53]. To date, numerous methods have been used to fabricate constructs to repair tissues *in vitro* and *in vivo*, ranging from bone to blood vessels [54–56]. Recently, increasing awareness of the influence of molecular composition and nanoscale architecture on cellular responses to materials has focused nanobiotechnology research on controlling these fabrication parameters in the development of better materials for tissue-engineering applications.

Present tissue-engineering methods consist primarily of porous three-dimensional scaffolds of various materials designed to imitate and replace the extracellular matrix (ECM) that supports all cell growth in the body [57]. Ideal tissue engineering would result from full implant integration, followed by its gradual degradation and replacement by natural cell-produced ECM as new tissue is generated. For this to occur, scaffolds should promote cellular migration, adhesion, proliferation, and differentiation, while supporting natural cell processes [58]. Development of more suitable scaffolds requires materials that exhibit not only appropriate mechanical characteristics, but also that elicit favorable cellular responses [59]. Fueled by a greater understanding of surface topography and composition on cell–scaffold interaction, research continues to improve nanoscale control of scaffold architecture in the design of more compatible engineered materials [60].

The term "biocompatible," often used to describe ceramic, metal, and polymers utilized in biological applications, is rather imprecise given the broadness of its accepted definition: "the ability of a material to perform with an appropriate host response in a specific application." Such an indefinite description fails to suggest what type of host response should be desired or considered acceptable [61]. Despite the range of physical and chemical properties exhibited by biomaterials, the body reacts in a similar fashion after implantation; a layer of proteins is randomly adsorbed onto the surface immediately, followed by macrophage attack, and finally encapsulation of the device through the classic foreign body reaction (Figure 24.2) [62,63]. This encapsulation segregates the device from the body, impeding normal wound-healing processes and preventing the device from functioning as intended.

The need to develop biomaterials capable of eliciting specific responses was recognized as early as 1993 [58]. The resulting emerging class of engineered biomaterials aims to design bioactive materials that control the host response to promote full tissue–scaffold integration and natural wound healing while simultaneously preventing the foreign body reaction [62]. The main techniques presently investigated for tissue engineering endeavor to address one or more of the following objectives [58]:

1. Control of the chemical environment through surface biomolecule immobilization or self-assembled systems.
2. Control of the nanostructure through surface modification or *de novo* fabrication.
3. Control of the biological environment of the surface through cell patterning.

Each of these seeks to reproduce an environment that successfully mimics *in vivo* conditions for successful tissue integration. It should be noted that very few examples of *in vivo* tissue replacement

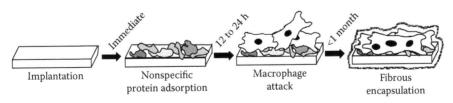

FIGURE 24.2 Implantation of a biomaterial ultimately leads to fibrous encapsulation of the device in the foreign body reaction, preventing natural tissue formation and device integration.

or repair have been reported. For this reason, promising *in vitro* results are discussed in the sections that follow, together with examples of preliminary tissue-engineering applications.

24.3.1 SURFACE MOLECULAR ENGINEERING FOR CONTROLLED PROTEIN INTERACTION

It is hypothesized that nonordered protein adsorption on materials begins the cascade leading to the foreign body reaction, because all normal biological events operate on a system based on specific recognition of proteins and polysaccharides [62,63]. To control nonspecific adsorption, several surface modification approaches have been explored: "stealth" materials are designed to decrease nonspecific protein adsorption and others promote adsorption of specific proteins in an effort to direct biological processes [63,64].

24.3.1.1 Stealth Materials

There are three main methods used in the development of protein-resistant stealth materials, also referred to as nonfouling or noninteractive surfaces [65]:

1. Hydrophilic surfaces can be achieved through thin-film deposition of hydrogels, phospholipids, or other suitable materials, such as PEG and PEO.
2. Chemical surface modification leads to surface expression of specific functional groups known to inhibit protein adsorption. For example, surfaces rich in carboxylic acid groups support cell growth, whereas methyl, hydroxyl, and carboxymethyl ester groups generally do not. Interestingly, surfaces modified by random mixtures of organic functional groups can exhibit significant bioactivity and specificity [63].
3. Surface immobilization of biomolecules, including proteins, seeks to recreate the natural *in vivo* environment.

The first two approaches address the problem of nonspecific adsorption but fail to address the coexisting need to promote the favorable cell–surface interactions required to encourage natural healing and tissue repair processes. Alternatively, the last method represents a tactic that addresses both issues by creating materials that manage biological interactions while resisting nonspecific protein adsorption.

24.3.1.2 Biomimetic Materials

Immobilization of proteins to control tissue–surface interaction represents a more biomimetic approach to material development, in contrast to stealth materials that prevent interaction. Natural biomolecules have the dual capability of triggering specific cellular responses while their presence can prevent nonspecific adsorption. The techniques used to append biologically active moieties on surfaces can be grouped into three main categories: physical micro-/nanofabrication, imprinting, and direct chemical immobilization.

Conventional microfabrication techniques used to pattern biomolecules on surfaces include photolithography, photochemistry, and microcontact printing [66]. Capable of immobilizing biomolecules on a variety of surfaces, the resolution of these techniques is generally limited to the microscale. Though nanometer-scale configurations are achievable through modifications of these techniques, including the application of lasers and finely focused ion beams (FFIBs) to photolithography, they remain unable to meet the desired resolution [66,67].

Recently developed nanofabrication techniques include dip-pen nanolithography (DPN), which uses functionalized atomic force microscopy (AFM) tips to deposit biomolecules on surfaces with pattern features as small as 10 nm [68]. Used for the creation of protein nanoarrays and virus arrays, this technique can create patterns with multiple components [68,69]. Enzymes have also been selectively deposited for biochemical modification of self-assembled monolayers [70]. While nanofabrication techniques are effective for patterning two-dimensional surfaces, these methods are quite limited in terms of processing time and are not suitable for three-dimensional scaffolds.

Imprinting is an approach used to confer biological recognition to surfaces through the creation of templates in synthetic polymers. This process yields surfaces with imprint accuracy at the nanometer scale, resulting in binding cavities that possess the correct conformation and functionality to allow selective binding to the appropriate molecules [71]. Despite the promising results, this technique is also limited to two dimensions, and multicomponent systems would be difficult to achieve.

Chemical immobilization, a more facile method of introducing biomolecules on materials, exploits functional groups present at the surface as binding sites. Successful covalent and noncovalent immobilization of biomolecules on synthetic polymers has been reported extensively, resulting in modified bioactivity and specificity [63]. Polymers requiring functionalization are modified by blending, copolymerization, and chemical and physical treatments, among others [64]. Although not amenable to specific patterning, multicomponent systems and three-dimensional surface modification are feasible. For example, the natural interaction between osteopontin and type I collagen can be exploited to functionalize a surface with oriented protein binding, leading to increased endothelial cell adhesion and proliferation [72].

In many cases, it is preferential to chemically immobilize peptide segments rather than whole proteins, as they retain the activity of full proteins while being less likely to invoke immune responses and are more resistant to degradation. RGD (arginine–glycine–aspartic acid), perhaps the most studied peptide sequence for tissue engineering, is well documented as a stimulant for cell adhesion on synthetic surfaces and its ability to interact with multiple cell adhesion receptors (for a review see [64]). Several *in vivo* studies have investigated the fate of peptide-covered surfaces in the body, including an RGD-functionalized hydrogel that demonstrates improved neural tissue adhesion, tissue regeneration, and host tissue infiltration when implanted into brain lesions, as compared to the nonmodified hydrogel [73]. A similar material also improves neurite repair and angiogenesis in spinal cord repair [74]. Bone tissue ingrowth and direct tissue-implant contact is also promoted on scaffolds with immobilized RGD-bearing peptides, whereas uncoated scaffolds become segregated by a fibrous tissue layer resulting from the foreign body reaction [75].

Control of protein–surface interaction is essential to the development of materials designed for biological application. Reduction of nonspecific protein adsorption with stealth materials is a promising first step to avoiding the foreign body reaction. Surface immobilization of biological molecules is a more sophisticated approach, leading to materials exhibiting favorable cell interactions and having the potential to influence biological processes. Continued advancements in this area will require the development of additional nanoscale techniques to allow greater control over surface characteristics. Such advancements must be supported by microbiological research to elucidate the nature of interactions between proteins and surfaces, as well as the protein spatial arrangements and orientations necessary for the desired application-specific response.

24.3.1.3 Nanostructured Surfaces

In addition to the importance of recognition events, advances in microbiology provide a greater understanding of the effect of surface micro- and nanostructure on cell behavior. *In vivo*, all cells live within the structural support of a complex nanoscale topology of pores, ridges, and fibers provided by the ECM [53]. Interactions between the substrate and cells are thought to activate the cytoskeleton in a way that mediates attachment, migration, growth, and differentiation, in addition to affecting cytokine and growth factor production [60,76]. As noted by Miller et al. [77], it is surprising that the design of optimal tissue engineering structures has not focused more on nanostructure, given that ECM has been known for some time to be crucial to cell and tissue growth. However, advanced techniques allowing nanostructured construction have renewed research interest, with the majority of work to date related to blood vessel and bone tissue regeneration.

Several methods are used to create nanoscale architecture in polymers, including phase separation [53,57,59,78], electrospinning [79], rubbing [80], chemical etching [77,81], colloidal lithography [82], AFM-assisted nanolithography [83], and template patterning [76,84]. Phase separation is particularly useful because it is amenable to the production of three-dimensional scaffolds. In addition to polymers, hydroxyapatite (HAP) and collagen are generally the materials of choice for nanostructured bone regeneration scaffolds. Two novel ways of coating nanoscale HAP onto three-dimensional scaffolds include chemical vapor deposition and growth directly from aqueous solutions [85,86]. Regardless of the method or material used to produce nanostructures, cell adhesion and proliferation increase compared to smooth or microstructured surfaces, with cell morphology more closely resembling the native state [59,76,77,82,87,88]. Further, protein adsorption, a necessary precursor to cell adhesion, is improved on nanostructured surfaces [88]. Additionally, submicron-sized grooves in surfaces can affect cell alignment for directed growth [87,89].

In addition to angiogenesis and osteogenesis, nanostructured materials demonstrate application for the growth of neurons and smooth muscle cells. Successful *in vitro* neurite growth is achievable with a biodegradable porous nanostructured PLA scaffold, where nanoscale topography promotes neuron adhesion and differentiation, including neurite outgrowth [53]. Similarly, bladder smooth muscle cell growth is improved on chemically treated PLGA and polyurethane (PU)-bearing nanostructures [81].

In a considerably different application, nanofabricated polyimide surfaces can be useful for the preparation of cell spheroids, roughly spherical masses composed of cells and associated ECM that demonstrate tissue-like morphological and physiological functions. Cell culture on nanostructured fluorinated polyimides results in fibroblast cell spheroids with a density comparable to tissue *in vivo*, fostering interest in their development for tissue-engineering applications [90].

Research in this field demonstrates the importance of nanoscale architecture on cell adhesion, growth, differentiation, and phenotype expression. Furthermore, it promotes the development of techniques for producing nanostructured surfaces, which are certain to be an important component in the future of successful tissue-engineering scaffolds.

24.3.1.4 Self-Assembled Systems

The techniques previously discussed represent approaches based on modifications to materials already employed in biomedical applications. Conversely, natural tissues consist of hierarchical organizations of molecules, giving rise to a multitude of nanostructures, microstructures, and macrostructures [91]. This has led researchers to explore the "bottom-up" approach of self-assembling systems, using biological systems as a design base to create biomaterials with improved properties.

Biomimetic approaches to self-assembly are used to apply surface functionality onto materials to direct specific cell responses, while also displaying nanostructure resembling natural tissue ECM. Advantages of this approach include ease with which functional groups can be incorporated; well-defined structures and intrinsic stability; and a lack of defects in the surfaces, which reduces nonspecific interactions [63,92]. The two main methods of self-assembly include the formation of self-assembled monolayers (SAMs) on surfaces and the formation of nano-, micro-, and macroscopic structures from natural self-assembly. To date, the majority of research into SAMs uses molecularly flat surfaces as templates. Multicomponent SAMs with controllable nanoscale features are attainable with a variety of patterning techniques (for a review see [93]). Though largely dependent on the functional groups of the surface on which they form and the nature of the molecules to be assembled [94], successful biomedical SAM application would require application to three-dimensional structures and to materials used for medical applications; such reports are limited [92].

In contrast to two-dimensional self-assembly, self-assembly of three-dimensional structures holds much promise and is of greater interest for tissue engineering. The challenge in this field remains the design of molecules that spontaneously self-organize into stable structures with desirable characteristics [95]. Peptides are particularly well suited as they are capable of self-assembly, are sufficiently robust, can confer biological activity to a surface, and degrade into nontoxic constituents. A particular advantage is that peptides may be designed to incorporate specific chains and functional groups to confer cell adhesive properties or cell differentiation signaling abilities to a surface. Several examples of synthetic and natural peptides demonstrate self-assembling abilities and form favorable structures, with amphiphilic peptides proving exceptionally useful in this domain [95].

In vitro, self-assembled peptide scaffolds facilitate cell attachment, migration, proliferation, and differentiation for a number of cell lines; cells are also found to produce components of natural ECM [95]. Primary rat neuronal cells project lengthy axons following the contours of a peptide nanofiber scaffold and form active and functional synapses. Chondrocytes encapsulated in the same scaffold produce components of natural ECM, including collagen and glycosaminoglycans. As well, liver progenitor cells differentiate and demonstrate natural enzyme activity. *In vivo*, these scaffolds promote repair of brain lesions [96]. In a separate application, an engineered amphiphilic self-assembling peptide forms collagen-like cylindrical structures that guide HAP crystal formation in orientations and sizes similar to natural bone [97]. Finally, an interesting combination of self-assembly and cell patterning techniques developed by Auger et al. employs a cohesive sheet of self-assembled human vascular cells as a template for fibroblast adhesion on the exterior surface. The interior surface then acts as a scaffold for endothelial cell growth, resulting in the formation of tissue-engineered blood vessels exhibiting appropriate structural characteristics and *in vitro* hemocompatibility [98,99].

24.3.2 SUMMARY

It has long been recognized that the materials and constructs used to replace damaged tissues are vastly inferior to their natural counterparts. There is little doubt that an ability to induce healing processes to produce natural tissue would bestow an enormous wealth of medical treatment options. Expanded understanding of the processes involved in healing and tissue growth processes, provided by progress in nanobiotechnology and microbiology, has furthered insight into the interactions occurring between cells and substrates, leading to improved designs that demonstrate promising results for future *in vivo* tissue applications. Though nanomedicine has yet to benefit from the true promise held in tissue engineering, it is clear that further developments in this area, including fabrication processes amenable to three-dimensional constructs with modifiable bioactivity, possess the potential to provide achievable, application-specific, tissue-engineering methods for medical treatments.

24.4 DIAGNOSTIC IMAGING AND MONITORING

The principles and methods employed in the nanoscale development of materials for therapeutic delivery systems and tissue engineering serve equally well in the design of imaging and monitoring diagnostics. Nanoparticles, primarily of inorganic materials such as silica, gold, and silver, serve as imaging aids for a range of *in vitro* and *in vivo* investigations, while tissue engineering methods of controlled biomolecule immobilization aid in the creation of biosensors to detect a range of proteins, DNA, and pathogenic compounds. Many of the techniques that now fall within the realm of nanomedicine were originally developed for other applications, such as the polymerase chain reaction (PCR) used to amplify DNA samples prior to diagnostic genetic screening [100]. The broad range of nanoscale manipulations used in diagnostic nanomedicine are highly cross-application tools, rendering difficult the categorization of such methods as biomolecule detection using nanoparticles for intracellular sensing [101]. The following division of imaging and diagnostic monitoring tools is intended to simplify these multidisciplinary applications, though the scope of current studies certainly exceeds these boundaries.

24.4.1 Biophotonics

Imaging techniques have benefited from progress in nanotechnology, with the main advances incorporating nanoparticles and quantum dots. Although not yet applicable for diagnostic purposes, new optical techniques are also being designed to improve nanoscale imaging.

24.4.1.1 Nanoparticles in Imaging

Advances in imaging have occurred in conjunction with the development of nanoparticles, which are used to enhance existing imaging techniques by serving as contrast agents or as markers in various optical techniques. The myriad of diagnostic imaging techniques currently used includes radiography (x-ray), magnetic resonance imaging (MRI), computed tomography (CT), positron emission tomography (PET), and ultrasound. Each of these methods represents an invaluable medical tool permitting diagnosis and monitoring of numerous conditions; however, each suffers from limitations due to one or more tissues that are difficult to image. Nanoparticles have been designed to overcome these limitations and to expand functionality of the techniques.

Radionuclide-encapsulating liposomes are used to prolong the lifetime of positron emitters in the body for improved diagnostic PET imaging. Liposomal radionuclides are designed to enhance blood pool imaging and specific tissue observation, with leaky tumor vasculature and subsequent accumulation making them particularly useful for diagnostic tumor imaging [102]. Radioisotope-carrying polymeric nanoparticles are similarly used to target bone and bone marrow [35]. As with drug delivery systems, these particles must be designed to avoid RES clearance and with appropriate molecular markers to allow accumulation in the target tissue.

Analogous strategies for improved radiography include the use of liposome-encapsulated contrast enhancers. In soluble form, these contrast enhancers are rapidly cleared from the body and pose some toxicity risks. In contrast, the nontoxic liposomal iohexol formulation increases residence time to 3 hours, with applications in cardiac imaging and early tumor detection [103]. Similarly, improvements in MRI imaging result from the use of a gadolinium carrying polyamidoamine dendrimer as a contrast agent. This nanosized paramagnetic molecule allows noninvasive localization of sentinel lymph nodes and mestastases in breast cancer patients, which is crucial for treatment design [104].

Although nanoparticles provide improved imaging using existing techniques, image resolution remains limited by the system design. Therefore, nanoparticles are being designed to be used with systems capable of nanoscale resolution. New optical imaging techniques have emerged as supporting technology enables more precise imaging and labeling at the nanoscale. Fluorescence-based imaging systems are by far the most popular and most studied systems for cell analysis. These systems, based on bioconjugation of an optical dye to biological molecules, offer the same resolution as optical microscopy systems, but allow for precise localization and sensitive quantification of individual biomolecules. However, fluorescent dyes suffer from limitations, including potential toxicity, reduced sensitivity compared to radioactivity, and interference from other molecules.

To improve upon the sensitivity of fluorescent imaging modalities, several methods of signal enhancement are employed in optical systems. Signal enhancement through fluorescence resonance energy transfer (FRET, also called fluorescence *in situ* hybridization, or FISH), which results when two fluorescent molecules are brought into close proximity leading to a change in fluorescence intensity, permits the intracellular detection of specific sequences in nucleic acids, including mRNA and DNA [105]. Signal enhancement is also achieved with plasmon–plasmon resonance interactions, which are similar to FRET. Gold and silver plasmon-resonant particles (PRPs), as well as super-particles consisting of colloidal gold nanoparticle shells around silica cores, strongly scatter optical light, making them easily visible using conventional microscopy or surface-enhanced Raman scattering (SERS) [44,106]. Two PRPs in close proximity produce changes in their plasmon optical resonance, allowing their distinct and bright signals to be quantified; a single PRP is as bright as 5

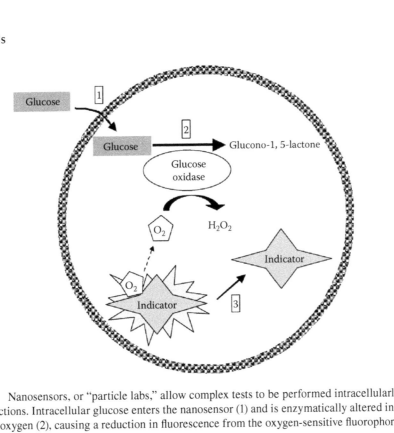

FIGURE 24.3 Nanosensors, or "particle labs," allow complex tests to be performed intracellularly through synergistic reactions. Intracellular glucose enters the nanosensor (1) and is enzymatically altered in a process that consumes oxygen (2), causing a reduction in fluorescence from the oxygen-sensitive fluorophore (3).

$\times 10^6$ fluorescein molecules or 10^5 quantum dots [107]. These are used for highly sensitive detection of antibodies in whole blood, as well as investigations of intracellular transport pathways [44,106].

24.4.1.2 Nanosensor Probes

Nanoparticles are also used to enhance and exploit fluorescent signals in the form of optical nanosensors. These sensors, which generally consist of an encapsulated fluorescent detection system, are designed to take advantage of fluorescent systems while improving their sensing abilities and reducing toxicity [108]. Dendrimers, in particular, allow the colocalization of several chromophores, increasing the signal and sensitivity to levels provided by less-favorable radioactive tags [109]. Nanosensors are more complex than simple fluorescent tags, usually consisting of a fluorophore that is activated or quenched by a particular analyte.

Nanosensors incorporating several different molecules, including fluorophores, enzymes, fluorionophores, and associated ionophores, are being designed for the detection of specific analytes, including intracellular pH, glucose, and potassium [108,110]. These show fast response time, reversible analyte detection, and high selectivity [111]. For example, quantitative glucose-sensing nanosensors, also called PEBBLEs, contain glucose oxidase and an oxygen-sensitive fluorescent indicator to detect intracellular glucose levels (Figure 24.3) [112]. The versatility of nanosensors allows encapsulation of a variety of molecules, leading to "particle labs" that perform complex tests intracellularly through synergistic reactions. They can be targeted to specific organelles, though they do suffer some cell entry limitations. Physical cell delivery methods are also viable, including injection and gene gun delivery, though liposomal and ultrasound-based delivery leads to better cell survival [106,108]. Advances in delivery vehicle design will also benefit nanosensors, enabling improved cellular entry.

The development and use of nanoparticles for diagnostic imaging is a relatively new endeavor that is likely to increase, owing to the promising results obtained to date. It can be expected that the development and use of nanosensors and image enhancers for common diagnostic techniques will predominate in the coming years.

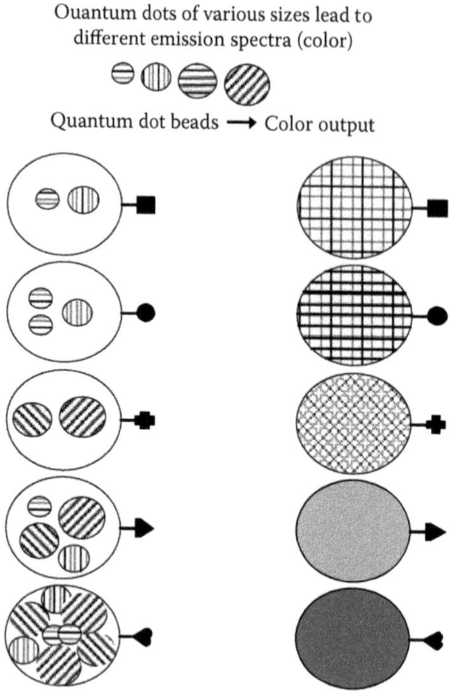

Quantum dots of various sizes lead to
different emission spectra (color)

Quantum dot beads → Color output

FIGURE 24.4 The versatility of quantum dots allows preparation of numerous distinctive tags using encapsulated combinations of uniquely colored dots.

24.4.1.3 Quantum Dots in Imaging

Although these nanosized devices could technically be included with nanoparticles, their unique properties place quantum dots (Qdots) in their own category. Qdots are semiconductor nanocrystals (diameter 2 to 10 nm) that exhibit broad excitation spectra and narrow emission spectra in the visible range [44]. Originally designed for information technology purposes, their application to biophotonic imaging through conjugation with biomolecules was quickly realized. They demonstrate significant advantages over classic fluorescent dyes, including size-tunable emission wavelength, photobleaching resistance, increased stability, reduced toxicity, and persistent residency in cells [113–115]; as well, simultaneous detection of multiple agents is possible due to their broad excitation and tunable narrow emission spectra.

Qdots are used in a number of biological applications, including visualization of DNA hybridization [116], immunoassays [117], receptor-mediated endocytosis [118], and *in vivo* cellular imaging [113]. *In vitro* and *in vivo* intracellular labeling with Qdots does not interfere with cell viability, growth, or differentiation over extended periods of time [119], and no signs of systemic toxicity result from intravenous administration [120]. Their persistence in cells makes them amenable for following extended tissue development, including embryo development [116].

The versatility of Qdots is best demonstrated through the design of multicolor optical coding systems. Hundreds of thousands of unique Qdots can be created through the encapsulation of combinations of zinc sulfide-capped cadmium selenide nanocrystals of slightly different sizes in polymeric beads expressing various biomolecules at the surface (Figure 24.4) [121]. This ability makes them ideal for high-throughput parallel analysis of biological molecules, such as gene expression studies and medical diagnostics. The use of Qdots in analysis is expected to replace planar DNA chips due to reduced costs, faster binding kinetics, and greater flexibility in target selection [121].

Even greater promise for Qdots lies with *in vivo* application. As with all nanoparticles, these must be designed with surface coatings to reduce RES clearance and to promote cell-specific targeting; they must also demonstrate sufficient lifetimes to allow visualization. Amphiphilic poly(acrylic acid) surface coatings allow noninvasive whole-body fluorescence imaging of targeted tissues up to 4 months postinjection [122]. Qdots are used for live imaging of tumors [123], capillaries, skin, and adipose tissue [120]. Further *in vivo* application will require Qdots with near-infrared (NIR) emission spectra, because transmission of these wavelengths is possible through tissue [44,113,122]. They may also find application in tracking of viral particles, drug molecules, and migratory tumor cells *in vivo* [113]. Qdots may also play a role in the development of spectroscopic and spectral imaging techniques for molecular analysis of pathologic tissue, leading to the identification of "disease fingerprints" and subsequent diagnostic techniques [124]. As interest and research in Qdots increases, it is likely that the consequential development of new noninvasive imaging techniques will ensue.

24.4.2 Diagnostic Biosensors

Though the majority of biorecognition techniques incorporate nanoscale manipulations, the majority of these methods would more classically fall within the domains of analytical chemistry than applied nanomedicine. However, the increasing sensitivity of these techniques greatly improves their utility as diagnostic aids and broadens the scope of applied and investigative nanomedicine.

24.4.2.1 Molecular Biointerfaces for Gene and Protein Biorecognition

Biomolecule detectors incorporate a biorecognition device capable of selectively recognizing the analyte of interest in connection with a signal transducer and a suitable output device. Transduction methods include a variety of optical (surface plasmon resonance [SPR], fluorescence), electrochemical (voltammetry, impedance, field effect), mechanical (cantilever, surface probe microscopy), and mass-based systems (quartz crystal microgravimetry [QCM], mass spectrometry). Selection of the appropriate transduction system is partially determined by the nature of information sought (quantitative or qualitative), the analyte (concentration, molecular weight), the sample size, and assay timeline.

The study of genomics promises to unravel the link between specific gene sequences and phenotype. Consequently, the majority of present diagnostic research focuses on the recognition of specific oligonucleotide (ON) sequences, DNA mutations, and single nucleotide polymorphisms (SNPs) in order to identify predisposition to genetic disorders or the presence of disease. Recognition using any of the above transduction methods requires the immobilization of biomolecules to a surface to form a biointerface in a manner similar to the design of tissue-engineering scaffolds. However, biosensors not intended for *in vivo* implantation avoid biocompatibility requirements and can employ inorganic and metallic substrates.

Common DNA recognition protocols rely on thiolated ONs adhered to gold surfaces as recognition ligands. Following PCR amplification of DNA to increase assay sensitivity, samples are exposed to the biointerface with the immobilized ON. Even though label-free assays are possible [125], the majority of hybridization and mutation assays use labels to obtain additional sensitivity through transduction dependent signal amplification. For example, silver and gold nanoparticles (PRPs) act as reporters for hybridization monitoring of cystic fibrosis genes by Raman scattering detection of plasmon resonance [126], and in the quantitative detection of SNPs in breast cancer genes using optical microscopy [127]. Monitoring of plasmon effects has the added advantage of multiple labeling capacity, where particles may generate different colors to label different sequences.

Nanoparticles are also used for fluorescent transduction methods, with single base mutation detection using 2.5 nm gold particles as scaffolds for fluorescently labeled ONs [128], or 2 to 100 nm fluorophore-loaded silica nanoparticles conjugated to unlabeled ONs [129]. For nonoptical transduction, charged liposomes amplify DNA recognition in electrochemical methods [130]. Recent studies also demonstrate amplified recognition of DNA hybridization using gold nanoparticles with QCM, which act as secondary ligands to increase the mass of recognized sequences [131], and provide sublayers to which the ligand ON is immobilized [132]. Continuing research has lowered the detection sensitivity of DNA hybridization events by three to four orders of magnitude in as many years, with current sensitivity in the subfemtomolar concentration range for fluorescence and QCM-based techniques [129,132]. For the more difficult task of identifying single base mutations, assays are approximately one order of magnitude less sensitive in terms of molar concentration [131].

The use of biosensors is not limited to genomic applications and DNA analysis. Proteomics, which seeks to identify and define the roles played by cellular proteins, has led to improved protein recognition sensors [133]. Traditional highly sensitive enzyme-linked immunosorbent assays (ELISA) are being replaced with microarray format protein assays similar to DNA analyses with the goal of increasing knowledge of associated disease biomarkers, protein functions, and drug target identification. Depending on the transduction method, detection relies on protein composition (mass spectroscopy), or on the immobilization of a monoclonal antibody to a biointerface for optical (SPR, fluorescence) [134,135] and mass (QCM)-based strategies [136,137]. Such methods require the oriented immobilization of a functional protein to allow correct binding with the target analyte.

To achieve proper antibody immobilization while preventing nonspecific protein adsorption, which is a particular necessity to maintain assay sensitivity and specificity for serum or unpurified samples, biointerfaces are designed using principles similar to those used in tissue engineering. The importance of such nanoscale design factors is clearly demonstrated by the tenfold increase in sensitivity obtained in one study upon a doubling of the distance at which the antibody was immobilized from the surface [138]. The sensitivity of immobilized ligands to their environment is an additional obstacle with protein arrays as individual antibodies require unique pH and ionic conditions to maintain functional conformation [139]. Despite the advances in this field, the sensitivity of specific protein biorecognition has yet to experience the enormous gain in sensitivity found with DNA assays, which benefit from the advantage of sample amplification.

The array formats popular for both protein and DNA detection are well adapted to genetic screening for a multitude of illnesses. By using automation to process large numbers of samples, databases generated from mass samplings are used to correlate genetic presence or susceptibility to illnesses such as breast cancer [140,141] and bacterial infection [142,143]. The genetic test for cystic fibrosis, which requires 25 genetic traits to be tested, highlights the utility of arrays for both exploratory and diagnostic studies [144]. Presently, the requirements of such broad analysis limit testing to optical methods. However, the prohibitive cost of specialized optical analyzers, in conjunction with improvements in electrochemical and QCM sensors, may soon popularize alternative diagnostic techniques. Ideally, these technologies will soon permit nanomedicine diagnostics to attain current sensitivity limits in a point-of-care device that requires no sample preparation, amplification, or labeling.

24.4.2.2 Pathogen Recognition

In addition to providing identification of disease susceptibility, advances in DNA hybridization techniques have greatly enhanced pathogen detection. These improvements permit the diagnosis timeline for sepsis, a rapidly advancing systemic infection for which timely treatment is critical, to narrow from the 24 to 48 hours required for culturing methods to 5 to 6 hours using PCR amplification methods [145]. Specific pathogenic nucleotide sequences correlate with a variety of pathogens, allowing detection using biorecognition techniques: the 16S rRNA gene is detectable in a range of Gram-negative and Gram-positive bacteria [146], the mecA gene is specific for the detection of methicillin-resistant staphylococci [147], and the 18S rRNA sequence is indicative of the fungus *Candida albicans* [148].

Viral genetic materials are also quantitatively detectable using PCR-amplified fluorescent techniques, the sensitivity of which is particularly important for measuring human immunodeficiency virus (HIV) loads in patients undergoing retroviral therapies that reduce viral loads to below the detection limit of commercial diagnostic methods [149]. Though the majority of pathogen detection relies on PCR amplification followed by electrophoretic or fluorescent identification, hybridization of viral DNA has recently been reported using alternate transduction methods. Thiolated ONs immobilized on gold nanoparticles provide a platform for optical Rayleigh scattering to detect hepatitis B and C viral DNA [150], and ester linkages to immobilize hepatitis B fragments allow for similar detection by voltammetry [151]. Silicon miniaturization technology also presents future possibilities for point-of-care viral detection using arrayed electrodes for amperometric techniques [152]. However, the true potential of bacterial and viral recognition techniques will not be achieved until assays are capable of detecting natural load levels, negating the necessity of PCR amplification, to provide more rapid and cost-effective diagnosis.

24.4.3 SUMMARY

Imaging and biorecognition systems used in diagnostic techniques rely predominantly on biophotonics, though alternate transduction methods are becoming more prevalent in biorecognition.

Both domains mutually benefit from rapid technological advances, fueled by an enormous research focus in the application of nanoparticles for enhanced imaging and DNA recognition techniques. Nanoparticles have been developed to improve contrast and allow imaging of target tissues using conventional medical diagnostic equipment, such as PET, MRI, and X-ray. Fluorescence-based nanoparticle systems are also generating much interest, particularly as nanosensors demonstrate the capability for detecting and quantifying intracellular processes. Greater interest still stems from advances in quantum dots, which have already demonstrated superiority over conventional fluorescent tags, and demonstrate a capacity for *in vivo* live imaging. Such interest is paralleled in biorecognition techniques, where similar particle-based tagging and amplification strategies have vaulted hybridization sensitivity to subfemtomolar levels. These techniques expand medical diagnoses to include abilities to genetically screen for disease susceptibility using microarrays that simultaneously analyze thousands of genes, as well as to detect disease and pathogen presence through DNA fingerprints. Continued development of nanobiotechnology techniques will increase the ease with which such diagnostic routines are performed, ensuring that analytical nanomedicine will prevail in the future of medical diagnostics.

24.5   ON THE HORIZONS OF NANOMEDICINE

Even though the methods discussed represent applied modalities that promise to be implemented within years, the scope of nanobiotechnology encompasses research that, though currently on the very frontiers of modern science, presents innumerable possibilities for the future of nanomedicine. The prophesied ability of nanoscale machines, or nanobots, to provide molecular-level construction and repair to exterminate disease and erase genetic defects represents the pinnacle of nanomedicine aspirations. Although the inherent difficulties in the design and manufacture of such devices raise questions as to their feasibility [153], research continues to explore nature's nanomachines and issues crucial to the development of nanobots. In cells, proteins are nanomachines that act as transporters, actuators, and motors, and are responsible for meticulous monitoring and repair processes [154,155].

Single molecule analysis can reveal the mechanistic details of protein function, with AFM and FRET imaging being the favored tools in such investigations. FRET imaging allows observation of biomolecular structure and intracellular motion [156,157], and AFM is used to explore molecular forces and energetics, and can be used to manipulate folding and structure [154]. A recently developed imaging technique, scanning near-field optical microscopy (SNOM), offers visual imaging with nanometer resolution, promising further developments in this field. Combining AFM and SNOM advances single molecule analysis by allowing simultaneous nanoscale topographic and fluorescence imaging, making nano-FRET analysis possible [158].

Through analysis of individual molecules, researchers aim to elucidate the principles of biomolecular machinery to understand the mechanisms governing gene activation, DNA repair, and motor proteins such as kinesin and myosin. These biological motors have the capability of precise molecular positioning in the construction of energetically unfavorable structures. Advances in this field could lead to the exploitation of nanomotors for molecular assembly; researchers have already begun to envisage the use of kinesin motors as molecular transporters in nanoscale syntheses [155]. Whether or not these discoveries ultimately contribute to the future development of nanorobots, the wealth of knowledge generated by single molecule analyses will undoubtedly expand the boundaries of treatment and diagnosis in nanomedicine.

24.6   CONCLUSIONS

The term "nanomedicine" presently incorporates a vast multitude of techniques that are roughly divisible into categories of therapeutic delivery systems, tissue engineering, and diagnostic imaging and biorecognition. As discoveries in various fields, such as nanofabrication, proteomics, and biophotonics, continue, and given the enormous funding presently directed toward nanotechnology by government and private-sector investors, this field will undoubtedly proliferate. Ultimately, nano-medicine treatments and diagnostics must bypass or control the host immune response to fulfill their function over a desired timeline. Though many of the treatments currently under development have yet to reach their envisioned performance at a research level, the potential clinical application of such interventions provides sufficient promise to ensure that nanomedicine represents the future of medical care.

REFERENCES

1. National Institutes of Health 2004.
2. NBTC 2004.
3. DeFrancesco L. 2003. *Nat. Biotechnol.* 21: 1127.
4. Kereiaks D.J., Choo J.K., and Young J.J. 2004. *Rev. Cardiovasc. Med.* 5: 9.
5. Panyam J. and Labhasetwar V. 2003. *Adv. Drug Deliv. Rev.* 55: 329.
6. Tanabe K., Regar E., Lee C.H. et al. 2004. *Curr. Pharm. Des.* 10: 357.
7. Thierry B., Winnik F.M., Merhi Y. et al. 2003. *Biomacromolecules* 4: 1564.
8. Cohen-Sacks H., Najajreh Y., Tchaikovski V. et al. 2002. *Gene Ther.* 9: 1607.
9. Kolodgie F.D., John M., Khurana C. et al. 2002. *Circulation* 106: 1195.
10. Tan P.H., Manunta M., Ardjomand N. et al. 2003. *J. Gene Med.* 5: 311.
11. Baker A.H. 2004. *Prog. Biophys. Mol. Biol.* 84: 279.
12. Vijayanathan V., Thomas T., and Thomas T.J. 2002. *Biochemistry* 41: 14085.
13. Cui Z. and Mumper R.J. 2002. *Pharm. Res.* 19: 939.
14. Ameller T., Marsaud V., Legrand P. et al. 2003. *Int. J. Cancer* 106: 446.
15. Cui Z. and Mumper R.J. 2002. *Bioconjug. Chem.* 13: 1319.
16. Dauty E., Behr J.P., and Remy J.S. 2002. *Gene Ther.* 9: 743.
17. Lindner L.H., Eichhorn M.E., Eibl H. et al. 2004. *Clin. Cancer Res.* 10: 2168.
18. Dong Y. and Feng S.S. 2004. *Biomaterials* 25: 2843.
19. Kumar N., Ravikumar M.N., and Domb A.J. 2001. *Adv. Drug Deliv. Rev.* 53: 23.
20. Rosler A., Vandermeulen G.W., and Klok H.A. 2001. *Adv. Drug Deliv. Rev.* 53: 95.
21. Nishiyama N., Okazaki S., Cabral H. et al. 2003. *Cancer Res.* 63: 8977.
22. Kim I.S. and Kim S.H. 2002. *Int. J. Pharm.* 245: 67.
23. Ibrahim N.K., Desai N., Legha S. et al. 2002. *Clin. Cancer Res.* 8: 1038.
24. Prabha S., Zhou W.Z., Panyam J. et al. 2002. *Int. J. Pharm.* 244: 105.
25. Douglas K.D. and Tabrizian M. 2005. *J. Biomater. Sci. Polym. Ed.* 16: 43–56.
26. Gonzalez F.M., Tillman L., Hardee G. et al. 2002. *Int. J. Pharm.* 239: 47.
27. Kim T.H., Ihm J.E., Choi Y.J. et al. 2003. *J. Control Release* 93: 389.
28. Park Y.J., Liang J.F., Ko K.S. et al. 2003. *J. Gene Med.* 5: 700.
29. Bremner K.H., Seymour L.W., Logan A. et al. 2004. *Bioconjug. Chem.* 15: 152.
30. Keller M., Harbottle R.P., Perouzel E. et al. 2003. *Chembiochem* 4: 286.
31. Rejman J., Oberle V., Zuhorn I.S. et al. 2004. *J. Biochem.* 377: 159.
32. Chesnoy S., Durand D., Doucet J. et al. 2001. *Pharm. Res.* 18: 1480.
33. Merdan T., Callahan J., Petersen H. et al. 2003. *Bioconjug. Chem.* 14: 989.
34. Itaka K., Yamauchi K., Harada A. et al. 2003. *Biomaterials* 24: 4495.
35. Park Y.J., Nah S.H., Lee J.Y. et al. 2003. *J. Biomed. Mater. Res.* 67A: 751.
36. LaVan D.A., McGuire T., and Langer R. 2003. *Nat. Biotechnol.* 21: 1184.
37. Murthy N., Xu M., Schuck S. et al. 2003. *Proc. Natl Acad. Sci. USA* 100: 4995.
38. Kim J.K., Choi S.H., Kim C.O. et al. 2003. *J. Pharm. Pharmacol.* 55: 453.
39. Tang G.P., Zeng J.M., Gao S.J. et al. 2003. *Biomaterials* 24: 2351.
40. Na K., Bum L.T., Park K.H. et al. 2003. *Eur. J. Pharm. Sci.* 18: 165.
41. Quintana A., Raczka E., Piehler L. et al. 2002. *Pharm. Res.* 19: 1310.
42. Gu H., Ho P.L., Tsang K.W. et al. 2003. *J. Am. Chem. Soc.* 125: 15702.
43. O'Neal D.P., Hirsch L.R., Halas N.J. et al. 2004. *Cancer Lett.* 209: 171.
44. West J.L. and Halas N.J. 2003. *Annu. Rev. Biomed. Eng.* 5: 285.
45. Cui Z. and Mumper R.J. 2003. *Eur. J. Pharm. Biopharm.* 55: 11.
46. Kim S.Y., Lee Y.M., Baik D.J. et al. 2003. *Biomaterials* 24: 55.
47. Fonseca C., Simoes S., and Gaspar R. 2002. *J. Control Release* 83: 273.

48. Fernandez-Lopez S., Kim H.S., Choi E.C. et al. 2001. *Nature* 412: 452.

49. Lee M., Rentz J., Han S.O. et al. 2003. *Gene Ther.* 10: 585.

50. Anderson D.G., Lynn D.M., and Langer R. 2003. *Angew. Chem. Int. Ed. Engl.* 42: 3153.

51. Liu G., Li D., Pasumarthy M.K. et al. 2003. *J. Biol. Chem.* 278: 32578.

52. Ziady A.G., Gedeon C.R., Miller T. et al. 2003. *Mol. Ther.* 8: 936.

53. Yang F., Murugan R., Ramakrishna S. et al. 2004. *Biomaterials* 25: 1891.

54. Matsuda T. 2004. *Artif. Organs* 28: 64.

55. Ochi M., Adachi N., Nobuto H. et al. 2004. *Artif. Organs* 28: 28.

56. Sharma B. and Elisseeff J.H. 2004. *Ann. Biomed. Eng.* 32: 148.

57. Chen V.J. and Ma P.X. 2004. *Biomaterials* 25: 2065.

58. Prokop A. 2001. *Ann. NY Acad. Sci.* 944: 472.

59. Dalby M.J., Riehle M.O., Johnstone H.J. et al. 2003. *J. Biomed. Mater. Res.* 67A: 1025.

60. Woo K.M., Chen V.J., and Ma P.X. 2003. *J. Biomed. Mater. Res.* 67A: 531.

61. Ratner B.D. 2002. *J. Control Release* 78: 211.

62. Garrison M.D., McDevitt T.C., Luginbuhl R. et al. 2000. *Ultramicroscopy* 82: 193.

63. Ratner B.D. 1996. *J. Mol. Recognit.* 9: 617.

64. Hersel U., Dahmen C., and Kessler H. 2003. *Biomaterials* 24: 4385.

65. Martin S.M., Ganapathy R., Kim T.K. et al. 2003. *J. Biomed. Mater. Res.* 67A: 334.

66. Lee K.B., Park S.J., Mirkin C.A. et al. 2002. *Science* 295: 1702.

67. Blawas A.S. and Reichert W.M. 1998. *Biomaterials* 19: 595.

68. Lee K.B., Lim J.H., and Mirkin C.A. 2003. *J. Am. Chem. Soc.* 125: 5588.

69. Cheung C.L., Camarero J.A., Woods B.W. et al. 2003. *J. Am. Chem. Soc.* 125: 6848.

70. Hyun J., Kim J., Craig S.L. et al. 2004. *J. Am. Chem. Soc.* 126: 4770.

71. Shi H. and Ratner B.D. 2000. *J. Biomed. Mater. Res.* 49: 1.

72. Martin S.M., Schwartz J.L., Giachelli C.M. et al. 2004. *J. Biomed. Mater. Res.* 70A: 10.

73. Plant G.W., Woerly S., and Harvey A.R. 1997. *Exp. Neurol.* 143: 287.

74. Woerly S., Pinet E., de Robertis L. et al. 2001. *Biomaterials* 22: 1095.

75. Kantlehner M., Schaffner P., Finsinger D. et al. 2000. *Chembiochem* 1: 107.

76. Goodman S.L., Sims P.A., and Albrecht R.M. 1996. *Biomaterials* 17: 2087.

77. Miller D.C., Thapa A., Haberstroh K.M. et al. 2004. *Biomaterials* 25: 53.

78. Ma P.X. and Zhang R. 1999. *J. Biomed. Mater. Res.* 46: 60.

79. Boland E.D., Matthews J.A., Pawlowski K.J. et al. 2004. *Front Biosci.* 9: 1422.

80. Nagaoka S., Ashiba K., and Kawakami H. 2002. *Artif. Organs* 26: 670.

81. Thapa A., Miller D.C., Webster T.J. et al. 2003. *Biomaterials* 24: 2915.

82. Dalby M.J., Berry C.C., Riehle M.O. et al. 2004. *Exp. Cell Res.* 295: 387.

83. Lyuksyutov S.F., Vaia R.A., Paramonov P.B. et al. 2003. *Nat. Mater.* 2: 468.

84. Guo C., Feng L., Zhai J. et al. 2004. *Chemphyschem* 5: 750.

85. Price R.L., Haberstroh K.M., and Webster T.J. 2003. *Med. Biol. Eng. Comput.* 41: 372.

86. Li P. 2003. *J. Biomed. Mater. Res.* 66A: 79.

87. Flemming R.G., Murphy C.J., Abrams G.A. et al. 1999. *Biomaterials* 20: 573.

88. Wei G. and Ma P.X. 2004. *Biomaterials* 25: 4749.

89. Suzuki I., Sugio Y., Moriguchi H. et al. 2004. *J. Nanobiotechnol.* 2: 7.

90. Nagaoka S., Ashiba K., Okuyama Y. et al. 2003. *Int. J. Artif. Organs* 26: 339.

91. Sarikaya M., Tamerler C., Jen A.K. et al. 2003. *Nat. Mater.* 2: 577.

92. Kwok C.S., Mourad P.D., Crum L.A. et al. 2000. *Biomacromolecules* 1: 139.

93. Smith R.K., Lewis P.A., and Weiss P.S. 2004. *Prog. Surf. Sci.* 75: 1.

94. Jang C.H., Stevens B.D., Carlier P.R. et al. 2002. *J. Am. Chem. Soc.* 124: 12114.

95. Zhang S., Marini D.M., Hwang W. et al. 2002. *Curr. Opin. Chem. Biol.* 6: 865.

96. Zhang S. 2003. *Nat. Biotechnol.* 21: 1171.

97. Hartgerink J.D., Beniash E., and Stupp S.I. 2001. *Science* 294: 1684.

98. L'Heureux N., Paquet S., Labbe R. et al. 1998. *FASEB J.* 12: 47.

99. Remy-Zolghadri M., Laganiere J., Oligny J.F. et al. 2004. *J. Vasc. Surg.* 39: 613.

100. Saiki R.K., Bugawan T.L., Horn G.T. et al. 1986. *Nature* 324: 163.

101. Koo Y.E., Cao Y., Kopelman R. et al. 2004. *Anal. Chem.* 76: 2498.

102. Oku N. 1999. *Adv. Drug Deliv. Rev.* 37: 53.

103. Kao C.Y., Hoffman E.A., Beck K.C. et al. 2003. *Acad. Radiol.* 10: 475.

104. Kobayashi H., Kawamoto S., Sakai Y. et al. 2004. *J. Natl Cancer Inst.* 96: 703.

105. Tsuji A., Sato Y., Hirano M. et al. 2001. *Biophys. J.* 81: 501.
106. Zhao Y., Sadtler B., Lin M. et al. 2004. *Chem. Commun. (Camb)* 7: 784–785.
107. Schultz S., Smith D.R., Mock J.J. et al. 2000. *Proc. Natl Acad. Sci. USA* 97: 996.
108. Aylott J.W. 2003. *Analyst* 128: 309.
109. Abdalla M.A., Bayer J., Radler J. et al. 2003. *Nucleosides Nucleotides Nucl. Acids* 22: 1399.
110. Brown J.Q. and McShane M.J. 2003. *IEEE Eng. Med. Biol. Mag.* 22: 118.
111. Lu J., Rosenzweig Z., and Fresenius J. 2000. *Anal. Chem.* 366: 569.
112. Xu H., Aylott J.W., and Kopelman R. 2002. *Analyst* 127: 1471.
113. Chan W.C., Maxwell D.J., Gao X. et al. 2002. *Curr. Opin. Biotechnol.* 13: 40.
114. Wu X., Liu H., Liu J. et al. 2003. *Nat. Biotechnol.* 21: 41.
115. Kaul Z., Yaguchi T., Kaul S.C. et al. 2003. *Cell Res.* 13: 503.
116. Dubertret B., Skourides P., Norris D.J. et al. 2002. *Science* 298: 1759.
117. Sun B., Xie W., Yi G. et al. 2001. *J. Immunol. Meth.* 249: 85.
118. Akerman M.E., Chan W.C., Laakkonen P. et al. 2002. *Proc. Natl Acad. Sci. USA* 99: 12617.
119. Jaiswal J.K., Mattoussi H., Mauro J.M. et al. 2003. *Nat. Biotechnol.* 21: 47.
120. Larson D.R., Zipfel W.R., Williams R.M. et al. 2003. *Science* 300: 1434.
121. Han M., Gao X., Su J.Z. et al. 2001. *Nat. Biotechnol.* 19: 631.
122. Ballou B., Lagerholm B.C., Ernst L.A. et al. 2004. *Bioconjug. Chem.* 15: 79.
123. Weissleder R., Tung C.H., Mahmood U. et al. 1999. *Nat. Biotechnol.* 17: 375.
124. Gao X. and Nie S. 2003. *Trends Biotechnol.* 21: 371.
125. Uslu F., Ingebrandt S., Mayer D. et al. 2004. *Biosens. Bioelectron.* 19: 1723.
126. Docherty F.T., Clark M., McNay G. et al. 2004. *Faraday Discuss.* 126: 281.
127. Oldenburg S.J., Genick C.C., Clark K.A. et al. 2002. *Anal. Biochem.* 309: 109.
128. Maxwell D.J., Taylor J.R., and Nie S. 2002. *J. Am. Chem. Soc.* 124: 9606.
129. Zhao X., Tapec-Dytioco R., and Tan W. 2003. *J. Am. Chem. Soc.* 125: 11474.
130. Patolsky F., Lichtenstein A., and Willner I. 2001. *J. Am. Chem. Soc.* 123: 5194.
131. Liu T., Tang J., Han M. et al. 2003. *Biochem. Biophys. Res. Commun.* 304: 98.
132. Liu T., Tang J., and Jiang L. 2004. *Biochem. Biophys. Res. Commun.* 313: 3.
133. Walter G., Bussow K., Lueking A. et al. 2002. *Trends Mol. Med.* 8: 250.
134. Brynda E., Houska M., Brandenburg A. et al. 2002. *Biosens. Bioelectron.* 17: 665.
135. Deckert F. and Legay F. 2000. *J. Pharm. Biomed. Anal.* 23: 403.
136. Liss M., Petersen B., Wolf H. et al. 2002. *Anal. Chem.* 74: 4488.
137. Zhang J., Su X., and O'Shea S. 2002. *Biophys. Chem.* 99: 31.
138. Grubor N.M., Shinar R., Jankowiak R. et al. 2004. *Biosens. Bioelectron.* 19: 547.
139. Washburn M.P. 2003. *Nat. Biotechnol.* 21: 1156.
140. Bertucci F., Viens P., Hingamp P. et al. 2003. *Int. J. Cancer* 103: 565.
141. Hao X., Sun B., Hu L. et al. 2004. *Cancer* 100: 1110.
142. Boldrick J.C., Alizadeh A.A., Diehn M. et al. 2002. *Proc. Natl Acad. Sci. USA* 99: 972.
143. Nau G.J., Richmond J.F., Schlesinger A. et al. 2002. *Proc. Natl Acad. Sci. USA* 99: 1503.
144. Grody W.W., Cutting G.R., Klinger K.W. et al. 2001. *Genet. Med.* 3: 149.
145. Carrigan S.D., Scott G., and Tabrizian M. 2004. *Clin. Chem.* 50: 1301.
146. Shang S., Chen Z., and Yu X. 2001. *Acta. Paediatr.* 90: 179.
147. Tarkin I.S., Henry T.J., Fey P.I. et al. 2003. *Clin. Orthop.* 89.
148. Tirodker U.H., Nataro J.P., Smith S. et al. 2003. *J. Perinatol.* 23: 117.
149. Gibellini D., Vitone F., Schiavone P. et al. 2004. *J. Clin. Virol.* 29: 282.
150. Wang Y.F., Pang D.W., Zhang Z.L. et al. 2003. *J. Med. Virol.* 70: 205.
151. Ye Y.K., Zhao J.H., Yan F. et al. 2003. *Biosens. Bioelectron.* 18: 1501.
152. Albers J., Grunwald T., Nebling E. et al. 2003. *Anal. Bioanal. Chem.* 377: 521.
153. Smalley R.E. 2001. *Sci. Am.* 285: 76.
154. Janicijevic A., Ristic D., and Wyman C. 2003. *J. Microsc.* 212: 264.
155. Hess H., Bachand G.D., and Vogel V. 2004. *Chemistry* 10: 2110.
156. Zhuang X. and Rief M. 2003. *Curr. Opin. Struct. Biol.* 13: 88.
157. Murakoshi H., Iino R., Kobayashi T. et al. 2004. *Proc. Natl Acad. Sci. USA* 101: 7317.
158. Yoshino T., Sugiyama S., Hagiwara S. et al. 2003. *Ultramicroscopy* 97: 81.

25 Biomedical Nanoengineering for Nanomedicine

Jie Han
NASA Ames Research Center, Moffett Federal Airfield,
California, USA

CONTENTS

25.1 INTRODUCTION

Nanomedicine is the medical application of nanotechnology in prevention, diagnostics, and treatment of diseases. In this handbook, the term "biomedical nanoengineering" is used to address the engineering issues in the biomedical applications of nanomaterials and nanodevices. In October 2003, the National Institutes of Health (NIH) announced the NanoMedicine Initiative (NMI) [1]. The NMI envisions, for example, the biomedical nanodevices or nanosystems to search out and destroy the very first cancer cells of a tumor developing in the body, the biological nanomachines to remove and replace the cell's broken part, and the molecule-sized pumps to deliver life-saving medicines precisely where they are needed in the human body.

Nanomedicine was mentioned in many early publications. For example, in two books [2,3], the nanomachines were proposed to monitor and repair the damaged cells and the intracellular structures, the nanorobots equipped with wireless transmitters to circulate in the blood and lymph systems and send out warnings when chemical imbalances occur or worsen, and at the extreme, these nanosystems to replicate themselves or correct genetic deficiencies by altering or replacing DNA molecules.

These scenarios may have sounded unbelievable years ago and may sound so even now, but the rapid, tremendous progress in nanotechnology is promising the formation of the nanomedicine through development of the biomedical nanoengineering. For example, nanostructures such as functional nanoparticles, dendrimers, fullerenes, carbon nanotubes, and semiconductor nanocrystals including quantum dots have been exploited for drug delivery, diagnostics, and treatment of diseases at the molecular level; the assembled nanostructured fibrous scaffolds reminiscent of extracellular matrix have been used for mimic properties of bone; and protein nanotubes based on self-assembly of unique cyclic peptides for novel antibiotics. Although most work is still in the laboratory research, some has found applications. For example, nanoparticles have been used in commercial products including drug-delivery systems and point-of-care diagnostics.

Tremendous medical benefits for health care from nanotechnology have been repeatedly described in many publications and media reports. Continued nanotechnology research in biomedicine is bringing up as much challenge as opportunities. Nanomedicine is a multidisciplinary field that needs the integrated teamwork and mutual understanding from professionals and public in the areas of medicine, biology, chemistry, physics, materials science, engineering, health care, law, and government.

Biomedical nanoengineering is a very broad yet deep multidisciplinary field and cannot be fully covered in this chapter. This chapter is only to offer a basic understanding of this emerging field for the professionals and public with different backgrounds. It will mainly discuss biologically functional nanomaterials such as dendrimers, single-crystal nanoparticles and nanowires, and fullerenes and carbon nanotubes, and their biomedical applications mainly in the prevention, diagnostics, and treatment of diseases.

25.2   NANOMATERIALS AND NANODEVICES

The NIH defines nanotechnology as "the creation of functional materials, devices and systems through control of matters at the scale of 1 to 100 nanometers, and exploitation of novel properties and phenomena at the same scale." Nanomaterials can be simply defined to have three features: 1 to 100 nm in one dimension, functional in applications, and producible in manufacturing. They have to be biologically engineered for biomedical applications.

Living systems are built upon from molecular materials or nanostructures such as nucleic acids (DNA and RNA) and protein. They are 2 and 5 to 50 nm wide, respectively. They can be produced from the self-assembly or self-organization processes in the living system or by chemical synthesis. DNA or RNA and associated enzymes and proteins or lipids can be self-assembled into 75 to 100 nm wide viruses. They can be further assembled into bacteria. Bacteria are 1 to 10 μm in size, with thin, rubbery cell membrane surrounding the fluid (cytoplasm) and all genetic information needed to make copies of its own DNA. Viruses and bacteria cause many diseases. A white blood cell is about 10 μm big, whereas all materials internalized by cells are smaller than 100 nm.

The size domain of nanomaterials is similar to that of the biological structures, as shown in Figure 25.1. For example, a single-wall carbon nanotube is as wide as a double-strand DNA; dendrimers and nanoparticles can be made similar to the sizes of proteins or viruses; and fullerene may present the smallest molecular nanostructures. These nanostructures have significantly different properties from bulk or microstructures and they are especially suitable for biomedical applications. For example,

- Nanoscale single-crystal or ordered structures are stronger, lighter, and less corrosive, yet cause less damage to cells or tissues.
- High specific surface allows to load the recognition molecules and drugs, enter the cells, and seek out specific nucleic acids and proteins or other molecular marks.
- Quantum confinement at nanoscale makes them more electrically conductive, superparamagnetic, and tunable optical emission for control of drug delivery and sensing at intra- and extracellular levels through external light and magnetic or electric field.
- Electricity or heat generated by external light, magnetic field, or electric field can destroy sick cells locally while leaving neighboring healthy cells intact.

The nanomaterials shown in Figure 25.1 are commercially available now. The biological functionalization to load drugs or recognition molecules chemically or physically is the critical step in their biomedical applications. A brief introduction to these nanomaterials and their properties and biomedical application is presented in Figure 25.2.

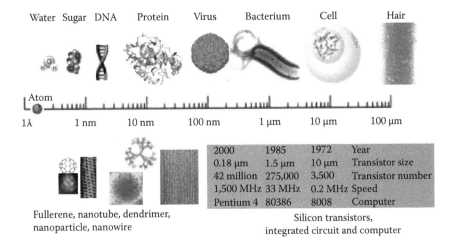

FIGURE 25.1   Nano- and microscale materials, devices, and systems.

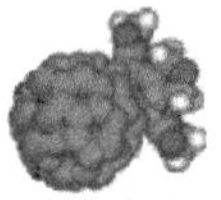	*Fullerenes*, C_{60}, 0.7 nm, functional drug carrier with linked antibodies or other targeting agents on the surface carbon atoms, and implanted medical devices.
	*Dendrimers* (5–50 nm), branched structure allows to link labels, probes, and drugs individually for drug carrier, implanted sensors, and medical devices.
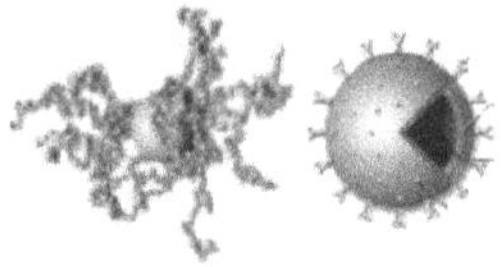	*Nanoparticles* (<100 nm, inorganic or organic) for implanted materials, nanoshells, and nanoemulsions for drug delivery; quantum dots (<8 nm) and magnetic nanoparticles for labeling in diagnostics and implanted sensors and medical devices.
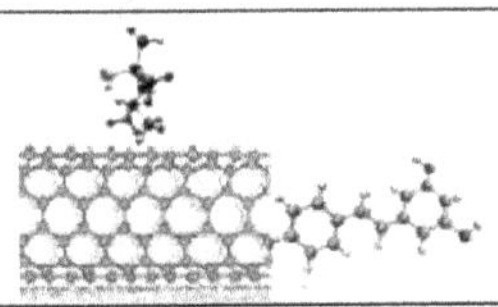	*Carbon nanotubes* (1 nm for single wall and 10–100 nm for multi-wall), one dimensional fullerene nanoelectrode arrays for *in vitro, in vivo,* and implanted sensors and medical devices; capable for diagnostic handheld systems for multiplexing without need of labeling and PCR.
	Single crystal nanowires (5–100 nm), one-dimensional nanoparticles (magnetic, electrical, and optical), capable of doing what nanoparticles and carbon nanotubes can do.

FIGURE 25.2   Biologically functional nanomaterials and biomedical applications.

25.2.1   Fullerenes and Carbon Nanotubes

Fullerenes (C_{60} and C_{70}) were discovered in 1985 by Smalley, Curl, and Kroto, which won the 1997 Nobel Prize in Chemistry [4]. Fullerenes are roughly spherical in shape and approximately 1 nm big. A carbon atom sits at each vertex of a buckyball, bonding with three of its neighbors. The strained sp2 configuration allows them to be chemically or biologically modified with small or large molecules for the biomedical applications. Fullerenes are transparent over a wide spectral range extending from the mid-infrared throughout the visible. They possess a high thermal and oxidative stability compared to many other organic materials, and they are extremely resilient and relatively impervious to damage. They do not react with corrosive compounds and are capable of absorbing and releasing electrons without being harmed or without changing. Fullerenes are mainly used for drug delivery and implanted sensing and treatment devices. They allow active pharmacopheres to be grafted to its surface in three-dimensional orientations for precise control in matching biological targets, in entrapping atoms within the fullerene cage, and for attaching fullerene derivatives to targeting agents.

Carbon nanotubes (CNTs) are one-dimensional fullerenes with a cylinder shape, discovered by Sumio Iijima in 1992 [4]. These cylinders can be closed or open in the ends, have single or multiple walls, can be metallic or semiconducting, and can be tens of nanometers to tens of micrometers long. They are about 1 nm wide for single-wall structures and 5 to 100 nm wide for multiwall structures. While retaining the properties of the fullerenes, the one-dimensional extension and the quantum confinement in the circumference make nanotubes intrinsically highly electrical and thermal conductive, mechanically strong, but very gentle to biological structures. Therefore, carbon nanotubes can be made stand-alone functional devices such as probe tips for biological images and chemical force detection, and vertically aligned nanoelectrode arrays for diagnostics and implanted medical sensing and treatment devices. The chemical functionalization is relatively easier in the open end than in the sidewall or closed ends.

25.2.2 DENDRIMER

Dendrimer, discovered by Don Tomalia in 1992 [5], is precisely constructed molecules built on the nanoscale in a multistep process through up to ten generations (5 to 50 nm). Each step doubles the complexity at the branching end. Drugs and recognition molecules can be attached to their ends or placed inside cavities within them. Dendrimers are versatile, with discrete numbers and high local densities of surface functionalities in one molecule, very attractive for biomedical applications, especially in cancer therapy. The dendritic multifunctional platform is ideal to combine various functions like imaging, targeting, and drug transfer into the cell.

25.2.2.1 Nanoparticles and Nanowires

Nanoparticles might be the earliest nanomaterials [6,7]. Their research and applications started at least two decade ago. They can be made of almost all known materials, whereas the biomedical applications have been mainly highlighted for organic and several types of inorganic nanoparticles:

- Polymer nanoparticles have been widely used for drug delivery and implantable materials. They can be made into nanoshells for drug encapsulation, and hydrophilic and hydrophobic for the expected biocompatibility. Their diameter ranges from 10 to 100 nm. The first nanoparticles were reported in 1976 with protein molecules entrapped inside 80 nm hydrophilic polymers.
- Magnetic nanoparticles can be made from most of the bulk magnetic materials such as cobalt, iron, and nickel, and their alloys, ferrite, nitride, or oxide. The superparamagnetic properties of smaller nanoparticles (<10 nm) can provide control for drug delivery, implanted sensing, and heating treatment to destroy sick cells through an external electromagnetic field.
- Quantum dot nanoparticles can be made of the semiconductors and metals when the particle size is less than 8 nm. They are capable of confining a single electron, or a few, and in which the electrons occupy discrete energy states just as they would in an atom (quantum dots have been called artificial atoms). These particles show optical gain and stimulated emission at room temperature. They are suitable for biological markers, drug delivery, and implanted sensing and heating devices through external lighting.

25.2.2.1.1 Single-Crystal Nanowires

Single-crystal nanowires (SNWs) are one-dimensional single-crystal nanoparticles like fullerenes versus carbon nanotubes. Although retaining the properties of nanoparticles, SNWs have been made into functional devices such as transistors, nanoelectrode arrays, and probes for biological sensing. Depending on the material type and diameter, SNWs and devices can be made electrically, optically, or magnetically functional.

Nanomaterials and nanodevices can be fabricated basically using two approaches. The bottom-up approach builds up devices or systems from atoms or molecules through chemical synthesis, self-assembly, or self-organization processes. This has been a natural way in life science to build up viruses, bacteria, cells, and living systems from molecular materials such as DNA and proteins. The top-down approach works from bulky materials through machining, etching, and lithography processes. There has been a drive in the semiconductor industry to make faster and more powerful computers by scaling down silicon transistor devices from 10 μm in 1972 to about 0.1 μm or 90 nm in 2004, as shown in Figure 25.1. In this chapter, we mainly discuss the bottom-up nanomaterials and nanodevices.

The scaling in top-down approaches is reaching the limit (or 100 nm barriers) beyond which the previous working principle and optical lithography may no longer work or may work at an extremely high cost. On the other hand, 100 nm is also a scale boundary in biotechnology and medicine between cellular or larger scale and molecular scale (DNA and protein level). Thus, we can understand why nanotechnology defines the scale below 100 nm.

The impact of the semiconductor technology is felt far beyond laptop and desktop computers, accounting for personal electronics, telecommunications, medical devices, automotive applications, and almost every aspect of our daily lives, all reaping healthy benefits from the increasing power of semiconductor chips. It has enabled the micro-electrical-mechanical-systems (MEMS), micro-fluidics, DNA microarrays or gene chips, protein chips, and all kinds of micromedical devices. These technologies will be further brought down to nanoscale by using the bottom-up biologically functional nanomaterials.

Nanomaterials offer many unique, novel features in biomedical applications, which other scale materials and technologies may not reach. For example,

> *Intracellular delivery capability*: The size of virus, bacterium, and cell is approximately 100 nm, 1 and 10 μm or larger, respectively (which, interestingly, corresponds to the feature size of transistors in 2003, 1985, and 1972) as shown in Figure 25.1. The biologically functional nanostructures with the size of less than 100 nm can enter cells and the organelles inside them to interact with DNA and proteins or stick on the surface of specific cells or organelles. This enables the earliest prevention and treatment of diseases. For example, detection of cancer at early stages is a critical step in improving cancer treatment. Currently, detection and diagnosis of cancer usually depend on changes in cells and tissues that are detected by a doctor's physical touch or imaging expertise. In many cases, it was too late to treat when it was diagnosed at microscale cellular or larger image scale level. The best way is to detect the earliest molecular changes, long before a physical exam or imaging technology is effective or even a tumor is formed.
>
> *PCR- and label-free detection capability*: Nanomaterials can readily be chemically attached with specific probes such as cDNA and antibody, which can seek out specific target DNA, RNA, or proteins for diagnostics and treatment. For example, a specific tag can be attached onto the nanotube ends to look for the specific mutations responsible for the diseases of interest. Nanoparticles or dendrimers can be attached with cDNA for gene expression analysis. Conventionally, gene expression analysis or mutation detection has to be carried out after or during the tedious sample amplification and optical labeling. However, the utilization of nanodevices as biosensor can eliminate these tedious laboratory steps for the sample preparations. For example, the carbon nanotube electrodes attached with DNA probes have been developed for label- and PCR-free detection for DNA samples.
>
> *Drug- and surgery-free intracellular treatment capability*: The detection or diagnosis is carried out with optical, magnetic, and electrical response of the nanodevices to the interaction

or binding. In addition, the treatment can be carried out using the magnetic, optical, or electrical properties of nanostructures. For example, heat generated by the light-absorbing, electrical thermal emission, and magnetic thermal generation can kill tumor cells while leaving neighboring cells intact.

In the following section, we will briefly introduce the biomedical applications associated with these unique features.

25.3 BIOMEDICAL APPLICATIONS

Biomedical applications of nanotechnology or nanomedicine have been reviewed by the nanotechnology alliance in Canada [8] and classified into the following category, as summarized in Table 25.1. In this chapter, we will only briefly introduce some of the applications listed in Table 25.1, which we believe will illustrate the unique features of nanomaterials and nanodevices in biomedical applications that conventional technologies may not reach. In addition, we will discuss these applications based on the prevention, diagnostics, and treatment of diseases. Nanomedicine and biomedical nanoengineering are only emerging, and much information has been appearing only at conferences, and not much has come from referred journals. Therefore, the references in this chapter are mainly from the Internet and Reference [8].

TABLE 25.1

Nanomedicine Taxonomy

Biopharmaceutics

Drug delivery
 Drug encapsulation
 Functional drug carriers
Drug discovery

Implantable Materials

Tissue repair and replacement
 Implant coatings
 Tissue regeneration scaffolds

Structural Implant Materials

Bone repair
Bioresorbable materials
Smart materials

Implantable Devices

Assessment and treatment devices
 Implantable sensors
 Implantable medical devices
Sensory aids
 Retina implants
 Cochlear implants

Surgical Aids

Operating tools
 Smart instruments
 Surgical robots

Diagnostic Tools

Genetic testing
 Ultrasensitive labeling and detection technologies
 High-throughput arrays and multiple analyses
Imaging
 Nanoparticle labels
 Imaging devices

25.3.1 Prevention

The best health care or medicine is preventive medicine. Common diseases to many people include diverticulitis, kidney failure, dialysis, gallstones, diabetes, osteoporosis, hypertension, coronary artery disease, stroke, aging, and numerous cancers. Virtually all our major diseases are associated with the consumption of (1) smoke and pollutants through the respiratory system; (2) alcohol, colas, and caffeine; and (3) an abundance of fats, animal proteins, and sweets, and the neglect of exercise. In addition to air cleaning, exercise, and vegetarian program, nutritional approaches for personal care including skincare will be the best for multiple disease prevention. Although it has not been greatly addressed in publications, the author believes that nanomaterials are promising the most effective preventive medicines. This is because of the intracellular delivery features of the nanomaterials.

Infection control is very important in the prevention of diseases. Conventional disinfectants for infection control have not been safe enough in applications. Nanoengineered delivery system technology is now offering a solution. For example, scientists at the University of Michigan and EnviroSystems, Inc., located in northern California have independently demonstrated safe and effective use of nanoemulsions as antimicrobial solutions. Nanoemulsions, which are suspensions of nanoparticles in water, have the emergent property of killing bacteria and inactivating enveloped viruses. EnviroSystems developed a nanoemulsion as a targeted delivery device for the biocide PCMX. This PCMX-loaded nanoemulsion has a broad spectrum of activity against bacteria including tuberculosis (TB), both enveloped and nonenveloped viruses, and fungi. Because the nanoemulsion is targeted, it has no toxicity to higher animal cells. Other respective benefits of the new nanoemulsion disinfectants are contributions to institutional productivity. For example, because nanoemulsion disinfectants are not considered hazardous materials, they do not require any special compliance with the Occupational Safety and Health Administration (OSHA), the Environmental Protection Agency (EPA), or local water systems. Many of the widely used disinfectants have been implicated in occupational dermatitis and respiratory illness. A reduction in the hazardous waste stream also contributes to a healthier local environment.

Another example of application in the marketplace is skin-care products using nanoparticles [11]. Nanoparticles encapsulating different agents of cosmetic and pharmaceutical interest have been developed for novel skin-care applications. The nanoparticles were found to show unique additional physical properties and offer new application possibilities that conventional technology cannot reach. The nanoparticles were also found to be very stable and have a high affinity to the stratum corneum. In addition, nanoparticle delivery systems have been developed to target the vesicles to hair and for that purpose they have dotted the nanoparticle shell with cationic molecules, thus producing a positively charged surface.

Nanoparticles in skin-care products include various types of delivery systems and can be subdivided on the basis of the encapsulating membrane structure into liposomes, nanoemulsions, nanosomes, and nanotopes. They can carry many actives to penetrate into skin quickly and into intracellular structures while conventional skin-care products usually do not penetrate the skin and release the active by diffusion or by capsule destruction. Nanoparticles also bring up many other new applications. For example, skin whitening or lightening is a more recent application in which actives carried by nanoparticles penetrate beyond the skin barrier, and more active reaches the necessary site of action in the skin, resulting in improved performance.

In addition to novel drug-delivery systems, biomedical nanoengineering also offers new approaches for the earlier detection of diseases or pathogens. This is also very important for the purpose of screening or diagnostics testing in preventive medicine as well as treatment, as introduced below.

25.3.2 DIAGNOSTICS

Nanoparticles, especially gold nanoparticles, have been used for diagnostics applications. For example, Quantum Dot Corporation uses quantum dots to detect biological material [11]. Because their color can be tailored by changing the size of the dot, the potential for multiple colors increases the number of biological molecules that can be tracked simultaneously. In addition, quantum dots do not fade when exposed to ultraviolet light, and the stability of their fluorescence allows longer periods of observation. These technologies are expected to be more sensitive than fluorescent dyes and could more effectively detect low abundance and low-level expressioning genes. They may also make use of smaller and less-expensive equipment to light and detect the samples. Without the need for gene amplification, they can also provide results in less time.

In addition to optical detection approaches, new effort has been made in electrical detection. Chad Mirkin and Nanosphere Company [12] use gold nanoparticle probes coated with a string of nucleotides that complement one end of a target sequence in the sample. Another set of nucleotides, complementing the other end, is attached to a surface between two electrodes. If the target sequence is present, it anchors the nanoprobes to the surface like little balloons, and when treated with a silver solution, they create a bridge between the electrodes and produce a current. The nanotechnology group at NASA Ames Research Center has developed carbon nanotube electrodes with attached DNA probes for DNA diagnostics applications [13]. They claim their technology can detect 1000 DNA molecules and therefore does not need the sample amplification methods such as PCR. In addition, they use nanoelectrodes for guanine oxidation in the DNA sample and therefore do not need any sample labeling.

Nanomaterials also offer new biomedical imaging that provides high quality not possible with current technologies, along with new methods of treatment. Researchers at the University of Michigan [14] are developing magnetic nanoparticles attached to a cancer antibody and a dye that is highly visible on a magnetic resonance image (MRI). When these nanoparticles latch onto cancer cells, the cancer cells will be detected on the MRI and then destroyed by laser or low-dosage killing agents that attack only the diseased cells. Another group at Washington University [15] is using nanoparticles to attract to proteins emitted from newly forming capillaries that deliver blood to solid tumors. The nanoparticles circulate through the bloodstream and attach to blood vessels, containing their complementary protein. Once attached, chemotherapy is released into the capillary membrane. The nanoparticles traveling in the bloodstream would be able to locate additional cancer sites that may have spread to other parts of the body.

Miniature wireless nanodevices are being developed for providing high-quality images not possible with traditional devices. Given Imaging has developed a pill containing a miniature video system. When the pill is swallowed, it moves through the digestive system and takes pictures every few seconds. The entire digestive system can be assessed for tumors, bleeding, and diseases in areas not accessible with colonoscopies and endoscopies. A company, MediRad, is trying to develop a miniature x-ray device that can be inserted into the body. They are attempting to make carbon nanotubes into a needle-shaped cathode. The cathode would generate electron emissions to create extremely small x-ray doses directly at a target area without damaging surrounding healthy tissue.

Implantable or wearable nanosensors are being developed for providing continuous and extremely accurate medical information, incorporated with complementary microprocessors to diagnose disease, transmit information, and administer treatment automatically if required. For example, researchers at Texas A&M and the Pennsylvania State University use polyethylene glycol beads coated with fluorescent molecules to monitor diabetes blood sugar levels. The beads are injected under the skin and stay in the interstitial fluid. When glucose in the interstitial fluid drops to dangerous levels, glucose displaces the fluorescent molecules and creates a glow. This glow is seen on a tattoo placed on the arm.

Researchers at the University of Michigan [16] are using dendrimers attached with fluorescent tags to sense premalignant and cancerous changes inside living cells. The dendrimers are administered transdermally and pass through membranes into white blood cells to detect early signs of biochemical changes from radiation or infection. Radiation changes the flow of calcium ions within the white blood cells and eventually triggers apoptosis, or programmed cell death (PCD) due to the radiation or infection. The fluorescent tags attached to the dendrimers will glow in the presence of the death cells when passed through a retinal scanning device using a laser capable of detecting the fluorescence.

25.3.3 Treatment

Biomedical nanoengineering is showing great potentials in drug delivery nanosystems and biomedical nanodevices for safe and effective treatment.

For example, Advectus Life Sciences is developing a nanoparticle-based drug-delivery system for the treatment of brain tumors. The antitumor drug doxorubicin is adhered to a poly butylcyano acrylate (PBCA) nanoparticle and coated with polysorbate 80. The drug is injected intravenously and circulates through the bloodstream. The polysorbate 80 attracts plasma apolipoproteins and is used by the bloodstream to carry lipids. This is to create a camouflage effect similar to low-density lipoprotein (LDL) cholesterol, allowing the drug to pass through the blood–brain barrier. Neurotech

Company is developing a nanoencapsulated cell therapy to treat eye diseases. It uses a semiperme-able membrane to encapsulate cells, which also permits therapeutic agents produced by the cells to diffuse through the membrane. The membrane isolates the cells from the local environment and minimizes immune rejection. The encapsulated cells are administered by a device implanted in the eye to permit the continuous release of therapeutic molecules from living cells. This avoids direct injections into the eye, which may not be practical for regular administrations.

C Sixty [17] is developing fullerene-based drug-delivery platforms that link fullerenes with antibodies and other targeting agents. Their drug-delivery systems include fullerene-decorated che-motherapeutic constructs, fullerene-radiopharmaceuticals, and fullerene-based liposome systems for the delivery of single drug loads or multiple drug cocktails. Employing rational drug design, C Sixty has produced several drug candidates using its fullerene platform technology in the areas of HIV/AIDS, neurodegenerative disorders, and cancer. Nanospectra Company is developing an interesting drug-delivery nanosystem in which a gold exterior layer covers interior layers of silica and drugs. This nanoshell structure can be made to absorb light energy and then convert it to heat. As a result, when nanoshells are placed next to a target area such as a tumor cell, they can release tumor-specific antibodies when infrared light is administered.

Perhaps the most exciting biomedical nanoengineering research is implantable nanodevices that can integrate sensing, monitoring, and treatment functionalities. For example, researchers at Aalborg University in Denmark are applying nanostructures to the electrode surfaces to improve biocompatibility and acceptance in the neural/muscle tissue. When placed within a cell membrane, the nanostructures form a bioelectric interface with the neuron or muscle cell that enables the intra-cellular potential of the cell to be observed and manipulated. They are further using nanostructures to activate denervated muscles caused by injuries to the lower motor neurons located in the spinal cord. This can result from traumatic spinal cord injury, strokes in the spinal cord, repeated verte-bral subluxation, brachial plexus injuries, and peripheral myopathies such as polio, which destroy the nerve cells controlling muscle (i.e., denervated muscle). Similar nanodevices can be also used to restore lost vision and hearing functions. The devices collect and transform data into precise electrical signals that are delivered directly to the human nervous system. Degenerative diseases of the retina, such as retinitis pigmentosa or age-related macular degeneration, decrease night vision and can progress to diminishing peripheral vision and blindness. These retinal diseases may lead to blindness due to a progressive loss of photoreceptors (rods and cones), the light-sensitive cells of the eye.

Retinal implants are being developed to restore vision by electrically stimulating functional neurons in the retina. One approach being developed by various groups including a project at Argonne National Laboratory is an artificial retina implanted in the back of the retina. The artifi-cial retina uses a miniature video camera attached to a blind person's eyeglasses to capture visual signals. The signals are processed by a microcomputer worn on the belt and transmitted to an array of electrodes placed in the eye. The array stimulates optical nerves, which then carry a signal to the brain. Optobionics Company makes use of a subretinal implant designed to replace photoreceptors in the retina. The visual system is activated when the membrane potential of overlying neurons is altered by current generated by the implant in response to light stimulation. The implant makes use of a microelectrode array powered by as many as 3500 microscopic solar cells.

A new generation of smaller and more powerful cochlear implants is intended to be more pre-cise and offer greater sound quality. An implanted transducer is pressure-fitted onto the incus bone in the inner ear. The transducer causes the bones to vibrate and move the fluid in the inner ear, which stimulates the auditory nerve. An array at the tip of the device makes use of up to 128 elec-trodes, which is five times higher than current devices. The higher number of electrodes provides more precision about where and how nerve fibers are stimulated. This can simulate a fuller range of sounds. The implant is connected to a small microprocessor and a microphone, which are built into a wearable device that clips behind the ear. This captures and translates sounds into electric pulses which are sent down a connecting wire through a tiny hole made in the middle ear. Implant electrodes are continuously decreasing in size and in time could enter the nanoscale. The nanotechnology group at NASA Ames Research Center is developing the carbon nanotube electrode arrays for this purpose [13].

25.4 CONCLUSION

Biomedical nanoengineering is an offshoot of biomedical engineering at nanoscale. It will be further defined, as the engineering issues at nanoscale in a biological system are being addressed during the application, research, and practice of nanotechnology to biomedicine. This chapter is intended for researchers to pay attention to many engineering issues in nanomedicine research.

For example, many engineering issues have not been addressed. They include fluid mechanics, diffusions, interactions, and self-assembling of nanomaterials and nanodevices in inter- and intracellular structures. Others to be studied include the characterization for the exact quantities and variations, location, timescales, interactions, affinities, force generation, flexibility, and internal motion of the nanomaterials and nanodevices. These engineering studies will help identify and define the design principles and operational parameters of nanodevices.

Biocompatibility is another important issue to be addressed. The electrically, magnetically functional carbon nanotubes, inorganic nanoparticles, and nanowires present the most existing nanomaterials, but may suffer from poor biocompatibility. If so, can these materials be modified while retaining the expected properties or new nanomaterials be developed to be used *in vivo*?

It surely takes great effort and time to answer these and many other questions for nanomedicine practice. NIH Roadmap's Nanomedicine initiative anticipates nanomedicine will yield medical benefits as early as 10 years from now.

REFERENCES

1. "NIH Nanomedicine Overview," http://nihroadmap.nih.gov/nanomedicine/index.asp.
2. *"Unbounding the Future—Nanotechnology Revolution,"* K.E. Drexler, C. Peterson, and G. Permagit, William Morrow and Company, New York, 1991.
3. *"Nanomedicine,"* R.A. Freitas Jr., Landes Bioscience, Georgetown, TX, 1999.
4. *"The Science of Fullerenes and Carbon Nanotubes,"* M.S. Dresselhaus, G. Dresselhaus, and P. Eklund (Eds.), Academic Press, San Diego, 1996.
5. *"Dendrimers III: Design, Dimension, Function Series : Topics in Current Chemistry,"* Vol. 212, Fritz Vögtle (Ed.), Springer, New York, 2001.
6. *"Nanoparticles: From Theory to Practice"* Gunter Smidt, John Wiley & Sons, New York, 2004.
7. *"Nanowires and Nanobelts: Materials, Properties and Devices,"* Zhong-lin Wang (Ed.), Springer, New York, 2004.
8. *"Nanomedicine Taxonomy,"* www.regenerativemedicine.ca/nanomed/Nanomedicine%20Taxonomy%20(Feb%202003).PDF.
9. *"Nanotechnology Enters Infection Control,"* www.infectioncontroltoday.com/articles/391feat3.html.
10. *"Preparation and Properties of Small Nanoparticles for Skin and Hair Care,"* F. Zuelli and F. Suter of Mibelle AG— Biochemistry, *SOEFW J.* 123, 1997, 880–885.
11. *"Quantum Dots,"* www.qdots.com/new/corporate/team.html
12. Chad Mirkin Northwestern, www.chem.nwu.edu/~mkngrp/pictmenu.html.
13. NASA Ames Center for Nanotechnology, www.ipt.arc.nasa.gov.
14. Martin Philbert, University of Michigan, www.sph.umich.edu/faculty/philbert.html.
15. Buddy Ratner, University of Washington, www.avs.org/dls/ratner.html.
16. James Baker, University of Michigan, http://nano.med.umich.edu/Personnel.html#Baker.
17. Uri Sagman, C Sixty, www.asapsites.com/asap/db/tss/home.nsf/pages/s14-sagman.

26 Physiogenomics: Integrating Systems Engineering and Nanotechnology for Personalized Medicine

Gualberto Ruaño, Andreas Windemuth,
and Theodore R. Holford
Yale University School of Medicine, New Haven, Connecticut, USA

CONTENTS

26.1 PHYSIOGENOMICS AND NANOTECHNOLOGY

26.1.1 INTRODUCTION

A revolution in our understanding of human health and disease has been launched by the sequencing of a prototypical human genome. To a large degree, this achievement represents the pinnacle of reductionist scientific thought, as having all genes dissected, one could in principle allow reconstitution of the organism. In contrast, the classical discipline of physiology has been dealing with systems from its very outset. Although clinically extraordinarily relevant, physiology remained an engineering embodiment of scientific thought distant from the molecular basis of function. Physiogenomics bridges the gap between the systems approach and the reductionist approach by using human variability in physiological process, either in health or disease, to drive their understanding at the genome level. Physiogenomics is particularly relevant to the phenotypes of complex diseases and the clustering of phenotypes into domains according to measurement technique, ranging from functional imaging and clinical scales to protein serology and gene expression.

Nanotechnology probes can serve as ultrasensitive reporters of dynamic cellular phenomena and protein interactions, allowing precise physiological phenotypes to be coupled to genomics, the new discipline of physiogenomics. Nanotechnologies allow delivery of nutrients, supplements, and growth factors in localized and compartmentalized cellular environments. Nanotechnology allows veering into cellular function with materials and constructs of the same scale as biological organelles and even macromolecules. With the level of specificity and individualization achievable with nanotechnology, the emerging paradigm is "genomically guided nanotechnology." The coupling of physiogenomics and nanotechnology provides key underpinnings to personalized health preservation as well as disease management and treatment [9].

Although single gene effects are the basis of "genetic diseases," partial penetrance is the rule in common clinical care. The pathways of physiology in personalized medicine are multigenetic, as they rely on networks of genes and not on single receptors and enzymes. With the advent of gene nanotechnology-based arrays, parallel processing of gene variability is practically possible at the level of physiological systems.

The concept of "average response" and "deviation from the mean" are ingrained in biomedical science. Learning from variability in response, and translating that into predictive diagnostics for personalized medicine is the challenge confronting physiogenomics and nanotechnology. Positioning each individual along a continuum of response to environmental inputs is the goal of array technology. There is a major need to couple the engineering advances in highly parallel genomic screens with statistical tools to derive valid information from pattern-recognizing algorithms. The practical consequence is that by learning from variability, and not depending on means and standard deviations, we can expect reduced sample sizes in clinical studies, and most importantly the ability to discover the markers and implement them in practice for prototyping and clinical validation. In this chapter, we introduce physiogenomics theory and application through nanotechnology capabilities for the understanding of disease etiology and treatment and for the advancement of personalized medicine.

26.1.2 FUNDAMENTALS OF PHYSIOGENOMICS

Physiogenomics utilizes an integrated approach composed of genotypes and phenotypes and a population approach deriving signals from functional variability among individuals. In physiogenomics, genotype markers of gene variation or "alleles" (single nucleotide polymorphisms [SNPs], haplotypes, insertions/deletions) are analyzed to discover statistical associations to physiological characteristics (phenotypes). The phenotypes are measured in populations of individuals either at baseline or after they have been similarly exposed or challenged to environmental triggers. These environmental interactions span the gamut from exercise and diet to drugs and toxins, and from extremes of temperature, pressure, and altitude to radiation. In the case of complex diseases, we are likely to find both baseline characteristics and response phenotypes to as yet undetermined environmental triggers. Variability in a genomic marker among individuals that tracks with the variability in physiological characteristics establishes associations and mechanistic links with specific genes (Figure 26.1).

Physiogenomics integrates the engineering systems approach with molecular probes stemming from genomic markers available from nanotechnologies that have altered the face of life sciences research. Physiogenomics is a biomedical application of sensitivity analysis, an engineering discipline concerned with how variation in the input of a system leads to changes in output quantities [8].

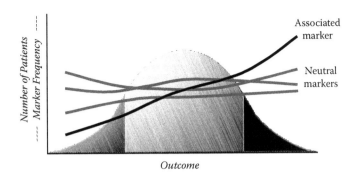

FIGURE 26.1   Physiogenomic analysis of gene marker frequency as a function of phenotype variability. An associated gene marker is one whose frequency in the cohort tracks with the variability in physiological responses. A neutral marker is one whose frequency is unchanged along the spectrum of response in the cohort.

Physiogenomics marks the entry of genomics into systems biology, and requires novel analytical platforms to integrate the data and derive the most robust associations. Once physiological systems are under scrutiny, the industrial tools of high-throughput genomics do not suffice, as fundamental processes such as signal amplification, functional reserve, and feedback loops of homeostasis must be incorporated. Physiogenomics includes marker discovery and model building. We will describe each of these interrelated components in a generalized fashion.

26.1.3   PHYSIOTYPE MODELS

We term the diagnostic models derived from physiogenomic diagnostics as "physiotypes." Physiotypes have several unique features. Physiotypes are predictive models incorporating genotypes from various genes and any covariates (e.g., baseline levels). Physiotypes are thus multigenetic in nature, and also include clinical information routinely gathered in medical care. Physiotypes harness the combined power of genotypes ("nature") and phenotypes ("nurture") to predict drug responses and the responses to other environmental challenges. Physiotypes are multimodular, and each individual module is derived from whether a significant association is found by univariate testing of the respective end point. The overall operational features of physiotypes are specificity and sensitivity each of 80% or more. Even though each component has individual characteristics, physiotypes reflect combined features of the various modules. Physiotypes provide answers to clinical management questions with high reliability and impact and can be used to address issues such as the risk of side effects from a medication.

Various specific genetic features of physiotypes are attractive for studying environmental interactions in prevention and treatment of disease. The genotype component does not change and is not confounded with environment. Some genotypes associated with a phenotype can become surrogate markers for the actual measurement of the phenotype. This capability may be particularly useful when measurement of the phenotype is difficult, expensive, or confounded by environmental conditions. Most importantly, genotyping technologies are rapidly decreasing in cost and are becoming increasingly automated. To this end, multiple genotypes from different genes coding for proteins in interacting pathways allow sampling the genetic variability in entire physiological networks quite economically.

Physiogenomics requires the highly multiplexed parallel processing capabilities available from nanotechnology. At Genomas we have employed automated, high-throughput genotype analysis with state-of-the-art fiber optic systems [3]. The genotyping analysis has an initial capacity of 300,000 SNPs per day and a multiplexing capability of 1500 SNPs from each genomic DNA sample. Our approach is to analyze each gene for variation at the SNP and haplotype level [11]. SNPs are available in the public domain from the National Center for Biotechnology Information (www.ncbi.nlm.nih.gov/SNP), and most have been confirmed for physical location in the chromosome locus of the gene, and validated as highly heterozygous in various populations by the International Haplotype Map Consortium [6]. Genomas also has access to Illumina's "SNP Knowledge Resource," a large, rapidly expanding SNP database of currently more than 1,000,000 high-confidence, mapped, and annotated SNP markers.

26.2 PHYSIOGENOMIC MARKER DISCOVERY

26.2.1 ASSOCIATION SCREENING

The purpose in association screening is to identify any of a large set of genetic markers (SNPs, haplotypes) and physiological characteristics that have an influence on the disease status of the patient, or the progression to disease. A single association test will proceed initially according to accepted multivariate methods in epidemiology [5]. The objective of the statistical analysis is to find a set of physiogenomic factors that together provide a way of predicting the outcome of interest. The association of an individual factor with the outcome may not have sufficient discrimination ability to provide the necessary sensitivity and specificity, but by combining the effect of several such factors, this objective may be achieved.

The purpose of the analysis at first is to identify significant covariates among demographic data and the other phenotypes and delineate correlated phenotypes by principal component analysis. Covariates are determined by generating a covariance matrix for all markers and selecting each significantly correlated marker for use as a covariate in the association test of each marker. Serological markers and baseline outcomes are tested using linear regression.

Next in the analysis is performing unadjusted association tests between genotypes and linear regression for serum levels and baselines. Tests will be performed on each marker, and markers that clear a significance threshold of $p < 0.05$ are selected for permutation testing.

26.2.2 PHYSIOGENOMIC CONTROL AND NEGATIVE RESULTS

Each gene not associated with a particular outcome effectively serves as a negative control and demonstrates neutral segregation of nonrelated markers. The negative controls altogether constitute a "genomic control" for the positive associations, where segregation of alleles tracks segregation of outcomes by requiring the representation of the least common allele for each gene to be at least 5% of the population, and then one can ascertain associations clearly driven by statistical outliers. Negative results are particularly useful in physiogenomics because one can still gain mechanistic understanding of complex systems from those, especially for sorting out the influences of the various candidate genes among the various phenotypes.

26.3 PHYSIOGENOMIC MODELING

26.3.1 MODEL BUILDING

Once the associated markers have been determined, a model is built for the dependence of response on the markers. In phase I, linear regression models will be used of the following form:

$$R = R_0 + \sum_i \alpha_i M_i + \sum_j \beta_j D_j + \varepsilon$$

where R is the respective phenotype variable, M_i are the marker variables, D_j are demographic covariates, and ε is the residual unexplained variation. The model parameters that are to be estimated from the data are R_0, α_i, and β_j.

The association between each physiogenomic factor and the outcome is calculated using a regression model, controlling for the other factors that have been found to be relevant. The magnitude of these associations is measured with the odds ratio, and statistical significance of these associations will be determined by constructing 95% confidence intervals. Multivariate analyses include all factors that have been found to be important based on univariate analyses.

26.3.2 OVERALL RATIONALE

The objective of these analyses is to search for genetic markers that modify the effect produced by a particular type of intervention, which epidemiologists refer to as an effect modifier. These markers are parameterized in our models as gene-intervention interactions. For example, if M_i is a 0 or 1 indicator of the presence of at least one recessive allele of gene i, and X_j represents the level of intervention, then the entire contribution to the outcome is given by the contribution of not only the gene and intervention main effects, but their interaction, as well—that is, $M_i \alpha_i + X_j \beta_j + M_i X_j (\alpha\beta)_{ij}$. Under this model, when the allele is absent ($M_i = 0$), the effect of a unit change in the intervention is described by the slope, β_j, but when the allele is present ($M_i = 1$), the effect of a unit change in the intervention is $\beta_j + (\alpha\beta)_{ij}$. Thus, the gene-intervention interaction parameter, $(\alpha\beta)_{ij}$, represents the difference in the effect of the intervention seen when the allele is present.

In the usual modeling framework, the response is assumed to be a continuous variable in which the error distribution is normal with mean 0 and a constant variance. However, it is not uncommon for the outcomes to have an alternative distribution that may be skewed, such as the gamma, or it may even be categorical. In these circumstances, we will make use of a generalized linear model, which includes a component of the model that is linear, referred to as the linear predictor, thus enabling us to still consider the concept of a gene-intervention interaction, as described earlier. The advantage of this broader framework is that it allows for considerable flexibility in formulating the model through the specification of the link function that describes the relationship between the mean and the linear predictor. It also provides considerable flexibility in the specification of the error distribution [7]. The S-Plus statistical software provides functions that calculate the maximum likelihood estimates of the model parameters.

To this point, we have described an analysis in which the effect of the intervention is assumed to be linear, but in practice the effect may not take place until a threshold is past, or it may even change directions. Thus, an important component of our exploration of the intervention effect on a particular response will involve the form for the relationship. In this case we will make use of generalized additive models (GAMs) [4]. In GAMs, the contribution of the marker and intervention is given by $M_i \alpha_i + \beta(X_j) + M_i \beta_\alpha (X_j)$. In this case, the effect when the allele is absent ($M_i = 0$) is $\beta(X_j)$ which is an unspecified function of the level of the intervention. In subject in which the allele is present ($M_i = 1$), the effect is given by the function $\beta(X_j) + M_i \beta_\alpha (X_j)$. In practice, we can estimate these functions through the use of cubic regression splines [1].

Predictive models may be sought by starting out with a hypothesis (which may be the null model of no marker dependence) and then adding each one out of a specified set of markers to the model in turn. The marker that most improves the p-value of the model is kept, and the process is repeated with the remaining set of markers until the model can no longer be improved by adding a marker. The p-value of a model is defined as the probability of observing a data set as consistent with the model as the actual data when in fact the null model holds. The resulting model is then checked for any markers with coefficients that are not significantly (at $p < .05$) different from zero. Such markers are removed from the model.

26.3.3 MODEL PARAMETERIZATION

The models built in the previous steps can be parameterized based on physiogenomic data. The maximum likelihood method is used, which is a well-established method for obtaining optimal estimates of parameters. S-plus provides very good support for algorithms that provide these estimates for the initial linear regression models, as well as other generalized linear models that we may use when the error in distribution is not normal.

In addition to optimizing the parameters, model refinement is performed by analyzing a set of simplified models and eliminating each variable in turn followed by reoptimizing the likelihood function. The ratio between the two maximum likelihoods of the original versus the simplified model then provides a significance measure for the contribution of each variable to the model. In a probabilistic network model, the approach is the same, except that instead of removing variables, dependency links are removed.

26.3.4 Model Validation

A cross-validation approach is used to evaluate the performance of models by separating the data used for parameterization (training set) from the data used for testing (test set). A model to be evaluated is reparameterized using all data except for one patient. The likelihood of the outcome for this patient is calculated using the outcome distribution from the model. The procedure is repeated for each patient, and the product of all likelihoods is computed. The resulting likelihood is compared with the likelihood of the data under the null model (no markers, predicted distribution equal to general distribution). If the likelihood ratio is less than the critical value for $p = .05$, the model will be evaluated as providing a significant improvement of the null model. If this threshold is not reached, the model is not sufficiently supported by the data, which could mean either that there is not enough data, or that the model does not reflect actual dependencies between the variables.

26.3.5 Multiple Comparison Corrections

Because the number of possible comparisons is very large in physiogenomics, we routinely include in our analysis a random permutation test for the null hypothesis of no effect for two to five combinations of genes. The permutation test is accomplished by randomly assigning the outcome to each individual in the study, which is implied by the null distribution of no genetic effect, and estimating the test statistic that corresponds to the null hypothesis of the gene combination effect. Repeating this process provides an empirical estimate of the distribution for the test statistic, and hence a p-value that takes into account the process that gave rise to the multiple comparisons.

For permutation testing, 1000 permuted data sets are generated, and each candidate marker will be retested on each of those 1000 sets. A p-value is assigned according to the ranking of the original test result within the 1000 control results. A marker is selected for model building when the original test ranks within the top 50 of the 1000 ($p < .05$).

The purpose of multiple comparison corrections is to generate a nonparametric and marker complexity adjusted p-value by permutation testing. This procedure is important because the p-value is used for identifying a few significant markers out of the large number of candidates. Model-based p-values are unsuitable for such selection, because the multiple testing of every potential serological marker and every genetic marker will be likely to yield some results that appear to be statistically significant even though they occurred by chance alone. If not corrected, such differences will lead to spurious markers being picked as the most significant. The correction is made by permutation testing—that is, the same tests are performed on a large number of data sets that differ from the original by having the response variable permuted at random with respect to the marker, thereby providing a nonparametric estimate of the null distribution of the test statistics. The ranking of the unpermuted test result in the distribution of permuted test results provides a nonparametric and statistically rigorous estimate of the false-positive rate for this marker. We also consider hierarchical regression analysis to generate estimates incorporating prior information about the biological activity of the gene variants. In this type of analysis, multiple genotypes and other risk factors can be considered simultaneously as a set, and estimates are adjusted based on prior information and the observed covariance, theoretically improving the accuracy and precision of effect estimates [10].

TABLE 26.1

Example of Physiogenomics Data Analysis and Screening for Gene Marker/Phenotype Associations

	Biochemical						Physiological			
A	B	C	D	E	F	G	H	I	J	Genotypes
					SNP Hits					
4	0	3	23	2	5	1	27	30	0	G1
4	3	1	5	3	16	17	25	23	3	G2
0	3	4	6	0	27	0	7	3	11	G3
0	0	3	0	3	2	4	1	5	16	G4
21	32	21	0	1	2	11	2	3	6	G5
9	5	0	0	23	5	3	9	12	11	G6
					Genomic Controls					
4	6	5	2	1	1	0	3	1	2	G7
1	2	0	1	5	8	9	1	0	0	G8
2	2	3	4	6	0	4	0	2	2	G9

26.3.6 SUMMARY OF ASSOCIATION RESULTS

The basis of the physiogenomics informatics platform is a parallel search for associations between multiple phenotypes and genetic markers for several candidate genes. The summary in Table 26.1 depicts the data set gathered from a hypothetical application of physiogenomics to a complex environmental exposure such as exercise or diet. In the top panel, each column represents a single phenotype measurement. Among the ten phenotypes in this example, various biochemical markers (denoted A–G) and physiological parameters (denoted H–J) are shown. Each row represents alleles for a given gene, and quantitatively renders associations of specific alleles to the variability in the phenotype.

The various numbers in Table 26.1 refer to the negative logarithms of p-value times 10. These p-values are adjusted for multiple comparisons using the nonparametric permutation test described earlier. For example, 30 refers to a p-value of less than 10^{-3}, or $p < .001$. The interface currently under development will also include the capability of interacting with the cell in the table to generate a further detailed analysis. As already noted, the p-value displayed in a cell is generated under the assumption of a linear trend for the effect of an intervention. The interface will be constructed to also generate a nonparametric estimate of the dose–response relationship for the different alleles using cubic splines [1], as well as other summary statistics such as the distribution of different alleles.

The interface allows visual recognition of highly significant association domains to both biochemical and physiological responses. Note the phenotype associations of gene 1 to both biochemical and physiological values. The biochemical and physiological outputs are associated by their relationship to the same gene. Noteworthy also are high-ranking associations of gene 2 to biochemical markers F–G and physiological phenotypes H and I. A summary table could list in order of significance the "hits" of positive association between genes and phenotypes.

There are also clearly negative fields. The same gene is associated to some phenotypes but not to others. Similarly, a given phenotype may have associations to some genes, but not others. Each negative result lends power to the positive associations. Some genes (G7 to G9) denoted as "genomic controls" have no association to any phenotype. Had a phenotype arisen from unsuspected population founder effects, most genes would have had specific founder allele frequencies reflected in the overall analysis. Such population stratification would give rise to a disproportionate number of genes associated to the phenotype.

The physiogenomics analysis includes the number of individuals with and without the associated allele. The individuals' counts among various phenotypes may be different depending on genotype sampling during a study. Physiogenomics can derive information from both well-represented distributions among the "in" and "out" groups and also those potentially driven by outliers. Thus, an association given an "in" group of 33% is interesting in most analytical platforms, whereas another with an "in" group of only 6% is potentially tractable only with physiogenomics with well-represented genomic controls. If the phenotype represents an undesirable outcome, the power of physiogenomics to detect trends among a small group of outliers is particularly valuable. In the case of side effects, the outliers may actually represent the susceptible population associated with a lower-frequency predictive marker.

26.4 FUTURE RESEARCH AND PROSPECTS

26.4.1 FUTURE RESEARCH

In the near future, more sophisticated models can be built using probabilistic networks. A probabilistic network is a factorization of the joint probability function over all the considered variables (markers, interventions, and outcomes) based on knowledge about the dependencies and independencies between the variables. Such knowledge is naturally provided by the hits coming out of the association screen, where each association can be interpreted as a dependency, and the absence of an association as an independency between variables. The model can then be parameterized by fitting to the data, similarly to the linear and logistic regression models, which are in fact special cases of probabilistic network models.

We also envision the informatics analysis to incorporate systems modules that could attribute various weights to each gene according to their relevance to the physiological mechanisms and predictive power statistically. The interface will include a framework for the expert physiologist to query the data set for mechanistic hypothesis, including feedback inhibition and signal amplification.

We know a priori that any list of candidate genes will miss some known key genes, and will certainly lack those genes not discovered so far or not identified yet as relevant. Physiogenomics posits the gene representation question in the following logic: representative genes are selected based on various functional criteria, and the associated genes are assembled into a predictive model. Through clinical research the predictive power of the model is ascertained. The hypothesis is that the genes in the panel together will explain a useful fraction of the variability in response among individuals. If the answer is affirmative, then the hypothesis is accepted, and the model is used. If the answer is inconclusive, the roster of genes will be modified until the multigene model's predictive level is clinically useful. Failure to establish predictive models with genes selected by either gene expression or pathway analysis can be circumvented with a supplementary analysis with other gene candidates.

An alternative physiogenomics strategy is genome-wide association study. The ultimate goal of The International Haplotype Map Consortium [6] is to render such an approach feasible, and the high multiplexing capacity of the various nanotechnology array platforms would be perfectly suited for such analysis. Such a strategy has been successful in family study of genetic traits but remains unproven for population studies.

26.4.2 PROSPECTS AND CONCLUSIONS

The wealth of interesting associations, some certainly unexpected, points to the power of physiogenomics. The above-described analytical tools permit the extension of physiogenomics to several thousand additional genes with the modern array nanotechnologies. Phenotypes could be added as well—for example, inflammatory and neuroendocrine markers are an area of intense interest in clinical medicine. The ability to measure changes in these markers for disease prevention strategies provides us a unique opportunity to examine genes determining a path to personalized health. The research can now utilize saved blood plasma and DNA for each patient to measure the appropriate genotypes and biochemical markers in blood, thus opening the possibility of retrospective analysis from archived clinical samples.

We conclude describing a prophetic embodiment of physiogenomics relevant to disease prevention. Figure 26.2 provides an example of personalized health care by customizing treatment intervention for the metabolic syndrome of obesity [2]. In the table, the choices are to recommend a given kind of exercise, drug, or diet regimen. If one of the options is high scoring, it can be used on its own. Thus, in the example, diet is high scoring in the first patient, a drug in the second, and exercise in the third. If the options are midrange, they can be used in combination (for example, if exercise and diet each have a positive effect but are unlikely to be sufficient independently). If none of the options is high or at least midscoring, the physiotype analysis suggests that the patient requires another option not yet in the menu.

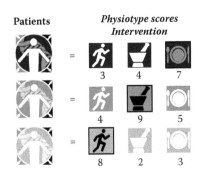

Patients   *Physiotype scores Intervention*

3   4   7

4   9   5

8   2   3

FIGURE 26.2 Physiotypes for personalized health.

As more environmental responses are characterized through physiogenomics, the chance that all patients will be served at increased precision of intervention and with optimal outcome will be greater. That we may even contemplate this scenario is a testament to the combined power of physiogenomics and nanotechnology. The highly parallel genome probing possible with nanotechnology and the systems engineering approach underlying physiogenomics in a real sense allow us to reconstitute the organism and its environmental response from the individual components. The specificity and individualization afforded by nanotechnology and physiogenomics are ushering medicine into the era of personalized health.

REFERENCES

1. Durrleman, S. and Simon, R. Flexible regression models with cubic splines. *Stat. Med.* 1989; 8: 551–561.
2. Ford, E.S., Giles, W.H., and Dietz, W.H. Prevalence of the metabolic syndrome among US adults: Findings from the third National Health and Nutrition Examination Survey. *JAMA* 2002; 287: 356–359.
3. Gunderson, K.L., Semyon, K., Graige, M.S. et al. Decoding randomly ordered DNA arrays. *Genome Res.* 2004; 14: 870–877.
4. Hastie, T. and Tibshirani, R. Generalized additive models. *Stat. Sci.* 1986; 1: 297–318.
5. Holford, T.R. *Multivariate Methods in Epidemiology.* New York: Oxford University Press, 2002.
6. International Haplotype Map Consortium. *Nature* 2003; 426: 789–796.
7. McCullagh, P. and Nelder, J.A. *Generalized Linear Models.* London: Chapman & Hall, 1989.
8. Saltelli, A., Chan, K., and Scott, E.M. *Sensitivity Analysis.* Chichester: John Wiley & Sons, 2000.
9. Ruano, G. Personalized Medicine Quo Vadis, *Personalized Medicine*, 2004 (in press).
10. Steenland, K., Bray, I., Greenland, S., and Boffetta, P. Empirical Bayes adjustments for multiple results in hypothesis-generating or surveillance studies. *Cancer Epidemiol. Biomarkers Prev.* 2000; 9: 895–903.
11. Stephens, J.C., Schneider, J.A., and Tanguay, D.A. Haplotype variation and linkage disequilibrium in 313 human genes. *Science* 2001; 293: 1048.

27 Patenting Inventions in Bionanotechnology: A Guide for Scientists and Lawyers*

Raj Bawa
Biology Department and Office of Tech Commercialization,
Rensselaer Polytechnic Institute, Troy, New York, USA

CONTENTS

* This chapter reflects the current views of the author which are likely to evolve. Furthermore, they should not be attributed, in whole or in part, to the author's affiliation listed above, nor should they be considered as expressing an opinion with regard to the merits of any particular company or product discussed herein. Nothing contained herein is to be considered as the rendering of legal advice. This chapter will focus on U.S. patents and the U.S. patent system. The author may be contacted at: bawabio@aol.com.

27.1 INTRODUCTION

The high-risk, high-payoff global nanotechnology phenomenon is in full swing as innovations at the intersection of engineering, biotechnology, medicine, physical sciences, and information technology are spurring new directions in research, education, commercialization, and technology transfer. In fact, the future of nanotechnology is likely to continue in this interdisciplinary manner. One of the major impacts of nanotechnology is taking place in the context of biology, biotechnology, and medicine.* This arena of nanotechnology is generally referred to as bionanotechnology, and sometimes nanomedicine. These two terms will be interchangeably used in this chapter.

There are quite a few bionanotechnology-related products in the market with numerous other potential applications under consideration and development. Generally speaking, commercial bionanotechnology is at a nascent stage of development and its full potential is years or decades away. However, there are a few bright spots where development is progressing more rapidly. For instance, consider the recent advances in bionanotechnology-related drug delivery, diagnostics, and drug development which are beginning to alter the landscape of medicine and health care. In the future, significant technologic advances across multiple scientific disciplines will continue under the nanotech banner to be proposed, validated, patented, and commercialized. This will likely accelerate in the coming years.

There is enormous excitement regarding bionanotechnology's potential impact. However, one thing is critical in this regard. Securing valid and defensible patent protection is a must for any player interested in bionanotechnology commercialization. Although early forecasts for bionanotechnology commercialization are encouraging, there are a few bottlenecks as well. Some formidable challenges include legal, environmental, safety, ethical, and regulatory questions as well as emerging thickets of overlapping patent claims. In fact, patent systems are under great scrutiny and strain, with patent offices around the world continuing to struggle with evaluating the swarm of bionanotechnology-related patent applications. The emerging thickets of patent claims has primarily resulted from patent proliferation and secondarily from the continued issuance of surprisingly broad patents in bionanotechnology by the U.S. Patent and Trademark Office (PTO) (Section 27.9). Adding to this confusion is the fact that the U.S. National Nanotechnology Initiative's (NNI) widely cited definition of nanotechnology is inaccurate and irrelevant, especially with regard to bionanotechnology (Section 27.2). This has resulted in the inadequate preliminary patent classification system that was unveiled in 2006 by the PTO (Section 27.7).

These challenges are creating a chaotic, tangled patent landscape in various sectors of bionanotechnology where the competing players are unsure as to the validity and enforceability of numerous issued patents. If this trend continues, it could stifle competition, limit access to some inventions, or simply cause commercialization efforts in certain sectors of bionanotechnology to grind to a halt.

---

* Though nanotechnology is characterized by distinctively new technologies such as scanning probe microscopy and nanolithography, many experts refute the popular notion that it is a distinct industry or sector. It may be more accurate to consider it to represent a set of tools (e.g., scanning probe microscopy) and processes (e.g., nanolithography) for manipulating matter that can be applied to virtually all manufactured goods. Similarly, I caution against envisioning a "nanotechnology market," per se. Instead, one should focus on how nanotech is being exploited across industry value chains, from basic materials to intermediate products to final goods. In fact, I believe that most nanotech-related products developed in the next few years will remain within existing markets or established sectors, and thus will not be marketed as "nanoproducts."

Given this backdrop, it is hard to predict whether bionanotechnology will make small but valuable contributions to medicine and biotechnology or whether it will act as a catalyst for a vast technological and health-care revolution. One thing is certain: bionanotechnology is here to stay, and it will definitely generate both evolutionary and revolutionary products in the future.* However, if the full potential of this "revolution" is to be fully realized, certain reforms are urgently needed at the PTO to address problems ranging from poor patent quality and questionable examination practices to inadequate search capabilities, rising attrition, poor employee morale, and a skyrocketing patent application backlog. All players involved in bionanotechnology agree that a robust patent system is essential for stimulating the development of commercially viable products.

In spite of all these challenges and bottlenecks, governments around the world are impressed by nanotechnology's potential and are staking their claims by doling out billions of dollars, euros, and yen for research.† International rivalries are growing.‡ Political alliances are forming, and battle lines are being drawn.§

Governments, corporations, and venture capitalists in 2006 spent $12.4 billion on nanotechnology research and development (R&D) globally, up 13% from 2005.¶ In 2006, globally, government spending grew to $6.4 billion, up 19% from 2005.** Global spending in 2006 on nanotech products ($50 billion) far surpassed that spent on nanotech R&D ($12 billion).†† In 2005, nanotechnology was incorporated into more than $30 billion worth of manufactured goods.‡‡ U.S. federal funds are supplemented by state investments in nanotechnology (approximately 40 cents per U.S. dollar). The president's budget for 2008 allocated 1.44 billion U.S. dollars for nanotech as compared to 1.35 billion U.S. dollars in 2007.

Numerous nanotechnology market reports and economic forecasts are available, each varying in their statistics. Often, poor assumptions underlie the analyses, rendering the results highly questionable or largely irrelevant. Also, most market reports on nanotechnology rely on the flawed NNI definition of nanotechnology (see Section 27.2) to draw their conclusions. Therefore, the data reflected in these reports should be taken as indicating trends rather than reflecting absolute numbers. For example, according to the National Science Foundation (NSF), by 2015 the annual global market for nano-related goods and services will top $1 trillion, making it one of the fastest-growing industries in history. On the other hand, Lux Research, Inc., predicts that by 2014, $2.6 trillion in global manufactured goods may incorporate nanotechnology (about 15% of total global manufacturing output).§§

---

* Although many sought-after innovations are decades away, a recent study claims that there are over 600 nanotech-based consumer products in the marketplace today (*See* A nanotechnology consumer products inventory—www.nanotechproject.org/index.php?id=44 [last accessed April 10, 2008]). Some examples of commercially available nanotech products include wrinkle-resistant khakis, super-light tennis rackets, and quick-drying paint.
† The passage of the 21st Century Nanotechnology Research and Development Act (Pub. L. No. 108-153) in 2003, which authorized $3.7 billion in federal funding from 2005 through 2008 for the support of nanotechnology R&D, is fueling the fervor over nanotechnology in the United States. This legislation resulted in the creation of R&D centers in academia and government. At present, there are over 50 institutes and centers dedicated to nanotechnology R&D. For example, the NSF has established the National Nanotechnology Infrastructure Network—composed of university sites that form an integrated, nationwide system of user facilities to support research and education in nanoscale science, engineering, and technology. Similarly, there are currently numerous government agencies with R&D budgets dedicated to nanotechnology.
‡ Edwards SA: *The Nanotech Pioneers—Where Are They Taking US?* Wiley-VCH Verlag GmbH & Co. KGaA, Weinheim, Germany (2006); Van Lente MA: Building the new world of nanotechnology. *Case W. Res. J. Int. Law* 38(1), 173–215 (2006).
§ I am optimistic that current fears about self-replicating nanobots, the potential toxic effects of nanoparticles, and the focus on strict regulations or a nanotech moratorium, will eventually give way to intelligent public dialogue on the realistic impact of bionanotechnology.
¶ Reisch MS: Nano goes big time. *Chemical & Engineering News* 85(4), 22–25 (2007).
** *The Nanotech Report* (4th Edition). Lux Research, Inc., New York (2006).
†† *See supra Note **.*
‡‡ *See supra Note **.*
§§ Report: *Sizing Nanotechnology's Value Chain*, Lux Research, New York (2004).

27.2   CURRENT DEFINITIONS OF NANOTECHNOLOGY AND NANOMEDICINE—A SOURCE OF CONFUSION

The term *nanotechnology* is very much in vogue. However, one of the problems it faces is the confusion, hype, and disagreement among experts about its definition.* Nanotechnology is an umbrella term used to define the products, processes, and properties at the nano/micro scale. By manipulating atoms, scientists can create stronger, lighter, more efficient materials ("nanomaterials") with tailored properties. In addition to numerous advantages provided by this scale of miniaturization (over their conventional "bulk" counterparts), quantum effects at this scale impart additional novel properties. Many of the properties of nanomaterials are fundamentally different from those of their macroscopic/bulk analogues due to an increased surface area and quantum effects. As a particle's size decreases, a greater proportion of its atoms are located on the surface relative to the interior (core), generally rendering it more reactive. In fact, quantum effects coupled with surface area effects can affect optical, electrical, chemical, and magnetic properties of nanomaterials, which in turn can affect their *in vivo* behavior.

One of the most quoted, yet incorrect, definitions of nanotechnology is that used by the NNI†: "[n]anotechnology is the understanding and control of matter at dimensions of roughly 1 to 100 nanometers, where unique phenomena enable novel applications."‡ Clearly, this definition excludes numerous devices and materials of micrometer dimensions, a scale that is included within the definition of nanotechnology by many nanoscientists.§ Therefore, experts have cautioned against an overly rigid definition based on a sub-100 nm size, emphasizing instead the continuum of scale from the "nano" to "micro."¶

Various federal agencies are also grappling with the definition of nanotechnology. For example, both the U.S. Food and Drug Administration (FDA) and the PTO use the NNI definition based on a sub-100 nm scale. This definition continues to present difficulties not only for understanding nanopatent statistics,** but also for the proper assessment of nanotechnology's scientific, legal, environmental, regulatory, and ethical implications. This problem persists because nanotechnology represents a cluster of technologies, each possessing different characteristics and applications. Although the sub-100-nm size range may be critical for a nanophotonic company where size-dependent quantum effects are particularly important (e.g., a quantum dot's size dictates the color of light emitted therefrom), this size limitation is not critical to a drug company from a formulation, delivery, or efficacy perspective because the desired or ideal property (e.g., improved bioavailability, reduced toxicity, lower dose, enhanced water solubility, etc.) may be achieved in a size range greater than 100 nm. Numerous examples from the pharmaceutical industry highlight this important point

---

* Editors: Nanotechnology. *Nature Nanotechnology* 1(1), 8–10 (2006).

† What is Nanotechnology? The National Nanotechnology Initiative. www.nano.gov/html/facts/whatIsNano.html [last accessed April 10, 2008].

‡ A nanometer is one billionth of a meter, or 1/75,000th the size of a human hair. An atom is about one third of a nanometer in width. A nanometer is one-billionth of a meter. Here is the size of some common objects expressed in nanometers: a basketball is about 24 centimeters (240 million nanometers); a flea is about 1 millimeter (1 million nanometers); an anthrax bacterium is 1 micrometer (1000 nanometers); and a DNA molecule measures around 2.5 nanometers wide. A sugar granule is about 1 millimeter, and a single sugar molecule is about 1 nanometer.

§ Bawa R: Nanotechnology patenting in the U.S. *Nanotechnology Law & Business* 1, 31–50 (2004); Bawa R: Patents and nanomedicine. *Nanomedicine* 2(3), 351–374 (2007); Bawa R, Bawa SR, Maebius SB, Iyer C: Bionanotechnology patents: Challenges and opportunities. In: *The Biomedical Engineering Handbook* (3rd Edition). Bronzino JD (Ed.), CRC Press, Boca Raton, FL; 29-1–29-16 (2006); Morrow KL, Bawa R, Wei C: Recent advances in basic and clinical nanomedicine. In: *Medical Clinics of North America*, 91(5), 805–843 (2007). Bawa R, Bawa SR: Protecting new inventions in nanomedicine. In: *Foresight, Innovation, and Strategy: Towards a Wiser Future*. Wagner CG (Ed.), World Future Society Press, Bethesda, MD, 31–44 (2005); Bawa R, Bawa SR, Maebius SB, Flynn T, Wei C. Protecting new ideas and inventions in nanomedicine with patents. *Nanomedicine: Nanotechnology, Biology and Medicine* 1(2), 150–158 (2005).

¶ National Heart, Lung, and Blood Institute Programs of Excellence in Nanotechnology. http://grants.nih.gov/grants/guide/rfa-files/RFA-HL-04-020.htm.

** Regalando A: Nanotechnology patents surge as companies vie to stake claim. *Wall Street Journal* June 18 Issue, A1 (2004).

(e.g., Abraxane's albumin-paclitaxel nanoparticles; Elan Pharma International's nanoparticles, and Kereos's anticancer particles).

Clearly, a definition based on physical limits tends to be an unorthodox way of defining a technology. Other technologies tend to be defined by a key technology or breakthrough: genetic engineering technology is based upon recombinant DNA, and the Internet is a collection of "bulletin boards" networked in a World Wide Web.

Furthermore, nanotechnology is nothing new. For example, nanoscale carbon particles ("hightech soot nanoparticles") have been used as a reinforcing additive in tires for more than a century. Another example is that of protein vaccines that squarely fall within the definition of nanotechnology.* In fact, the scale of many biologic structures is similar to various "nanocomponents." For example, peptides are similar in size to quantum dots (around 10 nm), and some viruses are the same size as drug delivery nanoparticles (around 100 nm). Hence, most of molecular medicine and biotechnology can be classified as nanotechnology. In fact, biologists had been studying all these nanoscale biological structures long before the term "nanotechnology" became fashionable.† Given this confusion, I recently proposed a more practical definition of nanotechnology that is unconstrained by an arbitrary size limitation‡:

> The design, characterization, production, and application of structures, devices, and systems by controlled manipulation of size and shape at the nanometer scale (atomic, molecular, and macromolecular scale) that produces structures, devices, and systems with at least one novel/superior characteristic or property.

Naturally, disagreements over the definition of nanotechnology carry over to the definition of nanomedicine. At present, there is no uniform, internationally accepted definition for nanomedicine either. One definition, unconstrained by size while correctly emphasizing that controlled manipulation at the nanoscale results in medical improvements and significant medical changes, comes from the European Science Foundation.§

> The science and technology of diagnosing, treating and preventing disease and traumatic injury, of relieving pain, and of preserving and improving human health, using molecular tools and molecular knowledge of the human body.

---

* Nanotechnology also aims to learn from nature—to understand the structure and function of biological nanodevices and to use nature's solutions to advance science and engineering. Evolution has produced an overwhelming number and variety of biologic devices, compounds, and processes that function at the nanometer or molecular level and that provide performance that is unsurpassed by synthetic technologies. When nanotechnology is combined with molecular biology, the potential applications at this frontier are widespread and sound like the stuff of science fiction. Given the complex biological machinery that exists in nature, it is hard not to conclude that complex machines on the nanoscale may be possible someday (Jones R: What can biology teach us? *Nature Nanotechnology* 1, 85–86 [2006]). The construction principles used in nanotechnology often originate in biology, and the goals are often biomimetic or aimed at the solution of long-standing research problems. At the heart of the approaches in this field is the concept of self-assembly. In fact, self-assembly of ordered elements is a defining property of life. Bionanotechnologists attempt to exploit the self-assembly and ordered proximity of nanoscale structures found in biology.

† However, note that the U.S. National Institutes of Health (NIH) emphasize that: "While much of biology is grounded in nanoscale phenomena, NIH has not re-classified most of its basic research portfolio as nanotechnology" (*See* www.becon.nih.gov/nano.htm [last accessed April 10, 2008]). The NIH identifies three broad areas that qualify as nanotechnology: "studies that use nanotechnology tools and concepts to study biology; that propose to engineer biological molecules toward functions very different from those they have in nature; or that manipulate biological systems by methods more precise than can be done by using molecular biological, synthetic chemical, or biochemical approaches that have been used for years in the biology research community."

‡ Bawa R: Patents and Nanomedicine. *Nanomedicine* 2(3), 351–374 (2007).

§ Nanomedicine—An ESF-European Medical Research Councils (EMRC) forward look report. European Science Foundation, Strasbourg, France (2004). Nanomedicine is, in a broad sense, the application of nanoscale technologies to the practice of medicine—namely, for diagnosis, prevention, and treatment of disease and to gain an increased understanding of the complex underlying disease mechanisms. The creation of nanodevices such as nanobots capable of performing real-time therapeutic functions *in vivo* is one eventual long-term goal here. Advances in delivering nanotherapies, miniaturization of analytic tools, improved computational and memory capabilities, and developments in remote communications will be integrated. These efforts will cross new frontiers in the understanding and practice of medicine.

Hence, I propose that the size limitation imposed in NNI's definition should be dropped, especially when it is applied to bionanotechnology nanomedicine. Furthermore, the phrase "small technology" may be more appropriate, because it accurately encompasses both nanotechnologies and microtechnologies. I believe an internationally acceptable definition and nomenclature of nanotechnology should be promptly developed in this context.

27.3   BIG PHARMA AND NANOTECHNOLOGY

Numerous market forces and challenges are dictating that new drug discovery, development, and delivery approaches be developed and implemented. There is a real and urgent need for drug companies to focus on technologies that support miniaturization and high throughput. Such approaches enable faster drug target discovery and drug development. They can also lead to a reduction in the cost of drug discovery, design, and development. Moreover, the current business model of drug companies (with their mammoth size and excessive reliance on blockbusters) is broken and is in need of repair.

Nanotechnology can enhance the drug discovery process via miniaturization, automation, speed, and reliability of assays. This is likely to result in a faster introduction of new cost-effective products to the market. For example, nanotechnology can be applied to current microarray technologies, exponentially increasing the hit rate for promising compounds that can be screened for each target in the pipeline. Inexpensive and higher-throughput DNA sequencers based on nanotechnology can reduce the time for both drug discovery and diagnostics. Although, these high-throughput screening technologies have led to an increase in the number of poorly water-soluble new chemical entities (NCEs), nanotechnology can also tackle such formulation problems. Nanoscience research has also resulted in a need for novel analytical technologies that can directly impact aspects of drug delivery, such as determining the efficacy of targeting, therapeutic outcome, and so forth.

Big pharma in today's global economy faces enormous pressure to deliver high-quality products to the consumer while maintaining profitability. It must constantly reassess how to improve the success rate of NCEs while reducing R&D costs as well as cycle time for producing new drugs, especially new blockbusters. In fact, the cost of developing and launching a new drug to the market, although widely variable,* may be upwards of $800 million. Typically, the drug appears on the market some 10 to 15 years after discovery.† Furthermore, only one out of five lead compounds makes it to final clinical use.‡ Annual R&D investment by drug companies has risen from $1 billion in 1975 to $40 billion in 2003, while new approvals have essentially remained flat, between 20 to 30 drugs per year.§ In fact, in recent years, NCEs accounted for only 25% of products approved by the FDA, with the majority of approvals being reformulations or combinations of already approved agents. Although the cost of drug R&D continues to rise, only 30% of drugs are able to recover their R&D costs. The weakened product pipeline issue is a global problem; the decreasing number of new drugs approved by the FDA and foreign drug agencies continue to haunt the drug industry. For example, FDA approvals have fallen by half since 1996, with only 20 approvals in 2005. Unique drug development models are being successfully developed by competitors to circumvent some of big pharma's patented branded drugs.¶

---

* DiMasi J, Hansen R, Grabowski H: The price of innovation: New estimates of drug development costs. *Journal of Health Economics* 22, 151–185 (2003); Adams C, Brantner V: Estimating the cost of new drug development: Is it really $802m? *Health Affairs* 25(2), 420–428 (2006).
† Anon R: Health Informatics into the 21st Century. *HealthCare Reports*. Reuters Business Insight February (1999).
‡ Erickson J: Translation research and drug development. *Science* 312, 997 (2006).
§ Sussman NL, Kelly JH: Saving time and money in drug discovery—A pre-emptive approach. In: *Business Briefings: Future Drug Discovery 2003*. Business Briefings Ltd, London, UK, 46–49 (2003).
¶ Owens J: Ethical medicine or IP dodge? *Nat. Rev. Drug Disc.* 6, 104 (2007).

In other words, while the above-mentioned trends and issues are creating novel challenges for the drug industry, they also represent an impetus for the drug industry as a whole to focus on nano-enabled R&D technologies.*

27.4 THE FUTURE OF BIONANOTECHNOLOGY

The potential future impact of bionanotechnology on society could be huge. Bionanotechnology could drastically improve a patient's quality of life, reduce societal and economic costs associated with health care, offer early detection of pathological conditions, reduce the severity of therapy, and result in an improved clinical outcome for the patient. The ultimate goal is obviously comprehensive monitoring, repair, and improvement of all human biologic systems: basically, an enhanced quality of life.

As of mid-2006, 130 nanotech-based drugs and delivery systems and 125 devices or diagnostic tests were in preclinical, clinical, or commercial development. However, the nano-pharma market is expected to significantly grow in the coming years. Analysts project that by the end of this year, the market for nanobiotechnology will exceed $3 billion, reflecting an annual growth rate of 28%.† One report predicts that the drug market for nanotech will pass $200 billion by 2015.‡ According to a 2007 report, the U.S. demand for nanotechnology-related medical products (nanomedicines, nanodiagnostics, nanodevices, and nanotech-based medical supplies) will increase over 17% per year to $53 billion in 2011 and $110 billion in 2016.§ This report predicts that the greatest short-term impact of nanomedicine will be in therapies and diagnostics for cancer and central nervous system disorders.

27.5 PROTECTING BIONANOTECHNOLOGY INVENTIONS: THE U.S. PATENT OFFICE AND CRITERIA FOR PATENTING

Globally, industries that produce and manage "knowledge" and "creativity" have replaced capital and raw materials as the new wealth of nations. Property, which has always been the essence of capitalism, is increasingly changing from tangible to intangible.¶ Intellectual property (IP) rights are a class of assets that accountants call intangible assets. These assets play an ever-increasing role in the development of emerging technologies like biotechnology, drug development, and nanotechnology. Modern IP consists of patents, trademarks, copyrights, and trade secrets. Patents are the most complex, tightly regulated, and expensive form of IP. They have the attributes of personal property—they may be assigned, bought, sold, or licensed.

---

* It is important to note that at this stage several obstacles beset nano-enabled drug R&D commercialization, including high production costs; the public's general reluctance to embrace innovative medical technologies without real safety guidelines; the relative scarcity of venture funds; few near-term commercially viable products; a general lack of knowledge regarding the interaction between nanomaterials and living cells (the issue of biocompatibility and toxicity of nanomaterials); big pharma's reluctance to accept and seriously invest in nanomedicine; production issues such as the lack of quality control, reproducibility, and scalability of most nanostructures of commercial interest; confusion and delay at the PTO (with respect to the burgeoning number of bionanotechnology-related patent applications) and FDA (with respect to a lack of clear regulatory/safety guidelines); pricing pressures due to high industry margins; a sharp decline in public confidence in the pharma industry generally; state and federal government's increased vigilance pertaining to hyperaggressive business practices (e.g., illegal drug marketing and improper drug pricing); and the media's continuing focus on the negative aspects of nanomaterials (environmental, health, and safety concerns are at the forefront).
† Report: *Nanobiotechnology Opportunities and Technical Analysis.* San Mateo, CA, Front Line Strategic Consulting (2003).
‡ Sesquehanna Financial Group LLP, Griffin Securities, Inc.
§ Report: *Nanotechnology in Healthcare.* The Freedonia Group, Inc. Cleveland, OH (2007).
¶ Intangible assets, as a portion of corporate market capital, are steadily rising.

Patent law is a subtle and esoteric area of law that has evolved in response to technological change. It has been modified numerous times since 1790, the year the first U.S. Patent Act was enacted.* This is due to new interpretations of existing laws by the PTO and by the courts, or by creation of new laws by Congress, often in response to new technology. Patent law, arguably one of the most obscure legal disciplines, is now at the forefront of bionanotechnology.†

Patentable inventions need not be pioneering breakthroughs; improvements of existing inventions or unique combinations or arrangements of old formulations may also be patented. In fact, majority of inventions are improvements on existing technologies. However, not every innovation is patentable. For example, abstract ideas, laws of nature, works of art, mathematical algorithms, and unique symbols and writings cannot be patented. Works of art and writings, however, may be copyrighted and symbols may be trademarked. Laws of the universe or discoveries in the natural world, even if revolutionary, cannot be patented. For instance, Einstein's Law of Relativity cannot be considered anyone's IP. For a U.S. patent to be granted, an invention must meet specific criteria as set forth in U.S. statutes (Table 27.1).

A U.S. patent‡ provides protection only in the United States, its territories, and its possessions for the term of the patent. It is estimated that 90% of the world's patents are issued through the three main patent offices—the United States, Europe, and Japan. Legally speaking, a U.S. patent is a document granted by the federal government (at the PTO§) whereby the recipient (or "patentee") is conferred the temporary right to exclude others from making, using, selling, offering for sale, or importing the patented invention into the United States for up to 20 years from the filing date. Similarly, if the invention is a process, then the products made by that process cannot be imported into the United States. All patented inventions eventually move "off" patent at the end of their patent term ("patent expiration") at which time they are dedicated to the public domain. This is the basis

---

* The Founding Fathers incorporated the concept of patents into the Constitution under Article 1, Section 8, Clause 8, whereby Congress was given the authority "[t]o promote the progress of science and the useful arts, by securing for limited times to authors and inventors the exclusive right to their respective writings and discoveries." President Washington signed the first U.S. Patent Act on April 10, 1790. Title 35 of the *United States Code* codified the Patent Act of 1952, the Act currently in use. Since the granting of the first U.S. patent in 1790, more than 7 million patents have been issued by the PTO, a bureau of the U.S. Department of Commerce. In fact, 1790 was the first year of operation for the PTO and it issued only three patents. On the other hand, in the 2006 fiscal year, 183,187 patents were issued. For the past few years, the PTO has received over 400,000 patent applications annually. In 2006, the average pendency ranged from 25.4 to 44 months. The number of patent applications filed has been increasing, on average, by over 10% per year since 1996. Currently, there is an astounding backlog of over 1 million unexamined U.S. patent applications.

† It seems that in the new millennium, patent issues are making headlines on a daily basis. As the line between academia and industry becomes fuzzier, the axiom for success in science, "publish or perish," is being replaced with "patent or perish" or "patent and prosper." Universities are straying away from their "education mission" by focusing on patents for potential license revenue. I believe that patents are as important, if not more so, as publications on *curriculum vitae*, and have a major impact in academia on hiring, tenure, and promotion.

‡ A patent is not a "hunting license"; it is merely a "no trespassing fence" that clearly marks the boundaries of an invention. (For example, see *Brenner v. Manson*, 383 U.S. 519, 536 [1966].) In other words, a patent grant is a negative grant; it prevents other parties from using the invention without prior permission of the patent holder (which can be in the form of a license). This does not imply that the patent holder can automatically publicly practice (i.e., commercialize) the invention. Often, appropriate government regulatory approval is required.

§ The PTO issues three types of patents as defined by U.S. statutes: (a) utility patents for "any new and useful process, machine, manufacture, or composition of matter, or any new and useful improvement thereof"; (b) plant patents for "any distinct and new variety of plant" (i.e., asexually reproduced non-tuber plant varieties); and (c) design patents for "any new, original and ornamental design for an article of manufacture." (i.e., ornamental designs of an article of manufacture).

TABLE 27.1
Legal Requirements to Obtain a U.S. Bionanotechnology Patent

U.S. Patent Statute	Brief Description of Statue
35 USC § 102 Novelty requirement	Invention must be novel (i.e., sufficiently new and unlike anything that has been previously patented, marketed, practiced, publicized, or published).
35 USC § 103 Nonobviousness requirement	Invention must be nonobvious to a person with knowledge in the field related to the invention, meaning that the person would not automatically arrive at the present invention from a review of existing ones (i.e., trivial variations that are readily apparent to a person with knowledge in the field related to the invention cannot be patented).
35 USC § 101 Utility requirement	Invention must have utility (i.e., the invention has some use and it actually works or accomplishes a useful task).
35 USC § 112(1) Written description requirement	Invention must be adequately described to the public to demonstrate "possession" of the invention at the time of filing.
35 USC § 112(1) Enablement requirement, part I	Invention must enable a person with knowledge in the field related to the invention to make or carry out the invention without "undue experimentation" (i.e., to make the claimed product or carry out the claimed process).
35 USC § 112(1) Enablement requirement, part II	Invention must enable a person with knowledge in the field related to the invention to use the invention.
35 USC § 112(2) Clarity requirement	Invention must be described in clear, unambiguous, and definite terms.
35 USC § 112(2) Best mode requirement	Invention must set forth the best mode of making or using the invention, contemplated by the inventor at the time of filing of the patent application.

Note: USC stands for U.S. Code.

for low-cost generic drugs* that appear in the marketplace following expiration of the costlier versions of the patented branded drug.†

The basic rationale underlying patent systems, both in the United States and abroad, is simple enough: an inventor is encouraged to apply for a patent by a grant from the government of legal monopoly of limited duration for the invention. This limited monopoly or proprietary right justifies R&D costs by assuring inventors the ability to derive economic benefit from their work. In exchange for this grant, the inventor publicly discloses the new technology that might have oth-

---

* Generic Drugs: Questions and Answers, U.S. Food and Drug Administration, Center for Drug Evaluation and Research, http://www.fda.gov/cder/consumerinfo/generics_q&a.htm (last accessed April 9, 2008).
† Current U.S. patent laws allow granting a patent on new drug formulations that have been created from old drugs, for instance, via novel drug delivery systems (DDS). Nanotechnology could also allow reformulation of existing and orphaned compounds. These new reformulations may qualify as NCEs at the FDA and for patents at the PTO. In other words, "nanoformulations" of older drugs may be patentable as long as they fulfill all the criteria for patentability. Furthermore, innovative DDS or platforms may be patented on their own under current U.S. patent statutes. Innovative DDS could enable drug companies to devise novel drug reformulations of off-patent or soon-to-be off-patent compounds. This strategy could delay or discourage generic competition during the most profitable years of an innovator drug's life cycle, especially if the reformulated drug is superior to its off-patent or soon-to-be off-patent counterpart. This approach, in effect, stretches the product life cycle of an existing, branded, patented drug. This strategy, commonly referred to as "product-line-extension," is broad in scope and includes any second-generation adaptation of an existing drug that offers improved safety, efficacy, or patient compliance. In fact, successful reformulation strategies should focus on how to add value through added ease and convenience for the consumer. If this approach is successful, the innovative DDS or platforms can maintain market share even after generics appear in the marketplace. Another often-employed approach is to develop and patent a novel polymorph of the innovative drug compound prior to patent expiration. Yet another strategy involves generating patent protection from a competitor's formulation (patented or off patent), by analyzing the competitor's existing patent claims, then tweaking them and filing patents that circumvent the competitor's specific use or DDS.

erwise remained secret (an "immediate benefit" to the public) and allows the public to freely use, make, sell, or import the invention once the patent expires (a "delayed benefit" to the public). Hence, the new technology that is brought to light in the form of valuable technical information provides a continuous incentive for future innovation. In this way, society obtains a *quid pro quo* from inventors in exchange for the temporary grant of exclusive rights. Such an advantageous exchange stimulates commerce (a "long-term benefit" to the public). Patent protection is the engine that drives industry and the incentive for it to invest in R&D to innovate. Clearly, without such protection, most companies would avoid costly R&D, and society would be deprived of the many benefits thereof. However, it is critical that the scope or breadth of the patent issued by the PTO be just right; it should neither be unduly broad nor should it be too limiting. In other words, the invention granted a patent should just fit within the boundaries of that patent. Unfortunately, this is not often the case (see Section 27.9).

Obtaining a patent for an invention is often a long, expensive, and tedious process that generally involves the inventor, patent counsel, or practitioner (i.e., patent agent or patent attorney) and PTO staff (especially a "patent examiner"). Patent examiners are PTO personnel who review the filed patent application to ensure that it fulfils all pertinent requirements of the law (Table 27.1). This review process is commonly referred to as an "examination." The exchange of documents between the PTO and the patent counsel is broadly known as "prosecution." If the examiner believes that all requirements for a patent are met, then a "notice of allowance" is issued to the applicant. Following this, a patent is issued once the applicant pays an "issuance fee." Upon issuance, the entire contents of the patent application ("the file wrapper" or "prosecution history") along with a copy of the patent and all future documents pertaining to the patent, are made available to the public. The entire patent examination process, starting with the filing of the patent application to its allowance or final rejection, may take anywhere from 1 to 5 years, or longer. This depends upon variables such as the specific technology area within the PTO where the patent is being reviewed by the patent examiner and the time to process the paperwork that accompanies the patent application by the PTO clerical staff. As part of the patent prosecution, all applications filed on or after November 29, 1999, are generally published 18 months after filing (see also Section 27.7).

Because, for most patents, the patent term commences on the date of filing and ends 20 years thereafter, most commercially valuable bionanotech inventions are, in reality, in the marketplace prior to the actual patent grant date (unless regulatory approval is sought). Generally, it is impossible to predict the future commercial success or commercial viability of an issued patent. In part, this is due to the fact that most patents are filed at the PTO without any clear idea of whether the invention is commercially valuable. For example, in bionanotech, patent applications are continuously being filed on a large number of drugs, therapies, and devices even before it is known that they will be ruled safe and effective by the FDA. If litigation rates (which range from 1.5 to 2% of the issued patents) are any indicator of commercial value, then only a tiny fraction of patents are commercially significant. Although obtaining a patent does not ensure commercial success, economists view patenting as an indicator of scientific activity.* They argue that this is the basis for providing a nation with a competitive advantage, fueling its economy.

In recent years, however, patents have become the subject of much debate and controversy. In fact, there are plenty of antipatent players in the field who feel that patent laws (and most international treaties) are unfairly providing an economic advantage to some over others.† It has even been suggested that patent laws and IP are the products of a new form of Western colonialism designed to deny the developing world access to common goods. Issues like biopiracy and IP theft have been proffered as reasons for the unavailability of essential drugs to the poorest and neediest people in

---

* Merges RP: Commercial success and patent standards: Economic perspectives on innovation. *Cal L. Rev.* 76, 803–845 (1988).

† McGrath M: The patent provisions in TRIPS: Protecting reasonable remuneration for services rendered—Or the latest development in Western colonialism. *European Intellectual Property Review* 18, 398–453 (1996).

the world. Not surprisingly, those in the developing world support patent protection but prefer a regime that suits their own national interests. In this regard, they highlight the fact that although Western drug companies continue to cite the need to reward innovation as a justification for stronger patent laws or patent enforcement, the industry, in reality, continues to spend more on reformulating preexisting drugs and on expensive litigation to protect their current patents than to innovate.* Future struggles over patents on the international stage are almost certain to focus on multinational drug patents when they are revoked or challenged.†

The PTO does not police or monitor patent infringement and it does not enforce issued patents against potential infringers. It is solely up to the patentee to protect or enforce the patent, all at the patentee's own cost. The patentee may enlist the U.S. government's help via the court system to prevent patent infringement. However, PTO decisions are subject to review by the courts, including the Court of Appeals for the Federal Circuit (CAFC),‡ and rarely, the U.S. Supreme Court. Sometimes Congress intervenes and changes or modifies some of the laws governing patents. If a court deems a patent to be invalid, the patent holder is unable to enforce it against any party. However, suing an alleged infringer is a risky business because when a patent holder sues an alleged infringer, in certain technologies, there is a 50% risk that his own patent will be found to be invalid.

Based on my review of seminal CAFC patent decisions from the past decade or so, it is my conclusion that the CAFC has fostered the following: (a) expanded what can be patented under the patent statues; (b) lowered the threshold to obtain a U.S. patent; and (c) tilted its decisions in favor of patent holders. Clearly, this stance has resulted in stronger patent protection for patent holders. As a result, since the creation of the CAFC, the number of patents granted has increased at an annual rate of 5.7% as compared to less than 1% from 1930 to 1982.§ According to some experts, if this trend continues, it could stifle competition and limit access to some inventions. Moreover, this is contrary to the *quid pro quo* discussed earlier: it disturbs the delicate balance between the patent holder's limited-time monopoly over the invention on the one hand and the public's interest in accessing the invention's disclosure (from the public domain) on the other hand. Certainly, this could be the very reason why the Supreme Court is increasingly stepping in to hear more and more patent appeals of CAFC decisions. It is important to note that the Supreme Court, which has rarely reviewed patent decisions in the past, has heard at least five important patent appeals of CAFC decisions in the last 4 years alone, reversing all of them. One of these recent landmark rulings,¶ that broadly impacts bionanotechnology, allows drug companies to infringe drug patents held by others as long as the infringement is during the R&D phase (i.e., preclinical phase) of drug development and generates data (on the compound being tested) that may (or may not) be ultimately submitted to the FDA as part of the drug approval process. By these and other recent decisions, the Supreme Court may be trying to reestablish the balance between the patent holder and the public's interest, a certain flexibility that it may have viewed as eroding under the CAFC. It is critical that the CAFC refocus its efforts to provide greater clarity to patent law and render patent decisions that are more consistent. After all, this is its true mission.

One highly controversial yet important statistic worth briefly discussing is the patent grant rate (i.e., the patent application allowance rate). Because the PTO is often not very forthcoming in providing accurate patent statistics and data on this issue,** several legal scholars have published studies to gauge this figure. One widely cited estimate places the average PTO grant rate at 77% to 95%

---

* Saini A: Making the poor pay. *NewScientist* 193(2597), 20 (2007).

† Tremblay JF: Drug patents struggles in Asia. *Chem. Eng. News* 85(6), 11 (2007).

‡ The CAFC was created by Congress in 1982 with the aim of creating uniformity in the patent law, especially with respect to unpredictable, evolving technologies like biotechnology and nanotechnology. In reality, it has sometimes failed in this role by rendering inconsistent and contradictory patent decisions.

§ Jaffe AB, Lerner J: *Innovation and its Discontents: How Our Broken Patent System Is Endangering Innovation and Progress.* Princeton University Press, Princeton, NJ (2004).

¶ *Merck KGaA v. Integra Lifesciences I. Ltd. et al.* 545 U.S., No. 03-1237 (2005).

** Wegner H: The USPTO's 54% allowance rate. IPFrontline.com, Dec. 30, 2006, http://www.ipfrontline.com/depts/article.asp?id=13796&deptid=5 (last accessed May 4, 2008).

of filed patent applications for the years 1981 to 2005.* However, I agree with legal scholars who consider this estimate to be artificially high, because it is based on an inappropriate legal framework and somewhat flawed numbers.† In any case, it is immaterial as to what the exact figures are; the crux of the matter is that the PTO grant rates are rather high and this may indirectly reflect a less rigorous review of patent applications as compared to the other major patent offices. In other words, these high allowance rates may be partly to blame for the granting of poor-quality patents by the PTO.‡ (Other concerns are discussed in detail in Section 27.8.) In this context, it is interesting to note that the time taken for one million patents to be granted has greatly declined since the grant of the first patent in 1836.§,¶

27.6   SIGNIFICANCE OF BIONANOTECHNOLOGY PATENTS

Patents are critical to the bionanotech "revolution." When investors in nanomedicine or drug companies consider the merits of their investment, patent issues are one of the most important items they review. There is also ample evidence that companies, start-ups, and universities are ascribing ever-greater value and importance to patents. Increasingly, they are willing to risk a larger part of their budgets to acquire and defend patents. The process of converting basic research in bionanotech into commercially viable products is proving to be long and difficult. The development of bionanotech-related technologies is extremely research intensive, and without the market exclusivity offered by a patent, development of these products and their commercial viability in the marketplace would be significantly hampered.

Patents are especially important for start-ups and smaller companies because they may help in negotiations over infringement of their patents during competitive posturing with larger corporations.** In fact, patents may also protect the clients of a patent owner because they prevent a competitor from infringing or replicating the client's products made under license from the patentee. Moreover, patents provide inventors credibility with their backers, shareholders, and venture capitalists—groups who may not fully understand the science behind the technology. Generally, patents precede funding from a venture capital firm. For a start-up company, patents are not only a means of attracting investment, but also serve to validate the company's foundational technology. Therefore, start-up companies aggressively seek patents as a source of significant revenue. They cite the potential for licensing patents and the power to control emerging sectors of nanotechnology as major rea-

---

* Quillen CD, Webster OH: Continuing patent applications and the USPTO—Updated. *Federal Circuit Bar Journal* 15, 635 (2006).

† Ebert LB: On patent quality and patent reform. *Journal of the Patent & Trademark Office Society* 88(12), 1068–1076 (2006).

‡ In light of this discussion regarding patent allowance rates and patent quality, it is rather interesting to note the PTO's recent announcement of a 54% allowance rate for the past fiscal year (October 1, 2005 to September 30, 2006). In this regard, it is further worth noting that while the number of patent applications has continued to climb (creating a steady backlog that threatens businesses), the number of issued patents has declined in recent years. The most dramatic decline was in 2005 when a drop of 11% in the allowance rate was reported from the previous fiscal year. Do these figures imply a vast improvement in patent quality over the earlier years when allowance rates were much higher? Most experts would disagree. If this is not the case, then is it possible that numerous high-profile patent cases (like the recent BlackBerry case) have oversensitized PTO upper management, who are now actively engaged in artificially suppressing the high patent grant rate? If this is indeed the case, all this tinkering with numbers will have disastrous consequences for the entire innovation process. Moreover, it is clearly contrary to the basic tenet of the U.S. patent system: "[t]o promote the Progress of Science and Useful Arts." (U.S. Constitution. Article I, Section 8, Clause 8)

§ The first U.S. patent was issued in 1790, and the numbering system was established in 1836.

¶ Press Release, USPTO, USPTO issues 7 millionth patent. Feb. 14, 2006, http://www.uspto.gov/web/offices/com/speeches/06-09.htm (last accessed May 4, 2008).

** Often, larger competitors employ frivolous lawsuits to pressure smaller companies or start-ups whose patents stand in their way, or which they wish to acquire. Frequently, the cost in executive time and corporate money for the smaller company or start-up becomes so onerous that it caves in to a licensing agreement. One viable strategy to avoid being taken over is to license the patent to the large competitor, in whose interest it then becomes to protect its position by protecting and defending the patent.

sons for seeking patents on bionanotech-related technologies.* Moreover, venture capitalists will not support a start-up that relies on trade secrets alone. In sum, investors are unlikely to invest in a start-up that has failed to construct adequate defenses around its IP via valid, enforceable patents. Numerous technologies and techniques pertaining to bionanotech can be protected via a patent (Table 27.2).

A company seeking a dominant position in a particular sector of bionanotech may wish to review patent citations (i.e., patents cited in other patents). Patent citations can serve as a useful indictor of licensing potential: patents that are repeatedly cited are generally considered more commercially valuable.† One quarter of all patents receive no citations, and a mere 0.01% earn greater than 100 citations.‡ According to one study, a patent cited 14 times in other patents is, on average, 100 times more valuable than a patent cited only 8 times.§

Millions of dollars may be lost if a company fails to take the necessary steps to protect its patent assets. Securing valid defensible patent protection is vital to the long-term viability of virtually any drug or biotechnology company, whether nanotechnology is the platform technology involved or not. Often, loss of these critical assets is a result of both the researcher's excitement with his or her research as well as general ignorance about IP. In fact, experts agree that "patent awareness" (i.e., the knowledge that patents are intangible property that can be obtained and lost) is central to any business plan or strategy.¶ Furthermore, it is essential that managers and patent practitioners implement certain proactive measures to "box out" the competition (Table 27.3).** In other words, taking the correct preventive steps is critical to realizing the full commercial potential of an invention. Because nanomedicine interfaces with fields such as biology, physics, chemistry, engineering, medicine,

TABLE 27.2

Various Bionanotechnologies That Can Be Protected via a U.S. Patent

Biopharmaceutics
    Drug delivery
        Drug encapsulation
        Functional drug carriers
    Drug discovery

Implantable materials
    Tissue repair and replacement
        Implant coatings
        Tissue regeneration scaffolds
    Structural implant materials
        Bone repair
        Bioresorbable materials
        Smart materials

Implantable devices
    Assessment and treatment devices
        Implantable sensors
        Implantable medical devices
    Sensory aids
        Retina implants
        Cochlear implants

Surgical aids
    Operating tools
        Smart instruments
        Surgical robots

Diagnostic tools
    Genetic testing
        Ultrasensitive labeling and detection technologies
        High-throughput arrays and multiple analyses
    Imaging
        Nanoparticle labels
        Imaging devices
Understanding basic life processes

---

* Regalando A: Nanotechnology patents surge as companies vie to stake claim. *Wall Street Journal* June 18 Issue, A1 (2004).

† McKie S: Innovation asset management: Don't bottle up creativity. *Intelligent Enterprise*, Dec. 1, 2006, http://www.intelligententerprise.com/showArticle.jhtml?articleID=194500328 (last accessed June 10, 2008).

‡ Farrell C: Follow the patents. *BusinessWeek* January 8 Issue, 78–79 (2007).

§ See *supra* Note ‡.

¶ Forman D: IP storm clouds build on horizon. *Small Times* 21–24 (2004).

**Bawa R, Bawa SR, Maebius SB, Iyer C: Bionanotechnology patents: Challenges and opportunities. In: *The Biomedical Engineering Handbook* (3rd Edition). Bronzino JD (Ed.), CRC Press, Boca Raton, FL; 29-1–29-16 (2006).

TABLE 27.3

Legal Tactics Available to Bionanotechnology Companies Dealing with Patent Disputes

Predispute strategies	*Strategic patenting* to clearly box out the competition (i.e., patent-drafting guidelines used to obtain broad, enforceable patents that preempts the field).
	Patent interference practice to attack a competitor's patent application, thereby preventing overlapping patents from being issued.
Postdispute strategies	*Patent reexamination,* a procedure for legally challenging a competitor's issued patent.
	Cross-licensing of patents to peacefully coexist with a competitor.
	Patent infringement litigation to invalidate a competitor's overlapping claims.

TABLE 27.4

Inventor's Reality Checklist and Complex Marketing Factors

- Does the invention offer a unique solution to a real problem?
- Does it offer a measurable improvement over previous attempts to solve the problem?
- Is it a stand-alone product or part of an existing product?
- Can it be easily manufactured or integrated into an existing product or system?
- How big is the potential market?
- Is the market growing or shrinking?
- Is the market global? Can the invention be expanded into new markets as they evolve?
- Will the invention become passé before a prototype is designed?
- Who are the prospective investors, partners, or licensees in the field?
- What price will consumers put on its value?
- What are the estimates for commercialization and marketing?
- What are the incentives for the consumer to buy the product?
- Is federal regulatory approval required?
- How long will it take to bring the product to market?

and computer science, filing a patent application (or conducting a patent search) in this field may require expertise in these diverse disciplines. Hence, employing a qualified patent counsel (a patent agent, patent attorney, or a multidisciplinary team of lawyers) who understands the legal and technical complexities is a critical step in obtaining quality patents. Additionally, issued patents and other prior art* should be carefully evaluated and effective patent-drafting strategies devised accordingly. In 2005, the PTO proposed several sweeping changes in patent practice that could significantly alter the way in which bionanotechnology companies file and prosecute patent applications.† Therefore, companies may need to rethink their patent strategies in order to maximize their patent rights, including taking appropriate proactive action on pending patent applications prior to the actual implementation of these new rules. Additionally, many complex marketing factors may need to be carefully evaluated (Table 27.4).

---

* The phrase "prior art" refers to various sources of information that the PTO uses to reject a patent application. In other words, it is the "knowledge" that exists in the public domain prior to the date of the invention. Prior art is often in the form of a printed document that contains a disclosure or description that is relevant to an invention for which a patent is being sought or enforced. It can include documentary material like publications, prior patents, Web sites, or other disclosures that suggest that the invention is not new. It can also include evidence of actual uses or sales of the technology within the United States. Typically, prior art is submitted by the inventor during prosecution of his or her patent application.

† Notice: Changes to practice for continuing applications, request for continued examination practice, and applications claiming patentably indistinct claims. *Federal Register* 71(1), 48 (2006); Notice: Changes to practice for the examination of claims in patent applications. *Federal Register* 71(1), 61 (2006).

The phrases "patent value" and "patent quality" are distinct concepts; however, they are somewhat related. Both of these largely determine a patent's potential for commercialization, licensing opportunities, investor interest, and enforceability:

1. Patent quality is generally assessed by determining the degree to which a patent examiner has made proper, timely decisions about the validity and scope of protection during the examination process that are consistent with the legal ruling a court would make after comprehensive review of the same application.
2. On the other hand, the patent value of an issued bionanotechnology patent is often measured in terms of other factors ("valuation metrics"):
   a. The breadth and scope of the issued patent claims that affect others freedom to operate (i.e., the patent's originality)
   b. The number of potential competitors in that particular sector of nanomedicine
   c. Government fees ("maintenance fees") paid in order to keep the patent enforceable*
   d. The patent's applicability to other fields
   e. Licensing and litigation activity surrounding the patent
   f. The frequency by which that patent is cited by others (discussed earlier)
   g. Other IP held by the patent holder in that particular technology, including any blocking, pioneering, or upstream patents

27.7 KEY STRATEGIES FOR BIONANOTECHNOLOGY INVENTORS

There are certain key considerations and strategies that a bionanotechnology inventor and his or her company must follow in order to adequately protect an invention even before a patent application is drafted or filed. Some of these are discussed briefly below.

27.7.1 AVOID ANY EARLY PUBLICATION OR ANY PUBLIC DISCLOSURE

The inventor should refrain from publishing a description of, publicly presenting, submitting grant proposals for, or offering the invention for sale prior to filing a patent application. Often a company releases information on a new product, or discusses details during negotiations prior to filing a patent application. All of these activities create prior art against the inventor and can trigger a 1-year "on-sale bar." Note that one of the criteria for patentability in the United States is that the invention must be "novel" (Table 27.1) and not appear in the public domain in the form of prior art. According to current U.S. patent law, the inventor has 1 year to file for an application from the date that invention is known of or offered for sale (meaning that any public disclosure triggers a 1-year deadline to file a patent application in the United States). On the other hand, because this 1-year grace period is not offered by foreign patent offices, any publication or public disclosure will prevent the inventor from obtaining a foreign patent altogether, or prevent the inventor from realizing the full range of potential applications for which a patent is being sought overseas. In summary, a patent application should be drafted and filed as soon as possible after the completion date of the invention to realize its full commercial potential.

27.7.2 CONSIDER OBTAINING A FOREIGN PATENT

Filing a bionanotechnology patent in a foreign country should be carefully considered and should largely depend upon commercial considerations. If there is an interest in expanding into foreign markets, then obtaining patents abroad should be seriously considered. Furthermore, even if the inventor does not plan to establish a market for the particular bionanotechnology invention in a foreign country, obtaining a patent there could be critical in securing licensing deals (and discouraging

---

* All utility patents that issue from a patent application filed on or after December 12, 1980, are subject to the payment of maintenance fees to the PTO to maintain them in force. Failure to pay maintenance fees on time results in the expiration of the patent.

unlicensed copying or use by foreign competitors). The danger of steering clear of foreign patent filing is that a competitor could commercialize the invention in a foreign country and capture valuable market share there. For example, an inventor who patents a novel dendrimer-based drug-delivery system only in the United States is essentially giving away the entire technology to other countries because the patent discloses the best method of making this novel system.

Most U.S. inventors seeking foreign bionanotechnology patents first file a U.S. patent application (known as the "national stage" application) and follow it with a patent application under the Patent Cooperation Treaty (PCT). The PCT is a multilateral treaty that was established in 1978. As of January 1, 2008, it has 138 member states. The PCT allows reciprocal patent rights among its signatory nations. In other words, it simplifies the patenting process when an inventor seeks to patent the same invention in more than one country. It should be emphasized that there is no "world patent." Inventors have a year after filing the national stage patent application to file in the foreign country under the PCT. Under PCT rules, inventors can specify particular foreign countries where they intend to seek patent protection for their bionanotechnology invention and may take 30 months (or more) from their original national stage application filing date (priority date) to complete all foreign application requirements under the PCT (international phase). This delay may provide the inventors with time to determine whether their bionanotechnology invention is commercially viable and merits patenting in several countries, thereby sparing them substantial effort and expense should they decide not to file overseas.

27.7.3 BEWARE OF PREGRANT PUBLICATION OF U.S. PATENT APPLICATIONS

Today, as part of the application process, all U.S. patent applications are published 18 months from the earliest filing date (up to that point, during prosecution, they are kept confidential), unless the applicant "opts out" and foregoes foreign patent filing. Traditionally, applications filed at the PTO were kept secret until they matured into a patent. However, because of the American Inventors Protection Act (AIPA) of 1999, an application filed on or after November 29, 1999, generally loses its secret status when it is published. In effect, this implies that almost always a patent application, as filed, will eventually appear in the public domain (whether or not it is patented) and will be available to competitors.

27.7.4 MAINTAIN PROPER LABORATORY NOTEBOOKS

Laboratory notebooks often contain valuable and critical information that may not be readily apparent to a company or its R&D facility. Therefore, it is critical that laboratory notebooks be maintained properly. This is especially important when working in research teams. Here, proper laboratory notes documenting the creative effort, maintaining confidentiality, and securing communication among the teams and filing for a patent promptly are essential steps that safeguard inadvertent or premature invention disclosure of one group's work by another group. Laboratory notebooks are also useful to patent practitioners to establish the date of an invention, especially in light of a competitor's challenge in court as to who invented first in what is known as "interference proceeding."

27.7.5 CONDUCT A PRIOR ART SEARCH AND A "FREEDOM-TO-OPERATE" SEARCH

It is highly recommended that a proper prior art search be conducted prior to filing a patent application. The purpose of this is to guage the competition. This may also assist the inventor to design around potential prior art. Moreover, because the patent owner does not automatically have the right to practice his/her invention, it may be wise to conduct a "freedom-to-operate search" of the issued bionanotech patent prior to investing in and commercializing it. Note that filing a patent application (or conducting a prior art or freedom-to-operate search) on novel bionanotechnology such as a nanoparticle drug delivery system may require expertise in diverse disciplines like biotechnology, physics, medicine, chemistry, and engineering. The quality and value of the issued patent (see Section 27.6 for details) will largely determine its potential for commercialization, licensing opportunities, investor interest, and enforceability. Hence, employing qualified patent practitioners who understand the legal standards and the complexities of the technology at hand is a critical step to obtaining quality patents.

27.7.6 EDUCATE EMPLOYEES AND RESEARCHERS

It is important that business and IP professionals within a company educate scientists to spot potential inventions during the R&D phase, as this may not always be apparent to them. In fact, a company should implement policies involving incentives where scientists are rewarded for reporting or submitting invention disclosures. This may be especially critical in a university setting where generating invention disclosures may be less of an incentive to researchers who are promoted or tenured based on their research grants. Scientists often overlook the fact that their inventions can be patented. Further, "patent awareness" may enable a researcher to pursue a particular research path that has a greater likelihood of leading to a patentable invention.

27.7.7 REQUIRE STRONG EMPLOYMENT AGREEMENTS TO SAFEGUARD IP

Companies must require all employees to sign agreements that clearly specify that all company inventions, intellectual property, and proprietary information is company property and cannot be disclosed or exploited by any employee at any time. This could become critical if a former employee joins a competitor company or research laboratory. Similar agreements should be required of consultants and visiting scholars where all rights are assigned to the company or university. Nondisclosure agreements should be required during negotiations for venture capital or licensing discussions. Furthermore, confidential materials should be properly labeled and safeguarded; otherwise, value associated with specific information or invention may be lost or reduced.

27.7.8 EMPLOY STANDARD TERMINOLOGY WHILE DRAFTING PATENT APPLICATIONS

The fact remains that bionanotechnology is an inherently difficult topic for discussion, in part due to the confusion surrounding its definition (see Sections 27.2 and 27.8). Although it is well recognized in patent law that a patent applicant can be his or her own lexicographer, it is recommended that an applicant should employ standard language in bionanotechnology patent applications whose meaning is well recognized in the pharmaceutical, medical, or biotechnology fields. Nondisclosure agreements should be required during negotiations for venture capital or licensing discussions. Furthermore, the language should be precise and the use of terms consistent throughout the claims and specification (avoid synonyms and be repetitive in the use of phrases when appropriate). This will prevent confusion at the PTO as well as prevent possible prosecution delay.

This can be especially advantageous later on if litigation arises. Note that it is possible that the patent will be the subject of litigation in the future (e.g., an infringement suit initiated by the patentee against a competitor or a suit for declaratory relief initiated by an accused competitor/infringer, asking a court to declare a patent invalid). The success of the litigation may hinge on how the patent was drafted. A poorly drafted patent will give an advantage to the competitor, causing significant aggravation and resulting in substantial litigation fees for the inventor. Therefore, while drafting patent applications, the drafter should anticipate that the patent might have to be defended in court. Moreover, poorly drafted patents can adversely affect patent issues like licensing potential, validity, and enforceability.

27.7.9 RELATIVE EASE OF OBTAINING "BROAD" PATENTS IN BIONANOTECHNOLOGY

Broad patents continue to be issued by the PTO in bionanotechnology.* The overburdened PTO faces new challenges and problems as it attempts to handle the enormous backlog in bionanotechnology

---

* O'Neill S, Hermann K, Klein M, Landes J, Bawa R: Broad claiming in nanotechnology patents: Is litigation inevitable? *Nanotechnology Law and Business* 4(1), 595–606 (2007).

applications filed and the torrent of improperly reviewed patents granted. At present, all these factors favor obtaining broad patents in bionanotechnology. (See Section 27.9 for details.)

27.8 SEARCHING BIONANOTECHNOLOGY PATENTS AND OTHER PRIOR ART—ISSUES AND CHALLENGES

27.8.1 New Class 977 for Nanotechnology Patents

Due to the burgeoning number of new patent applications filed at the PTO and continued pressure from industry, in 2004, the PTO finally created a cross-reference classification scheme (also referred to as a cross-reference digest or art collection) for nanotechnology (designated as Class 977/Digest 1). The purpose of this class was described by the PTO on its official Web site in 2004*:

> [E]stablishing this nanotechnology cross-reference digest is the first step in a multi-phase nanotechnology classification project and will serve the following purposes: facilitate the searching of prior art related to Nanotechnology, function as a collection of issued U.S. patents and published pre-grant patent applications relating to nanotechnology across the technology centers and assist in the development of an expanded, more comprehensive, nanotechnology cross-reference art collection classification schedule.

It is important to note that this digest should not be construed as an exhaustive collection of all patent documents that pertain to nanotechnology.

The PTO currently expanded Class 977 into 250 subclasses. As of January 2008, the PTO has placed over 5000 U.S. patents in Class 977. However, these figures should only be considered a rough underestimate of the total number of nanotech-related patents. This is because the PTO has copied the NNI's narrow definition of nanotechnology (see Section 27.2) for classification purposes. This has resulted in a skewed patent classification system, especially with respect to bionanotechnology inventions. Furthermore, this classification scheme is neither sufficiently descriptive enough to accommodate many of the unique properties that nanotechnology inventions exhibit, nor does it address the interdisciplinary nature and range of technologies encompassed.

In conclusion, the PTO's efforts to provide a home for a few thousand U.S. patents via a skewed classification system defeats the very purpose for the creation of Class 977, namely: (a) to gauge the number of nanotechnology patent applications filed and patents issued, and (b) to assist patent practitioners as well as patent examiners in searching nanotechnology patent documents.

27.8.2 Searching Bionanotechnology Prior Art

There are various issues pertaining to bionanotech patent searching that are of concern. For example, some experts state that the PTO lacks effective automation tools to search prior art pertaining to bionanotechnology. Moreover, their databases may not be exhaustive. This problem may be particularly acute regarding nonpatent prior art. Although there has been a dramatic rise in bionanotechnology patent activity, most of the prior art still exists in the form of journal publications, technical reports, and book articles. Web sites and pregrant patent publications provide an additional resource. I believe that a large amount of this wealth of nonpatent scientific literature directed to bionanotechnology or nanomedicine predates many of the patents that have been issued and are currently issuing. It is possible that patent examiners lack access to some of this critical nonpatent information. This is possible either because the PTO does not subscribe to all relevant commercial databases, or because not all patent examiners are experienced searchers. As a result, patent examiners may miss discovering prior art (we have highlighted this in the case of carbon nanotubes [§]). The problem of access to nonpatent information is not unique to bionanotechnology

* New Cross-Reference Digest for Nanotechnology. U.S. Patent & Trademark Office, August 2004, www.uspto.gov/web/patents/biochempharm/crossref.htm

TABLE 27.5

Selected Prior Art Search Databases for Bionanotechnology

Issuing authorities' Web sites	U.S. Patent and Trademark Office (www.uspto.gov), the European Patent Office (www.epo.org), and the Japanese Patent Office (www.jpo.go.jp)
Thomson databases	Derwent World Patents Index, Delphion, Dialog, and Thomson Pharma
IFI CLAIMS patents database	Data on U.S. patents and current patent legal status
STN chemical abstracts database	Chemistry bibliographic data available from Chemical Abstracts Service, including patents and patent families
IMSWorld drug patents international database	Patent family data for commercially significant drugs
INPADOC	European patent search database
JAPIO	Patent abstracts of Japan
Engineering, technology, and scientific databases	INSPEC, EiCompendex, SCISEARCH, and Chemical Abstracts Service
Markets and business databases	Factiva and PROMPT

patent examination; it is seen in most technology areas.* Furthermore, the Internet usage policies of the PTO may prevent patent examiners from accessing all relevant databases to access information.† The issue here may be one of security because prior art searching on the Internet can run the risk of being tracked externally. Given these inferior search capabilities, I agree with the conclusion‡ that "the informational burdens on the examiner are clearly heavy—even before the examiner engages in the heavy lifting of interpreting the prior art."

Given this backdrop, it seems that patent examiners are basing decisions about the grant of bionanotechnology patent on limited information. It is scary to envision that their faulty decision making will shape a nascent industry for years to come. It appears that this information deficit has rendered examination unfocused and inefficient, resulting in the issuance of numerous "unduly broad" patents (Section 27.9).

Add to this confused state of affairs the general difficulty in searching bionanotechnology prior art. Because of its broad and often overlapping definition, searching and retrieving patents and publications is complicated relative to other technology areas. Different terms can refer to the same nanomaterial. For example, nanofibers, nanotubes, nanocylinders, buckytubes, nanowires, and fibrils all refer to carbon nanotubes.§ Because of this particular point, accurately mapping the patent landscape is also a real challenge. Patents or publications that are truly bionanotech-based may not use any specific nano-related terminology. In fact, patents or publications are often written "not to be found" in order to keep potential competitors at bay. On the other hand, there are business-savvy inventors and assignees who use key words incorporating a "nano" prefix into their patents or publications to better market their invention or concept. Therefore, part of the challenge in finding "true" bionanotechnology prior art while searching patent and commercial databases (Table 27.5) is the judicious use of key terms, patent classification codes, and alternative phraseology. Coupling this strategy with additional filtering (via a subject expert) is probably the most reliable way to uncover prior art.

---

* In view of all this, commentators have questioned the validity of issued patents in general. They state that a "granted patent" cannot be equated with "official government approval and certification of validity": "There can hardly be a patent agent who, privately, will not readily admit that he or she has got lots of things 'past the office' on flimsy grounds. In the nature of things, a patent office, however hard it tries can only be a coarse filter. Patents that pass the filter cannot be taken as necessarily valid." (Blackman M: Editorial. *World Patent Information* 29, 4–7 [2007].)

† Notice: Internet Usage Policy, 64 Federal Register 33,056,33,061, Dept. of Commerce (1999).

‡ Petherbridge L: Positive examination. *IDEA* 173, 182–183 (2006).

§ Harris D, Bawa R: The carbon nanotube patent landscape in nanomedicine. *Expert Opinion on Therapeutic Patents* 17(9), 1165–1174 (2007).

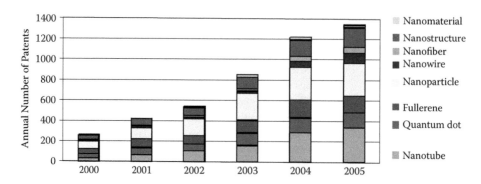

FIGURE 27.1   Nanomaterial-based U.S. patents issued.

27.9   BIONANOTECHNOLOGY PATENT PROLIFERATION AND PTO PROBLEMS—A RECIPE FOR DISASTER?

Federal agencies continue to grapple with nanotechnology.* The PTO is no exception. In fact, for more than a decade, all of the world's major patent offices have faced an onslaught of nanoscience patent applications.† The situation at the PTO is likely to worsen as more applications are filed and pendency rates further skyrocket. As companies develop products and processes and begin to seek commercial applications for their inventions, securing valid and defensible patent protection will be vital to their long-term survival. In the decades to come, with certain areas of bionanotechnology further maturing and promised breakthroughs accruing, patents will generate licensing revenue, provide leverage in deals and mergers and reduce the likelihood of infringement. The development of bionanotech-related products, which is extremely research intensive, will be significantly hampered in the absence of the market exclusivity offered by a patent. Due to the PTO's poor track record in handling issues like examination quality, skyrocketing patent pendency, out-of-control examiner attrition, and low morale, I note the following issues impacting bionanotechnology patenting.

27.9.1   A CHAOTIC NANOTECH PATENT LAND GRAB CONTINUES

Due to the potential market value of bionanotech products, every entity in the international race for technological dominance—researchers, executives, and patent practitioners—views the collection and exploitation of bionanotech patents as critical. In fact, these players are making an effort to obtain the broadest protection possible for new nanoscale polymers, devices, and systems that have applications in biotechnology and medicine. Therefore, a sort of "patent land grab" (Figure 27.1) is in full swing by "patent prospectors" as start-ups and corporations compete to secure broad patents in nanomedicine during these critical early days.‡ This land grab mentality is also fueled by the

---

* Weiss R: Nanotechnology regulation needed, critics say. *Washington Post*, Dec. 5, 2005, A08, http://www.washington-post.com/wpdyn/content/article/2005/12/04/AR2005120400729.html (last accessed May 4, 2008) (suggesting that most federal agencies have not created nanotech-specific safety regulations and discussing EPA's nanotech program, with which companies that manufacture nanotech products may voluntarily choose to comply).

† Bawa, R: Patents and nanomedicine. *Nanomedicine* 2(3), 351–374 (2007); Van Lente MA: Building the new world of nanotechnology. *Case W. Res. J. Int. Law* 38(1), 173–215 (2006); Bawa R: Nanotechnology patents and the U.S. Patent Office. *Small Times* 4, IP8 (2004); Huang Z et al: Longitudinal patent analysis for nanoscale science and engineering: country, institution and technology field. *Journal of Nanoparticle Research* 5, 333 (2003); Bawa R, Bawa SR, Maebius SB: The nanotechnology patent 'gold rush.' *Journal of Intellectual Property Rights* 10, 426–433 (2005); Lawrence S: Patently absurd: Too many patents could kill nanotechnology. *Red Herring* November 20 Issue, 119 (2002); Jaffe AB, Lerner J: Patent prescription: A radical cure for the ailing US patent system. *IEEE Spectrum* 42, 38–43 (2005); Sabety T: Nanotechnology innovation and the patent thicket: Which IP policies promote growth? *Nanotechnology Law & Business* 1(3), 262–283 (2004).

‡ See *supra* pg. 325 Note *.

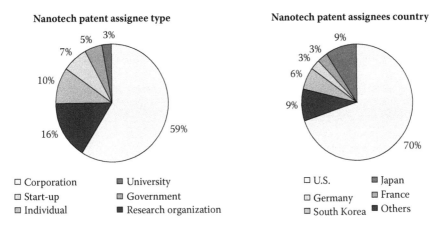

Nanotech patent assignee type

- □ Corporation
- □ Start-up
- □ Individual
- ■ University
- □ Government
- ■ Research organization

Nanotech patent assignees country

- □ U.S.
- □ Germany
- □ South Korea
- ■ Japan
- ■ France
- ■ Others

FIGURE 27.2   U.S. Nanotechnology patent demographics.

relative lack of products and processes in the marketplace. Companies feel that, to demonstrate confidence and sway venture capitalists, they must generate or claim IP. Some companies also feel pushed into claiming as much IP as possible due to fear that, if they lag behind in this effort, someone else will claim the broadest IP. Similarly, academic researchers feel this compulsion to file for bionanotechnology patents in order to bolster their reputation and *curriculum vitae*. Moreover, most inventors have quickly realized the opportunities of a disorganized PTO during these early days when they can secure broad patents on valuable upstream technologies with relative ease.

Certain general trends are being reported for nanopatents. With nanotechnology maturing further, the number of claims in the patent applications and the amount of scientific literature cited during patent prosecution is on the rise. This is significant because scientific publications are the most accurate indicator of scientific activity and productivity.* Another trend observed is that nanotechnology patent owners are eyeing commercial potential, and therefore, maintaining (i.e., paying maintenance fees) more of their patents.

However, these patent prospectors have to deal with an overburdened PTO, which historically has been slow to react to emerging technologies like biotechnology and software. In fact, the entire U.S. patent system is under enormous scrutiny and strain as the PTO continues to struggle with the evaluation of bionanotech-related patent applications. Some commentators have strongly voiced their concerns regarding the emerging patent thicket problem and its impact on global access to products.† Some of the concerns highlighted above are borne out by U.S. nanotechnology patent demographics from 2006 (Figure 27.2).

---

* Ilakovac V et al: Reliability of disclosure forms of authors' contributions. *Canadian Medical Association Journal* 176, 41 available at http://www.cmaj.ca/cgi/reprint/176/1/41 (last accessed June 10, 2008); Pololi L, Knight S, Dunn K: Facilitation scholarly writing in academic medicine. *Journal of General Internal Medicine* 19(1), 64-68 (2004).

† Report: Nanotech's "second nature" patents: Implications for the global south. ETC Group, Ottawa, Canada (2005). An extract from the report:

"Although industry analysts assert that nanotech is in its infancy, 'patent thickets' on fundamental nano-scale materials, building blocks and tools are already creating thorny barriers for would-be innovators. Industry analysts warn that, 'IP roadblocks could severely retard the development of nanotechnology.' Some insist that nano-scale technologies will address the most pressing needs of the [world's] marginalized peoples. But in a world dominated by proprietary science, it is the patent owners and those who can pay license fees who will determine access and price....The world's largest transnationals, leading academic labs and nanotech start-ups are all racing in the patent gold rush. Increasingly, universities are licensing on an exclusive basis. Nanotech's 'second nature patents' are positioning multinational matter moguls to own and control novel materials, devices and their manufacturing processes... Control and ownership of nanotechnology is a vital issue for all governments and civil society because nanomaterials and processes can be applied to virtually any manufactured good across all industry sectors... At stake is control over innovations that span multiple industry sectors ... companies that hold pioneering patents could potentially put up tolls on entire industries." [Citations omitted]

27.9.2   PROBLEMS PLAGUE THE PATENT EXAMINATION PROCESS

Although the PTO budget has bloated to its current $1.6 plus billion mark, various examination problems continue to haunt it. One patent expert recently summarized the current crisis at the PTO*:

> The U.S. Patent & Trademark Office is under siege for issuing patents that should never have issued, and for excessive delays. Congress is considering changes such as a new opposition system for challenging patents when they emerge from examination.

A law professor is blunter in her criticism†:

> The United States patent system is broken and desperately needs fixing.... Why are so many bad patents being issued?... Under our current system, granting an application with little scrutiny takes less time than subjecting it to rigorous review.... The examiners are unable to perform more than a cursory search of their own [due to time constraints and lack of expertise]...Third parties—competitors and consumers—are generally excluded from the patent examination process, even though these parties have the greatest incentive to discover the prior art and disclose it to the Patent Office in order to prevent bad patents from being issued.

Indeed, questionable patent examinations at the PTO seem to extend across technology areas. Some of the shortcomings that impact bionanotechnology patent examination are examined briefly below:

- At present, the PTO lacks a dedicated examining group ("technology center" or TC) to handle applications on nanotechnology or bionanotechnology. Moreover, few examiners have experience in the rapidly evolving field of nanotechnology. Because bionanotech is interdisciplinary in nature, patent applications that are searched, examined, and prosecuted in one TC could and should be examined more effectively by a coordinated review of more than one TC. In reality, there is no such collective review and applications continue to be examined differently within each TC. Obviously, this approach does not provide a cohesive and uniform examination of applications because examiners in each of its eight TCs may rely upon case law, legal standards, and prior art that may be unique to their specific TC.
- Many bionanotech patent applications may not receive adequate examination during prosecution because of the patent examiner's inability to locate applicable prior art, especially nonpatent prior art. Therefore, as discussed in detail earlier (Section 27.8.2), it is accurate to conclude that patent examiners may sometimes be basing decisions about the grant of a patent on limited information.
- The PTO continues to be understaffed in numerous TCs, and it is plagued by high attrition. The agency's inability to attract and retain a talented pool of patent examiners is creating havoc. At hearings on Capitol Hill and in its Annual Reports, the PTO brass proudly touts hiring hundreds of new patent examiners each year to alleviate the backlog that is clogging the patent system. In fact, in this context, the Commissioner for Patents continues to highlight that the PTO will hire 1200 new patent examiners in the current fiscal year to alleviate the backlog that is clogging the patent system.‡ However, it fails to focus on the critical issue of "brain drain" resulting from an exodus of so many experienced patent

---

* Mcdonald D: Fighting the modern patent wars. *Intellectual Property Today* 14(1), 7 (2007).

† The Press Register. http://ipbiz.blogspot.com/2007/01/lafrance-on-jaffelerner-on-patent.html (last accessed June 10, 2008).

‡ Marasco CA: Overlooked opportunities in government. *Chemical and Engineering News* 85(11), 47–50 (2007).

examiners and other senior-level officials.* It would be desirable for the PTO brass to focus on retaining more of its seasoned employees and not put all its efforts into hiring new ones. These attrition rates are likely to be further exacerbated by poor morale and work conditions. According to many experts, patent examiners are underpaid (relative to U.S. law firm salaries) and overworked (as compared to their colleagues at the European Patent Office). They also have to review applications under unreasonable time pressures and skyrocketing patent pendency (discussed later). Arguably, the internal quality review process that monitors quality of patents allowed by patent examiners is fraught with a general lack of legal and scientific expertise on the part of reviewers.

- The PTO's funding problems are legendary. Congress's long-standing practice of "diverting" PTO user fees collected from patent applicants to the general budget have always caused much consternation. Naturally, stopping this practice would alleviate some problems at the agency. In February 2006, a bill was signed by the President that allows the PTO to spend all its projected collected fees, thereby preventing funds from bring diverted to other government programs. I hope that because of this law, the damaging drain on the agency's financial resources will finally end. It is also hoped that the PTO will now temper its annual practice of hiking patent fees.

- Even today, with all the quality initiatives underway, examiners are still rewarded on the quantity of their work, not the quality. An antiquated quota system is firmly in place. The patent examiner's production goals (quota) have not been adjusted in decades in spite of the increased complexity of patent applications, not to mention the substantial increase in the amount of prior art that the examiner has to search and analyze. Quality continues to take a back seat. Although, year after year, the PTO Annual Reports paint a much rosier picture. According to recent PTO statistics, the allowance error rate has hovered around 4%. This could imply that the PTO's own conservative estimates indicate that thousands of U.S. patents were "wrongly" allowed.†

- The PTO has failed to effectively engage outside legal or technology experts. Only a handful of experts from industry or academia have lectured on legal or technical issues unique to bionanotech. This reluctance to use outside expertise has further added to the information deficit. It is clear that the PTO lacks internal expertise in these matters, and its isolationist policy only compounds the problem. Moreover, patent examiners are not required to have advanced degrees in science or engineering. Possessing advanced degrees or advanced training, by and large, goes unrecognized at the PTO.

- Few training modules or examination guidelines have been developed to educate patent examiners in the complexities and subtleties of bionanotech. Similarly, no written guidelines specific to bionanotech are available for patent practitioners.

Given all these challenges, it is hard to predict with certainty how all these issues and challenges will play out with respect to bionanotech patenting or commercialization. We will have to wait and see whether this industry thrives like the information technology industry, or becomes burdened like the radio patent deadlock.‡

Congress is continuing patent reform hearings in an effort to quell questionable patents as well as to provide adequate safeguards against abuses to the patent system. One of the proposals under serious consideration is a "post-grant review" of patent applications. However, I agree with some patent experts that "[s]erious doubts exist whether a politically controlled PTO can guarantee the promise

---

* Many reports have highlighted the fact that the federal government is vulnerable to "brain drain" both because baby boomers are retiring and their potential replacements (most notably graduate students) do not view the government as their first choice of work.

† Hultquist S: Statistical musings. *Intellectual Property Today* 14(5), 48 (2007).

‡ Sabety T: Nanotechnology innovation and the patent thicket: Which IP policies promote growth? *Nanotechnology Law & Business* 1(3), 262–283 (2004).

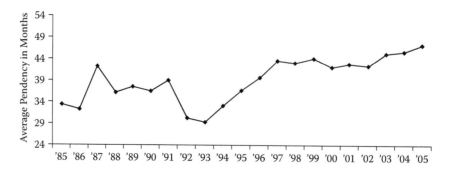

FIGURE 27.3 Nanotechnology patent pendency.

of the post-grant system that the patent community so desperately needs. ... [T]he patent community can hardly have confidence in a post-grant review system under the control of the PTO."*

27.9.3 The Nanotech Patent Onslaught Strains the U.S. Patent Office

For the past decade or so, there has been a dramatic increase in the number of new nanotechnology patent applications filed and patents granted (Figure 27.1) as well as an increase in published patent applications and scientific publications. This information overload has created numerous challenges for the PTO, an agency that traditionally struggles with this issue. Furthermore, this overburdened and inefficient agency has yet to implement a solid plan to handle the enormous growth in nanotechnology patent applications filed. This has resulted in added time to review patent applications (i.e., an increase in patent pendency) and concerns about the validity and enforceability of numerous issued patents (reflects a decrease in patent quality).

As stated earlier, the >1 million backlog of patent applications continues to build. A recent report puts the average nanotechnology patent pendency at 4 years† (Figure 27.3), a period that is simply too long for certain technologies that peak and are then obsolete in a few short years. This excessive delay has particularly serious business consequences for smaller companies and start-ups, as these entities rely heavily on venture funds for their success. Therefore, it was no surprise that these groups recently confronted the Undersecretary of Commerce regarding the high patent pendency of their nanotechnology inventions. Surprisingly, the Undersecretary blamed the excessive delays on nanotechnology companies, accusing them of poaching nanotech-trained examiners en masse from the PTO.‡ I find the Undersecretary's argument a sad excuse for inefficiency and incompetence.

Furthermore, surprisingly broad patents in bionanotech continue to be issued by the PTO.§ Obviously, this is partly the result of court decisions in the past two decades that have made it easier to secure broad patents. Laws have also tilted the table in favor of patent holders, no matter how broad or tenuous their claims. As a result, the PTO faces an uphill task as it attempts to handle the enormous backlog in applications filed. It also faces a torrent of improperly granted patents, many of which are likely to be "reexamined."

In this climate of patent proliferation, it is inevitable that in the near future there will be an increase in litigation. Most patent practitioners regularly highlight one or more of the following problems while discussing bionanotech patents:

- An improper rejection of a patent application due to an examiner's erroneous conclusion that the subject matter is not novel.

---

* Wegner H: Post grant review: is the PTO up to the task? www.ipfrontline.com/printtemplate.asp?id=15015 (last accessed June 10, 2008).

† *The Nanotech Report* (5th Edition). Lux Research, Inc., New York (2008).

‡ Editor: Poachers to blame for patent delays. *New Scientist* 194(2600), 23 (2007).

§ Bawa R: Patents and nanomedicine. *Nanomedicine* 2(3), 351–374 (2007).

- Issuance of an "overly broad" patent by an examiner that infringes previously issued patents and gives control over a broad swath of basic technology, allowing the patentee to unfairly exclude competition. This runs the risk of impeding important future downstream innovations.
- Issuance of a patent in spite of existing prior art that was overlooked by the examiner during patent examination.

Naturally, any of the above results is unacceptable to the nanotechnology community. Issuance of patents of poor quality, or too many invalid patents on early stage research (upstream technologies) is likely to cause enormous damage to commercialization efforts because it can result in one or more of the following:

- Suppressing market growth and innovation
- Causing loss of revenue, resources, and time
- Discouraging industry from conducting R&D and manufacturing
- Inducing unnecessary licensing
- Promoting a greater possibility of patent appeals and infringement lawsuits
- Stifling high-quality inventions (introducing noise into investment, valuation, and contracting decisions) and undermining the patent system
- Eroding public trust vis-à-vis bionanotech

One patent expert highlights the impact of poor-quality patents in economic terms*:

Questionable patents can harm competition and hinder innovation by forcing market participants to pay licensing royalties, incur substantial legal expense to defend against infringement claims, engage in design-around efforts that raise costs and/or hinder product performance. . . . [A] patent holder can have real power even without being a true inventor because the systems for patent issuance and patent litigation are tilted in favor of patent applicants and patent holders. The result is that the patent system, while intended to promote innovation, instead places sand in the gears of our innovation engine.

27.9.4 EMERGING BIONANOTECHNOLOGY PATENT THICKETS

Currently, there are too many players holding too many bionanotech patents; this has created the current fragmented, messy patent landscape. Most experts agree that this patent landscape is already producing "patent thickets" that have the potential of causing protracted legal battles. This is obviously an undesirable result and could easily stop bionanotech development in its tracks. Clearly, it poses one of the biggest threats to commercialization.

Patent thickets are broadly defined in academic discourse as "a dense web of overlapping intellectual property rights that a company must hack its way through in order to actually commercialize new technology."† Such patent thickets, a result of multiple blocking patents, naturally discourage and stifle innovation. Claims in such patent thickets have been characterized as broad, overlapping, and conflicting. Therefore, business planners and patent practitioners should steer company researchers away from such potential patent thickets. They may also need to analyze the patent landscape to gauge "white space" opportunities (i.e., no overlapping patents) prior to R&D efforts, patent filing, or commercialization activities (Figure 27.4). Classically, such an analysis of the number and quality of patents issued in a particular sector of bionanotech can highlight a particular

---

* Shapiro C: Patent system reform: Economic analysis and critique. *Berkeley Technology Law Journal* 19, 1017–1019 (2004).
† Shapiro C: Navigating the patent thicket: cross licenses, patent pools and standard-setting. *Innovation Policy Econ.* 1, 119–150 (2000).

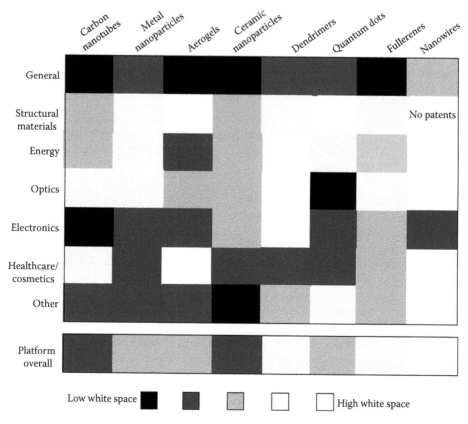

FIGURE 27.4 U.S. patent landscape analysis by nanomaterial platform.

technology trend, areas of high/low commercialization potential, and areas that indicate a high risk of market entry.

According to a widely circulated market study published in 2005,[*] nanoscience researchers around the world are steadily filing patents in the hope of creating "toll booths" that could slow future product development. Because there has been an explosion of overlapping and broad patent filing (and issuance of corresponding broad patents) on nanomaterials, it is likely that the companies that want to use these building blocks in products will be forced to license patents from many different players to implement their inventions. The report focused on five fundamental nanomaterials: carbon nanotubes, dendrimers, fullerenes, nanowires, and quantum dots. The study identified carbon nanotubes and quantum dots as of particular concern. The study noted that although fullerenes and nanowires are relatively free of overlapping patent claims, the other categories are quickly attracting patent applications. For example, the study found that a large number of patent claims for dendrimers have been assigned to recently acquired Dendritic Nano-Technologies, Inc. (Mount Pleasant, Michigan). It also noted that quantum dot patent claims tend to cover the materials themselves rather than specific applications and that the patent situation for using carbon nanotubes in electronics looks "messy." Although some dominant or pioneering patents on carbon nanotubes will expire in the near future, a classic patent thicket seems to be developing in the area of single-walled carbon nanotubes, where companies such as IBM (White Plains, New York), NEC Corpora-

---

[*] Nanotechnology gold rush yields crowded, entangled patents. Lux Research, www.nanotech-now.com/news.cgi?story_id=09134 (last accessed May 8, 2008).

tion (Tokyo, Japan), and Carbon Nanotechnologies, Inc. (Houston, Texas) are likely to aggressively stake out their claims.*

To analyze the perceived patent thicket in any bionanotech-related technology, a detailed legal review of the claim set from the patents in the thicket may be necessary before decisions regarding patent filings or substantial investment on commercialization are undertaken.

27.9.5 THE COMING BIONANOTECHNOLOGY PATENT WARS

Patent grants in bionanotechnology and nanomedicine-related inventions are likely to continue at a pace that is almost synchronous with funding. This is true on an international as well as national scale. The aggressive mentality described above has not only produced overlapping patents, but the race to patent anything "nano" has resulted in a flood of exceedingly broad upstream nanopatents. Although broad patents are generally awarded for pioneering inventions, they should never be allowed if prior art exists. Experts fear that bionanotech's constantly growing patent estate may actually retard innovation due to uncertainty over who is infringing on whose patent. Most experts directly blame the PTO for awarding numerous erroneously broad bionanotech patents.

Clearly, this proliferation of unduly broad patents and the resulting patent thickets will require litigation to sort out, especially if sectors of bionanotechnology become financially lucrative.† At the present time, it seems that nanotech companies in general are avoiding costly court battles. In fact, there is hardly any nanotech patent litigation underway in the United States. Companies sometimes avoid costly litigation to prevent exposing their own patents, some of which may be based on a poor review at the PTO and, thus, whose validity may be questionable. In any case, I believe that expensive litigation is as inevitable as it was with the biotechnology industry where extensive patent litigation resulted once products became commercially successful. The reason for this is simple: royalties may be collected by the patentee from potential infringers. However, in most patent battles, the larger entity with the deeper pockets will ultimately prevail even if the brightest innovative stars are on the other side. This situation is all too familiar to the business sector. It leads to higher costs to consumers (if and when products are commercialized), while deterring the innovation process.‡

Ultimately, companies introducing new products to the market will face considerable uncertainty regarding the validity of broad and potentially overlapping patents held by others. The ongoing land grab will definitely worsen the problem for companies striving to develop commercially viable products. In fact, bionanotech start-ups may soon find themselves in patent disputes with large, established companies, as well as between themselves. Start-ups may also become attractive acquisition targets for larger companies because takeover is generally a cost-effective alternative to litigation.

It is possible that companies may need to acquire costly licenses for patents from other companies in order to establish themselves. It is also possible that companies may use their patents to exclude rather than license out. Furthermore, those who do license may do so at an unreasonably high cost. However, I hope that none of these scenarios will come about. Instead, it is hoped that a more harmonious atmosphere will prevail, where (nonexclusive) cross-licensing agreements by start-ups and large corporations alike will become the norm. In my view, liberal patent licensing is another particularly effective strategy to maneuver through the patent thicket at this stage in the development of bionanotech, especially because the enforceability of so many patents is questionable. It should be noted, however, that when the total number of owners of conflicting intellectual

* Van Lente MA: Building the new world of nanotechnology. *Case W. Res. J. Int. Law* 38(1), 173–215 (2006); Harris D, Bawa R: The carbon nanotube patent landscape in nanomedicine. *Expert Opinion on Therapeutic Patents* 17(9), 1165–1174 (2007).

† O'Neill S, Hermann K, Klein M, Landes J, Bawa R: Broad claiming in nanotechnology patents: Is litigation inevitable? *Nanotechnology Law and Business* 4(1), 595–606 (2007).

‡ Lawrence S: Patently absurd: Too many patents could kill nanotechnology. *Red Herring* November 20 Issue, 119 (2002).

property is relatively small, cross-licensing has been the answer. But, when the number of owners of conflicting IP is relatively large, the transaction costs of cross-licensing may be too great for it to be effective. Also, critics consider cross-licensing as a settlement of a patent dispute that may not serve the public interest because cross-licensing (as compared to litigation) limits competition when it is between competitors, raising the spector of antitrust prosecution of these agreements as monopolies.

27.10 NAVIGATING THE BIONANOTECH PATENT THICKET

Following are some other proposals that may cut through the bionanotech patent gridlock and prevent widespread and wasteful litigation.

27.10.1 FORMATION OF PATENT POOLS

The multiple-party patent thicket problem may be solved by the cooperative formation of "patent pools" by technologically competing entities. Patent pools are defined as legally permissible cooperative agreements whereby the members of the pool have access to the patents of the entire pool in exchange for a set price. However, it is uncertain at this stage whether patent pools will be a lawful, desirable, or beneficial answer to the patent thicket problem.

27.10.2 GOVERNMENT ACTIONS TO ENCOURAGE NONEXCLUSIVE LICENSING

Some practitioners have also suggested that government must step in and use its existing authority under the Bayh-Dole Act to encourage nonexclusive licensing of foundational nanotechnology patents.* Under the Bayh-Dole Act of 1980, universities and small business entities may retain patent ownership rights if the research was funded by the U.S. government. The government retains a royalty-free license to any patented technology that is generated as a result of such funding. Naturally, the Bayh-Dole Act will assist bionanotech companies in the same way it helped biotechnology start-ups—by promoting the transfer of university-owned patents funded by government grants to the private sector, given that academia has become increasingly aggressive in patenting its bionanotech-related research.

27.10.3 OTHER GOVERNMENT ACTIONS

Government action, such as the imposition of compulsory licensing of upstream and/or foundational patents that have been financed by public funds, may assist in breaking up dominant patent monopolies. Enforcement of antitrust and competition laws by the government may encourage more cooperation between the various players as well as stimulate active cross-licensing and patent pooling.

There are, of course, other strategies available to prevent or navigate patent entanglements, both before and after the patent issues.† Companies could also focus on trade secrets as a supplement to patents. A greater willingness on the part of the patent applicant to provide prior art, particularly nonpatent prior art, would be helpful to the patent examiner.

---

* Sabety T: Nanotechnology innovation and the patent thicket: Which IP policies promote growth? *Nanotechnology Law & Business* 1(3), 262–283 (2004).
† Harris D, Miller J, Bawa R, Cleveland JT, O'Neill S: Strategies for resolving patent disputes over nanoparticle drug delivery systems. *Nanotechnology Law and Bus,* 1(4), 101–118 (2004); Bawa R, Bawa SR, Maebius SB, Iyer C: Bionanotechnology patents: Challenges and opportunities. In: *The Biomedical Engineering Handbook* (3rd Edition). Bronzino JD (Ed.), CRC Press, Boca Raton, FL; 29-1–29-16 (2006).

27.11 CONCLUSIONS, CAUTIONS, AND RECOMMENDATIONS

Securing valid and defensible patent protection is critical to the bionanotech "revolution." Although early forecasts for commercialization are promising, the emerging patent thicket in this arena of nanotech could be a major stumbling block.* It is almost certain that the enforceability of numerous U.S. bionanotechnology patents (like e-commerce patents previously) will be a major problem in the future. Furthermore, due to the substantial annual increase in costs associated with maintaining and enforcing issued patents, enforceability may also be a problem when the patent holder lacks the resources to maintain or enforce the patent against potential infringers. On the other hand, valid patents stimulate market growth and innovation, generate revenue, prevent unnecessary licensing, and greatly reduce infringement lawsuits.

The PTO continues to be under enormous strain and scrutiny. Reforms are urgently needed to address issues ranging from poor patent quality and questionable examination practices to inadequate search capabilities, rising attrition, and an enormous patent backlog. Numerous influential entities, ranging from government to nongovernment organizations, have recently become more vocal in their criticism of the PTO. They have produced authoritative reports with detailed recommendations regarding overhauling the PTO and the U.S. patent system.† These reforms are urgently needed in order to ensure a better balance between innovation and competition.‡ Without these reforms, the cursory patent examination that is currently in place coupled with patent proliferation and patent backlog that already exceeds 1 million will result in the issuance of too many invalid and unenforceable bionanotech patents. This will continue to generate a crowded, entangled patent landscape with few open-space opportunities for commercialization. For many companies, navigating this minefield will be an unattractive option.

Ownership of technology in the form of patents is one thing, deriving sufficient economic value therefrom is a different issue. Obtaining undeserving patents and profiting from the threat of litigation rather than providing beneficial bionanotech products runs counter to the foundations of our patent system. Therefore, if the current dense patent landscape becomes more entangled and the patent thicket problem worsens, it may prove to be *the* major bottleneck to viable commercialization,§ negatively impacting the entire bionanotech revolution. For investors, competing in this high-stakes patent game may prove too costly.

ACKNOWLEDGMENTS

I wish to thank Michael Burger of Nanowerk LLC (Honolulu, Hawaii) for kindly providing Figure 27.1. I thank Lux Research, Inc. (New York) and Foley & Lardner (Washington, DC) for kindly providing Figure 27.2, Figure 27.3, and Figure 27.4. Table 27.2 is courtesy of Neil Gordon of the Canadian NanoBusiness Alliance (Montreal, Canada).

---

* Bawa R: Patenting nanomedicine: A catalyst for commercialization? *Small Times* 5(8), 16 (2005); Bawa R: Will the nanomedicine "patent land grab" thwart commercialization? *Nanomedicine: Nanotechnology, Biology and Medicine* 1, 346–350 (2005); Bawa R: Patents and nanomedicine. *Nanomedicine* 2(3), 351–374 (2007).

† Report: *To Promote Innovation: The Proper Balance of Competition and Patent Law and Policy*. Washington, DC: Federal Trade Commission, (2003); Merrill SA, Levin RC, Myers MB (Eds): *A Patent System for the 21st Century*. National Academies Press, Washington, DC (2004); Mittal AK, Kootz LD: *Improvements Needed to Better Manage Patent Office Automation and Address Workforce Challenges*. Report GAO-05-1008T, U.S. Government Accountability Office, Washington, DC (2005); Report: *US Patent and Trademark Office: Transforming to Meet the Challenges of the 21st Century*. National Academy of Public Administration, Washington, DC (2005).

‡ Patent reform bills are currently pending in Congress. Similar measures have failed in the past 3 years as the information technology industry and big pharma have battled over the finer points of these bills. (See Sandburg B: Patent reform bill gets cool reception from Rx industry; damage cap is worry. *F-D-C Report* 69(17), 8–9 (2007); Hess G: Patent reform stalls in Senate. *Chemical and Engineering News* 86(22), 40–43 (2008).

§ Bawa R: Patenting nanomedicine: A catalyst for commercialization? *Small Times* 5(8), 16 (2005); Bawa R: Will the nanomedicine "patent land grab" thwart commercialization? *Nanomedicine, Nanotech. Biol. Med.* 1, 346–350 (2005).

Index

A

B

Milton Keynes UK
Ingram Content Group UK Ltd.
UKHW050454071024
449327UK00015B/371

9 780367 387051